Advances in Experimental Medicine and Biology

Volume 1180

Editorial Board

IRUN R. COHEN, *The Weizmann Institute of Science, Rehovot, Israel*
ABEL LAJTHA, *N.S. Kline Institute for Psychiatric Research, Orangeburg, NY, USA*
JOHN D. LAMBRIS, *University of Pennsylvania, Philadelphia, PA, USA*
RODOLFO PAOLETTI, *University of Milan, Milan, Italy*
NIMA REZAEI, *Children's Medical Center Hospital, Tehran University of Medical Sciences, Tehran, Iran*

AF729036

Advances in Experimental Medicine and Biology presents multidisciplinary and dynamic findings in the broad fields of experimental medicine and biology. The wide variety in topics it presents offers readers multiple perspectives on a variety of disciplines including neuroscience, microbiology, immunology, biochemistry, biomedical engineering and cancer research.

Advances in Experimental Medicine and Biology has been publishing exceptional works in the field for over 30 years and is indexed in Medline, Scopus, EMBASE, BIOSIS, Biological Abstracts, CSA, Biological Sciences and Living Resources (ASFA-1), and Biological Sciences. The series also provides scientists with up to date information on emerging topics and techniques. 2018 Impact Factor: 2.126.

More information about this series at http://www.springer.com/series/5584

Yiru Fang
Editor

Depressive Disorders: Mechanisms, Measurement and Management

Springer

Editor
Yiru Fang
Division of Mood Disorders
Shanghai Mental Health Center
Shanghai, China

The work was supported by the National Key Research and Development Program of China (2016YFC1307100), the National Natural Science Foundation of China (81771465, 81930033), the National Key Technologies R&D Program of China (2012BAI01B04); and also supported by the Innovative Research Team of High-level Local Universities in Shanghai.

ISSN 0065-2598 ISSN 2214-8019 (electronic)
Advances in Experimental Medicine and Biology
ISBN 978-981-32-9273-4 ISBN 978-981-32-9271-0 (eBook)
https://doi.org/10.1007/978-981-32-9271-0

© Springer Nature Singapore Pte Ltd. 2019, corrected publication 2020
This work is subject to copyright. All rights are reserved by the Publisher, whether the whole or part of the material is concerned, specifically the rights of translation, reprinting, reuse of illustrations, recitation, broadcasting, reproduction on microfilms or in any other physical way, and transmission or information storage and retrieval, electronic adaptation, computer software, or by similar or dissimilar methodology now known or hereafter developed.
The use of general descriptive names, registered names, trademarks, service marks, etc. in this publication does not imply, even in the absence of a specific statement, that such names are exempt from the relevant protective laws and regulations and therefore free for general use.
The publisher, the authors and the editors are safe to assume that the advice and information in this book are believed to be true and accurate at the date of publication. Neither the publisher nor the authors or the editors give a warranty, expressed or implied, with respect to the material contained herein or for any errors or omissions that may have been made. The publisher remains neutral with regard to jurisdictional claims in published maps and institutional affiliations.

This Springer imprint is published by the registered company Springer Nature Singapore Pte Ltd.
The registered company address is: 152 Beach Road, #21-01/04 Gateway East, Singapore 189721, Singapore

Contents

About the Editor

Dr. Yiru Fang is a Director and Chief Psychiatrist of the Clinical Research Center, and the National Key Clinical Disciplines at the Shanghai Mental Health Center. He is also a Director and Professor of the Management Center for Mood Disorders, and the Clinical Research Center for Mental Disorders at the Shanghai Jiao Tong University School of Medicine, and the Guest Professor of the CAS Center for Excellence in Brain Science and Intelligence Technology. Dr. Fang is the Chairman of Chinese Society of Neuroscience and Psychiatry, and the Vice Chairman of Chinese Society of Psychiatry. He also serves as the Editorial Board Member for Journal of Affective Disorders, Neuroscience Bulletin, International Journal of Bipolar Disorders, and General Psychiatry; and the Member of China Expert Advisory Group of the Lancet Psychiatry. Dr. Fang has worked in the field of mental health for 35 years, and devoted himself to practice, research, teaching and training on psychiatry; He is an editor of <Chinese Guideline for treating Bipolar Disorders>. As a principal investigator, Dr. Fang has conducted various biological, clinical and implementation researches on depression, bipolar disorders, and mood related psychiatric disorders, funded by national and local governments.

Contributors

Min Cai Department of Psychiatry, Xijing Hospital, Fourth Military Medical University, Xi'an, Shaanxi, China

Jun Chen Clinical Research Center and Division of Mood Disorders, Shanghai Mental Health Center, Shanghai Jiao Tong University School of Medicine, Shanghai, China

Peijun Chen VA Northeast Ohio Healthcare System, Cleveland VA Medical Center, Case Western Reserve University School of Medicine, Cleveland, USA

Lvchun Cui Division of Mood Disorders, Shanghai Mental Health Center, Shanghai Jiao Tong University School of Medicine, Shanghai, China

Jinxia Dai Bio-Medical Center and College of Veterinary Medicine, Huazhong Agricultural University, Wuhan, China

Yuqiang Ding State Key Laboratory of Medical Neurobiology, Department of Laboratory Animal Sciences, Institutes of Brain Science, Fudan University, Shanghai, China

Yiru Fang Clinical Research Center and Division of Mood Disorders, Shanghai Mental Health Center, Shanghai Jiao Tong University School of Medicine, Shanghai, China;
CAS Center for Excellence in Brain Science and Intelligence Technology, Shanghai, China;
Shanghai Key Laboratory of Psychotic Disorders, Shanghai, China

Wu Hong Division of Mood Disorders, Shanghai Mental Health Center, Shanghai Jiao Tong University School of Medicine, Shanghai, China

Hailan Hu Center for Neuroscience and Department of Psychiatry of First Affiliated Hospital, Zhejiang University School of Medicine, Hangzhou, China;
Interdisciplinary Institute of Neuroscience and Technology, Qiushi Academy for Advanced Studies, Zhejiang University, Hangzhou, China;

NHC and CAMS Key Laboratory of Medical Neurobiology, Mental Health Center, Zhejiang University, Hangzhou, China

Shaohua Hu Department of Psychiatry, First Affiliated Hospital, Zhejiang University School of Medicine, Hangzhou, China

Xiaobai Li The Department of Psychiatry, The First Hospital of China Medical University, Shenyang, China

Xiaohua Liu Division of Mood Disorders, Shanghai Mental Health Center, Shanghai Jiao Tong University School of Medicine, Shanghai, China

Xiancang Ma The First Affiliated Hospital of Xi'an Jiaotong University, Xi'an, People's Republic of China

Rongrong Mao Department of Pathology and Pathophysiology, School of Basic Medical Science, Kunming Medical University, Kunming, China

Ruizhi Mao Division of Mood Disorders, Shanghai Mental Health Center, Shanghai Jiao Tong University School of Medicine, Shanghai, China

Zhiang Niu Division of Mood Disorders, Shanghai Mental Health Center, Shanghai Jiao Tong University School of Medicine, Shanghai, China

Daihui Peng Division of Mood Disorder, Shanghai Mental Health Center, School of Medicine, Shanghai Jiao Tong University, Shanghai, China

Han Rong Department of Psychiatry, Shenzhen Kangning Hospital, Shenzhen, Guangdong, China

Huaning Wang Department of Psychiatry, Xijing Hospital, Fourth Military Medical University, Xi'an, Shaanxi, China

Qi Wang The Department of Psychiatry, The First Hospital of China Medical University, Shenyang, China

Zuowei Wang Hongkou District Mental Health Center, Shanghai, China

Zhiguo Wu Division of Mood Disorders, Shanghai Mental Health Center, Shanghai Jiao Tong University School of Medicine, Shanghai, China;
Department of Psychiatry and Neuropsychology, Shanghai Deji Hospital, Qingdao University, Shanghai, China

Chunlan Xiao Hongkou District Mental Health Center, Shanghai, China

Lin Xu Key Laboratory of Animal Models and Human Disease Mechanisms and Laboratory of Learning and Memory, Kunming Institute of Zoology, The Chinese Academy of Sciences, Kunming, China

Lu Yang Division of Mood Disorders, Shanghai Mental Health Center, Shanghai Jiao Tong University School of Medicine, Shanghai, China

Zhijian Yao Department of Psychiatry, The Affiliated Brain Hospital of Nanjing Medical University, Nanjing, China

Chen Zhang Division of Mood Disorders, Shanghai Mental Health Center, Shanghai Jiao Tong University School of Medicine, Shanghai, China

Qinting Zhang Shanghai Key Laboratory of Forensic Medicine, Shanghai Forensic Service Platform, Academy of Forensic Science, Shanghai, China

Xia Zhang Departments of Psychiatry and Cellular and Molecular Medicine, University of Ottawa, Institute of Mental Health Research at the Royal, Ottawa, ON, Canada

Guoqing Zhao Department of Psychology, Provincial Hospital Affiliated to Shandong University, Jinan, China

Chapter 1
Introduction

Yiru Fang and Ruizhi Mao

Abstract Depressive disorder is one of the most widespread forms of mental disorders which lead to a significant public health concern, such as disability, suicide, and so on. Its etiology remains vague but it is believed that depressive disorder is a multifactorial disease which is induced by the interaction of social, psychological, and biological factors. Thus, there is no clear and definite pathological theory could illustrate its mechanism independently until now, involving genetics, neuroimaging, neuroinflammation, neuroendocrine, and others. Comprehensive assessment to patients with depression is the starting point for a right diagnosis. History-taking of physical condition is as important as psychiatric interview and rational usage of scales would be beneficial for screening. There are many kinds of therapeutic measures for depressive patients nowadays, including general intervention, pharmacotherapy, psychotherapy, and physical therapy. For now, anti-depressants used in clinical practice is almost monoamine-based drugs while much more progress have been made in developing new antidepressant medications, like prototypical N-methyl-D-aspartate (NMDA) receptor antagonists, opioid agonists, gamma-aminobutyric acid ($GABA_A$) receptors, and psychedelics. Once these novel drugs are proved to be practicable, it will create a historical evolution in the field of psychiatry. In addition, we advocate that measurement-based care (MBC) should run through the whole duration of treatment and goals of MBC in every stage are different. As brain projects in many countries are conducting in inspiring ways, we believe that our understanding about depressive disorder, of course, and other neuropsychiatric disorders will be better in the future.

Y. Fang (✉)
Clinical Research Center and Division of Mood Disorders, Shanghai Mental Health Center, Shanghai Jiao Tong University School of Medicine, Shanghai, China
e-mail: yirufang@aliyun.com

CAS Center for Excellence in Brain Science and Intelligence Technology, Shanghai, China

Shanghai Key Laboratory of Psychotic Disorders, Shanghai, China

R. Mao
Division of Mood Disorders, Shanghai Mental Health Center, Shanghai Jiao Tong University School of Medicine, Shanghai, China
e-mail: maomaoaofei@163.com

© Springer Nature Singapore Pte Ltd. 2019
Y. Fang (ed.), *Depressive Disorders: Mechanisms, Measurement and Management*, Advances in Experimental Medicine and Biology 1180,
https://doi.org/10.1007/978-981-32-9271-0_1

Keywords Depressive disorder · Etiology · Diagnosis · Screening · Treatment

1.1 Introduction

Depressive disorder is characterized by the presence of sad, diminishing of interest or pleasure, irritable mood, along with somatic and cognitive changes (Battle 2013; Zhao et al. 2018). According to the data from WHO (World Health Organization) in 2015, more than 300 million people were suffering from depression, taking up 4.4% of the global population. Depression can result in the death of patients and has been the second leading cause of death among 15–29 years old (http://www.who.int/healthinfo/global_burden_disease/estimates). Sofar, there is no well-established theory that can illustrate the pathogenesis of depressive disorder completely. Efforts of different aspects has been made to get a better understanding about the mechanism of this mental disease, e.g., studies of genetics, neuroimaging, neuroinflammation, neuroendocrine, biological rhythm and studies from other perspectives. In the process, animal models become crucial tools for gaining more knowledge of the pathophysiology and testing new antidepressant medications (Czeh et al. 2016). When extrapolating findings of animal studies to human, we should be cautious that animal models of depression are not complete analogues of clinical patients. Lacking of effective and efficient biomarkers for identifying depressive disorder, clinical assessment still is the uppermost methods for diagnosis. Though the diagnostic criteria vary from country to country, there are few differences between them. To date, two main diagnostic systems are the Diagnostic and Statistical Manual of Mental Disorder (DSM) and the International Classification of Diseases (ICD) systems (Battle 2013; World Health Organization 1992). Once a patient diagnosed as depressive disorder, systematic and proper treatment strategies are necessary, e.g., pharmacotherapy, psychotherapy, physical therapy.

In this chapter, we provide a brief overview about depressive disorder with the current evidence, including pathological mechanism, diagnosis, screening, treatment, and measurement. In the end, we outline the conditions of brains projects which could provide the technological support for understanding depressive disorder in the future.

1.2 Etiology

Despite the fact that some breakthrough discoveries have been made in recent years, the underlying mechanism of depressive disorder remains elusive. Generally, the consensus had been formed that depressive disorder is a multifactorial disease and it is generated by the complicated interactions between social, psychological, and biological factors. As a result, it is inadvisable and unilateral to use one theory to explain the etiology of depressive disorder.

Genetics

Up to now, family and twin researches have drawn a firm conclusion that genetic factors play crucial roles in the pathogenetic process of depressive disorder (McGuffin et al. 2007; Sullivan et al. 2000). Many approaches have been used to make it clear about the relationship between the mental disease and genes, including candidate genes, genome-wide association analysis, whole-genome sequencing. For example, initial investigations of candidate genes focused on analyzing polymorphisms in genes regarding monoamine neurotransmission. These target genes include gene for serotonin transporter (SCL6A4), genes for serotonin (5-HTR1A, 5-HTR2A, 5-HTR1B, 5-HTR2C), and dopamine receptors (D3D4), genes for the enzymes monoamine oxidase A (MAOA) and so on. From these studies, it seems that these genes are related with depressive disorder to some extent but the association is not robust. Results of meta-analyses were controversial, highlighting the lack of replication studies (Clarke et al. 2010; Hu et al. 2017; Zhou et al. 2018). Genome-wide methods revealed that depressive disorder is highly polygenic and involves many hundreds of genes without a principal gene playing a main role (Shadrina et al. 2018). This leads to the heterogeneous phenotypes of the disease and drives up the number of sample size needed to find out significant genetic associations with depressive disorder in the future studies(Major Depressive Disorder Working Group of the Psychiatric 2013). Therefore, it needs a global coalition to overcome the tough obstacle and set up a database based on big data platform. At the same time, the intergenic interactions cannot be ignored and new methods are needed for further exploration. Another complex layer to the pathophysiology of mood disorder is the interaction between genes and environmental factors (GxE). Epigenetic modifications provide a possible mechanism to explain GxE interaction. These epigenetic changes, without inducing DNA sequence change, primarily involve DNA methylation, histone modifications, and non-coding RNA. The epigenetic studies in depression need replication in animal models and human as well. Animal researches are useful and relatively easy to control environmental exposure while the human trails allow insight into the human-specific biological processes. In addition, integrating epigenomic data with other data, like transcriptomic, proteomic and metabolomic data may enable a better understanding of depressive disorder (Dalton et al. 2014; Januar et al. 2015).

Neuroimaging

Recent advances in neuroimaging methods have made a considerable progress for confirming the brain structural and functional abnormalities of depressive disorder. Structural imaging technologies are computed tomography (CT), magnetic resonance imaging (MRI) while functional imaging technologies contain functional MRI (fMRI), positron emission tomography (PET), magnetic resonance spectroscope (MRS), and near-infrared spectroscopy (NIRS). MRI has the predominant superiority in visualizing brain morphology and function in vivo because of its high spatial resolution and non-invasion (Blamire 2008). Structural MRI studies have been conducted a lot in depressive population, mainly focusing on gray matter volume, cortical thickness, and integrity in white matter (Grieve et al. 2013; Han et al. 2014; Lake et al. 2016). Evidence from MRI studies supports that hippocampus, prefrontal cortex, amygdale, and insula are involved in the pathological alteration of major depressive disorder (MDD) (Hagan et al. 2015; Malykhin and Coupland 2015; Stratmann et al. 2014). Diffusion tensor imaging (DTI) is a MRI approach for detecting microstructure of white matter in neural tissue (Sexton et al. 2009). White-matter integrity of MDD patients reduces significantly in the corpus callosum, inferior fronto-occipital fasciculus, left superior longitudinal fasciculus, and anterior cingulate-limbic system (Korgaonkar et al. 2014; Ota et al. 2015). Abnormity in frontal and fronto-subcortical areas may account for some executive, behavioral, and emotional deficits (Sugimoto et al. 2018). There are also abnormal functional connections between regions, including default mode network, prefrontal-limbic-thalamic, anterior cingulate cortex (ACC)-thalamus, ACC-insula (Gong and He 2015). Either connectivity or activation changes of depressive patients involve in multiple brain networks, such as affective-salience circuit, the fronto-parietal cognitive control circuit, and the medial prefrontal-media parietal default mood network (Otte et al. 2016). Both structural and functional disruption exist in patients with depressive disorder, but the relationship between two of them remains unclear (Bi and He 2014; Wang et al. 2015). Further multimodal imaging studies are needed to clarify the topological association between the structural abnormality and dysfunction. Discoveries from PET and other functional studies reveal that there are changes in regional cerebral blood flow, aerobic glycolysis, and cerebral metabolic rate of glucose as well in brains of depressive patients (Bullmore and Sporns 2012; Liang et al. 2013; Tomasi et al. 2013). All of these results indicate that a complex network, involving neurochemical and physiological dysfunction, underlies the pathology of depressive disorder.

Neuroimmunology

Inflammatory activation is considered as a kind of survival mechanism for self-protection, which are self-defensive responses of our body to various traumatic stimuli. While the "cytokine hypothesis" of depression have been raised based on the observation that depressive individuals show elevated plasma cytokines compared with healthy subjects (Kohler et al. 2017; Mao et al. 2018; Zhang et al. 2016). Data supports that crosstalk between central nervous system and inflammatory pathways are related to the development of depression (Kiecolt-Glaser et al. 2015; Miller and Raison 2016). There are several pathways which inflammatory signals can be transferred from the periphery environment to the brain. For instance, peripheral cytokines could cross the weak regions of the blood–brain barrier (BBB), such as the circumventricular organs. Or they can be transferred by specific transporters or bind to peripheral afferent nerve fibers (Eyre and Baune 2012). Through interacting with neurotransmitters or neurocircuitry related to mood regulation, these inflammatory factors could drive the clinical symptoms of depressive disorders (Capuron et al. 2012; Eisenberger et al. 2010; Felger et al. 2013). Meanwhile, depression and cytokines may be a bi-directional association. Depression, as a self-sustaining stressor, may induce a systemic stress response to produce pro-inflammatory cytokines. Copelan et al. confirmed it and they found multiple depressive episodes exerted more robust effect on later C-reactive protein (CRP) levels (Copeland et al. 2012). CRP is a risk factor for several chronic diseases, such as metabolic or cardiovascular diseases (Emerging Risk Factors Collaboration 2010; Pradhan et al. 2001). Hence, they raised the question whether emotional dysfunction in childhood is related to the risk of middle-age chronic diseases.

Because anti-inflammatory treatment presents anti-depressant effects in patients with physical diseases who showed elevated peripheral inflammation, for example, individuals with osteoarthritis or psoriasis (Kohler et al. 2014; Raison et al. 2013). It holds promise that some classes of anti-inflammatory medications may serve as novel antidepressants. However, results from a study of mice and humans should be noted. Warner-Schmidt et al. reported that serotonergic antidepressant drugs may increase levels of certain cytokines (e.g., TNF-α and INF-γ) in frontal cortex and their antidepressant effects may be attenuated by anti-inflammatory agents. Then the results of mouse studies were confirmed in the Sequenced Treatment Alternatives to Relieve Depression (STAR*D) dataset that nonsteroidal anti-inflammatory drugs (NSAIDs) or other analgesic treatments reduced patients' response to citalopram (Warner-Schmidt et al. 2011). Given discoveries above, the use of anti-inflammatory medications on depressive patients still needs a prudent attitude. Clinicians should reflect on the therapeutic benefits of anti-inflammatory treatment versus its consequences of antagonizing the efficacy of antidepressants.

Biologicalrhythm

Biological rhythm patterns in human body consist of endogenous and exogenous rhythm. Endogenous clock is an internal time-keeping (approximately 24 h) to adapt to the light-dark cycle. The internal cycle interacts and synchronizes itself with the external stimuli, which lead to the exogenous rhythm (Orozco-Solis and Sassone-Corsi 2014). Sleep disturbance often exists in patients with depression, which presents as sleepiness during the day and insomnia at night (Srinivasan et al. 2009). The disruption of the sleep/wake rhythm may cause alterations of circadian rhythms in hormones, metabolites associated with sleep time. As the principal transducer of the photoperiodic signal, alteration in melatonin secretion may desynchronize endogenous rhythms. Nocturnal melatonin release often decreases in depressive individuals and it may be related to the patients' sleep disturbance (Soria and Urretavizcaya 2009). Depressive patients may also present increased plasma levels of cortisol, a hormone related to stressful conditions. The symptom of anhedonia and the metabolic alterations may be associated with the disturbed circadian rhythm of cortisol (Salgado-Delgado et al. 2011). Seasonal affective disorder (SAD) is a recurrent illness that patients experience depression during fall and winter most frequently and get remission or shift into hypo/mania during spring and summer. SAD is another evidence that circadian rhythm plays an important role in the pathology of depression. In fall and winter, day time is shorter than night. Light deficiency can cause negative mood, sleepiness during the day, and sleeplessness at night. Because of the low luminosity, some neurotransmitters, like dopamine and serotonin, are not secreted enough in some brain regions (Nutt 2008). To rectify the circadian rhythm disruption caused by light deficiency, bright light therapy (BLT) have been used to treat seasonal and non-seasonal depression with effect sizes of 0.84 and 0.53 (Golden et al. 2005).

Animal Models

Unlike plaques and tangles observed in Alzheimer's disease (AD), it is hard for depression to find specific clues in the brain to focus on. Plenty of preclinical and clinical trials have been performed in depressive disorder within these years. As for humans, findings of postmortem brains provide the direct evidence for probing neuropsychiatric diseases. Nevertheless, the source of postmortem brains is rare and it is difficult to establish a dynamic observation in postmortem samples. Clinical trials are good choices to produce dynamic pictures of the brain in humans but easy to raise ethical issues when encountering invasive manipulation or novel drug testing. Therefore, animal models may provide another source of insight into deciphering depressive disorders. Various animal models have been used to elucidate pathological mechanism that underlies depression and test new antidepressants. The most well-known genetic models are "knock-out" and "knock-in" mice which enable the observation of the functional consequences of disrupting a specific target gene. Of course, the traditional genetic models also have limitations. The effects of these genetic manipulations influence the entire body and the target proteins (e.g., receptor or transporter) are limited. To resolve the above problems, optogenetic tools become attractive in the neuroscience field for its high anatomical and temporal precision (Gautier et al. 2014; Packer et al. 2013). Besides genetic models, models based on different types of stress exposure are often used as surrogates of depressive disorder, such as social defeat stress models, chronic mild stress models, learned helpless paradigm, and many other models (Hollis and Kabbaj 2014; Slattery and Cryan 2014). Though animal models have contributed a lot in the field of neuroscience, but for now, there is no perfect models could mimic the complicated mental disease completely. For example, cognitive impairment is common in depressive patients, while it is not addressed in preclinical studies (Mathews and MacLeod 2005). As animal experiments cannot simulate the biological process in human body identically. Therefore, transferring the preclinical discoveries to clinical practice still need to be more cautious and put more efforts in the future.

1.3 Diagnosis and Measurement

Diagnosis

Diagnosis for depressive disorder depends on assessment of clinical manifestation, course of disease, physical examination, and laboratory testing. Explicit diagnostic criteria are important for defining patients who would be more likely to benefit from antidepressant therapy. Two diagnostic systems, DSM and ICD, are widely used throughout the world (Battle 2013; World Health Organization 1992). There are a few of differences about depressive disorders between these two diagnostic systems though they are similar generally. As for the description of depression, both of them consider "depressed mood" as the cardinal symptom. DSM-5 points out that the mood does not match the current plight while ICD-10 has no specific illustration. With regard to course of diseases, DSM-5 prescribes that every discrete episode lasts for more than 2 weeks, involving clear-cut changes in inter-episode remissions, affect, cognitions, and neurovegetative functions (Battle 2013). In ICD-10, it is characterized by the whole episode of at least 2 weeks' duration. For recurrent depressive patients, DSM-5 stipulates that remission experience is a period of 2 or more months with no symptoms, or one or two mild symptoms. ICD-10 also requires several months of remission between two depressive episodes but there is no clear description about the mental state in remission. Depressive disorders are divided into mild, moderate and severe degrees according to the severity of the disease in ICD-10. While the severity of symptom and social functional impairment is not included in the DSM-5 criteria.

Through the joint efforts of experts from WHO over a decade, the new ICD-11 was launched in June 18, 2018 which is an advance preview. It will come into effect on January 1, 2022 (http://www.who.int/news-room/detail/18-06-2018-who-releases-new-international-classification-of-diseases-(ICD-11). The ICD-11 is completely electronic and much more friendly for clinical users. There are some modifications in the ICD-11 draft about depressive disorder compared to ICD-10. For example, common expression of grief can include depressive symptoms but do not meet a depressive episode diagnosis. Nevertheless, a depressive episode could be superimposed on common grief (Bolton et al. 2016; Sampogna 2017). With the publication of DSM-5, there are also some changes in the new edition. "Depressive disorders" has been existing as an independent chapter, instead of being a part of "Mood Disorders" in the previous editions. Compared to DSM-IV, the content of depressive disorders becomes more detailed in DSM-5. They include disruptive mood dysregulation disorder, major depressive disorder (including major depressive episode), persistent depressive disorder (dysthymia), premenstrual dysphoric disorder, substance/medication-induced depressive disorder, depressive disorder due to another medical condition, other specified depressive disorder, and unspecified depressive disorder. In addition, major depressive disorder is supplemented with some specifiers, such as with anxious distress, with mixed features, with melancholic features, with atypical features and so on (Battle 2013). It may provide more narrowly defined entities for heterogeneous depressive disorders, which could help prompt the etiological exploring about the disease. Some studies have revealed that different subtypes of MDD present different pathological features, like melancholic versus atypical depression (Korte et al. 2015; Rudolf et al. 2014). Specific subtype of depressive disorders may correspond to a specific etiology and a specific treatment strategy. A great job is on the way which is supported by the National Key Research and Development Program of China (2016YFC1307100). Experts from China are devoting themselves to figure out multiple dimensional diagnosis and individual treatment for different types of depressive disorders (NCT01764867).

Screening and Measurement

Screening could help identify undiagnosed patients with depression, which could shorten around 4 years between depression onset and the first-time treatment. It may also be beneficial to find out patients with suicidal risk among those screening positive (O'Connor et al. 2016). Some self-rating scales can be used as screening tools and evaluating the severity of clinical symptoms for depressive disorder. These instruments include the Patient Health Questionnaire-9 (PHQ-9), Self-rating Depression Scale (SDS), Beck depression inventory (BDI), and Quick Inventory of Depressive Symptomatology (QIDS-SR) (Beck et al. 1961; Rush et al. 2003; Zuithoff et al. 2010; Zung et al. 1965). In particularly, PHQ-9 is a preferred tool used in most primary care. It is a multipurpose scale to assess the depressive symptoms within the last 14 days (Kroenke et al. 2001). Chibanda et al. reported that its sensitivity was 85% (95% CI: 78–90%) and specificity was 69% (95% CI: 59–77%) for identifying depressive and anxiety disorders (Chibanda et al. 2016). Some other-report questionnaires are also widely used in clinical practice, including Hamilton Depression Scales (HAMD) and Montgomery Depression Rating Scale (MADRS) (Hamilton 1960; Montgomery and Asberg 1979).

The concept of measurement-based care (MBC) has become much more popular nowadays in the field of psychiatry. It means that systematically using measurement tools to monitor progress and guide treatment choice (Crismon et al. 1999). It allows psychiatrists to make individual treatment decisions for every patient according to their response to antidepressant (Trivedi et al. 2006). Compared to standard treatment (clinicians' choice decision), it seems significantly easier and more efficient to achieve response and remission for patients treated along with MBC (Guo et al. 2015). The factors of MBC for depressive disorders contain the severity of depression, tolerance and dose of antidepressants, treatment compliance. Meanwhile, physical examination is as important as psychiatric test for depressive patients in the process of MBC. Because evidence from meta-analysis of some large-scale longitudinal studies suggested that MDD could increase the risk of somatic disorders, e.g., diabetes mellitus, hypertension, heart disease, stroke, and cancer (Penninx et al. 2013). Many patients with physical diseases may present depressive symptoms as well. For instance, clinically significant depression is present in one of every four people with type 2 diabetes mellitus (T2DM) (Semenkovich et al. 2015). As a result, comprehensive measurement is necessary for patients with depression, which is integration of biological, phycological, and sociological assessment. The goal of MBC varies from different stages. In acute stage, the key point of evaluation is changes and severity of depression, possibility of suicide and adverse effects of drugs. For patients in consolidate period, the main task is to assess patients' response to treatment, compliance, social functions systematically, and to monitor whether disease is likely to relapse. During maintenance phase, regular measurement needs to be carried out as well, especially for people at high risk of recurrence.

1.4 Treatment and Management

In fact, there are a wide range of options for treating depressive patients (Table 1.1). The National Institute for Health and Care Excellence (NICE) provides a stepped-care model for patients, caregivers, and clinical practitioners when they are making decision for treatment strategy (Fig. 1.1) (https://www.nice.org.uk/guidance/cg90/chapter/1-Guidance#stepped-care). Following this framework, patients with depression may be provided with the most efficient and least intrusive interventions first. If the outcome is unsatisfactory, it is better to offer the next step interventions. Currently, antidepressants drugs are still one of the main therapeutic measures for moderate-to-severe depressive episodes.

Close to 6 decades of efforts have not made them more efficient to lift patients out of depression. At least a third of patients with depression do not response to the monoaminergic medications, which has been proved by the STAR*D study (Rush et al. 2006). Emerging evidence revealed that antidepressant drugs do not just work in the roughly simple way, changing the concentration of serotonin, norepinephrine, or dopamine in the synaptic cleft. The effect of antidepressants on monoamine signaling pathway leads to a complicated sequence of neuronal adaptive changes on molecular and cellular levels. These changes can induce strengthened synaptic efficacy and connectivity by activating specific genes linked to neural plasticity (Conti et al. 2007; Gaska et al. 2012; Sillaber et al. 2008). The net effect of the changes may also switch neural networks into a more immature developmental state, which

Table 1.1 Treatment modalities for depressive disorders

General intervention	Spiritual support
	Good nutrition
	Quality sleep
	Regular exercise
	Relaxation training
Pharmacotherapy	Antidepressant medications
	Antidepressant medications augmentations
Psychotherapy	Supportive psychotherapy
	Cognitive behavior therapy
	Psychodynamic psychotherapy
	Interpersonal psychotherapy
	Marriage and family therapy
Physical therapy	Electric convulsive therapy
	Repetitive transcranial magnetic stimulation treatment
	Deep brain stimulation
	Vagus nerve stimulation

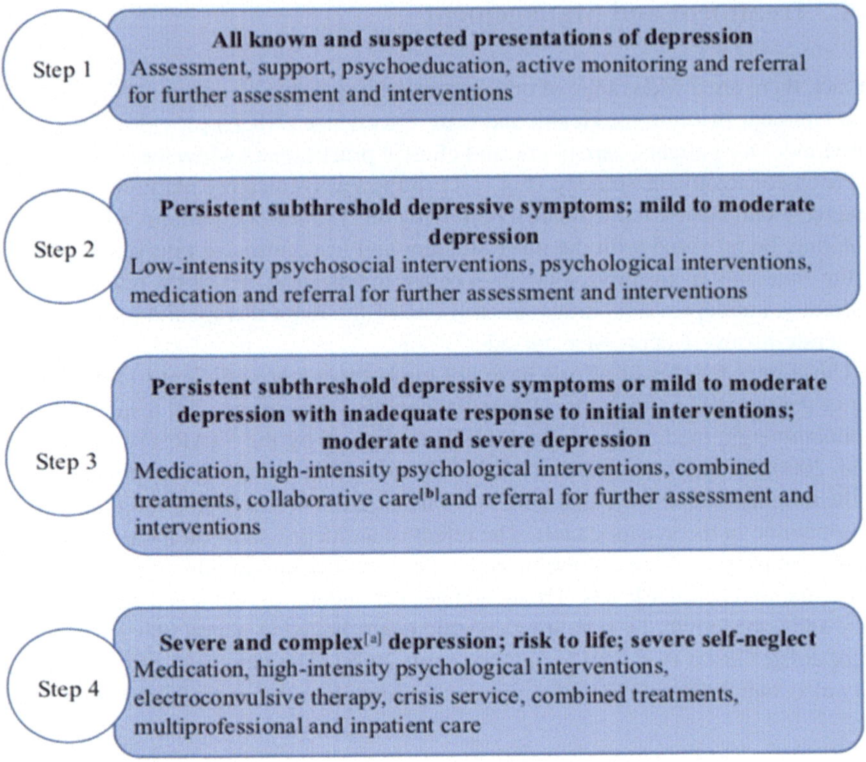

Fig. 1.1 The stepped-care model for adult patients with depression. This model provides a framework to guide patients, caregivers and clinical practitioners for their treatment decisions. **a** Complex depression includes depression that shows an inadequate response to multiple treatments, is complicated by psychotic symptoms, and/or is associated with significant psychiatric comorbidity or psychosocial factors. **b** Only for depression where the person also has a chronic physical health problem and associated functional impairment

could reactivate developmental-like plasticity in some crucial brain regions. The key circuits within the limbic system are more easily restructured by incoming emotional relevant stimulation (Karpova et al. 2011; Sharp 2013). During the last 3 decades, plenty of work on novel antidepressant medications that are outside of the monoamine hypothesis, have been done. The most remarkable potential "next-generation" antidepressant is ketamine which is one of the prototypical N-methyl-D-aspartate (NMDA) receptor antagonists (Schatzberg 2014). One continued frustration and clinical limitation of our current antidepressants that follow the monoamine mechanism is that it can take from 2 to 8 weeks to achieve an adequate antidepressant effect. Compared to monoamine-based drugs, ketamine shows a rapid antidepressant effects within 2–4 h (Machado-Vieira et al. 2010; Williams and Schatzberg 2016). Among treatment-resistant depression (TRD) patients specifically, multiple trials have illustrated that ketamine infusions could induce acute reductions in depression comparing with placebos (Cusin et al. 2017; Diamond et al. 2014; Singh et al. 2016). At present, the clinical use of ketamine is off-label. There is still much concern about its optimal dose, duration of treatment, adverse effects, and ethical issues. Besides NMDA antagonists, different novel paradigms for addressing TRD are in development, such as opioid agonists, gamma-aminobutyric acid ($GABA_A$) receptors, and psychedelics. However, without clear clues of how depression is produced in the brain, those drugs that may counteract the pathology cannot be sought out.

As the development of technology, intelligent electronic devices including mobile phones, tablets, and smart watches have become an essential part of daily life in modern society. Additionally, these devices are equipped with powerful sensing, computations, and other functional systems, specifically smartphones. These platforms have the potential to shift the current management patterns for depressive disorders, known as behavioral intervention technologies (BIT) (Mohr et al. 2013). They could offer the opportunity to monitor depressive patients' behaviors, such as reduced activity, psychomotor retardation, sleeping conditions continuously (Saeb et al. 2015). They allow patients to input information about their inter states anywhere and anytime to meet a person's immediate needs (Burns et al. 2011). These devices are also mediums carrying intervention paradigm such as web-based CBT. As BITs can store large amounts of data, analysis and visualization of big data make sense. And how to make the full use of personal digital instruments to help clinical treatments needs deeper explorations. We believe that patients' lives may evolve through a rapid stream of updates, as they are using their smartphones or other devices.

1.5 Outlook

In the field of natural science, mysteries about human brains raise a grand challenge as well as those of the universe. As one of the most severe and common mental disorders throughout the world, depressive disorder still remains a secret to people, but we keep an optimistic attitude to its future. Because many countries are proceeding their ambitious neuroscience projects with great enthusiasm and they are trying to find clues about human brains.

The Brain Research through Advancing Innovative Neurotechnologies (BRAIN) Initiative was announced by US president in 2013 (https://blog.csdn.net/gaohanggaolegao/article/details/51867316). In the past 5 years, the BRAIN Initiative focused on developing methods for monitoring, manipulating and computing neural activities. For example, the Cell Census Consortium, one of the ten pilot projects in the U.S. BRAIN Initiative are putting large and long-term effort to create whole-brain cell atlases in different species including humans. Understanding the cell types that comprise the neural circuits could help to elucidate how neural circuits process and integrate signals to induce behavior, emotion, cognition (Ecker et al. 2017; Lake et al. 2016). As it is mentioned above that animal models could not represent human completely, patient-derived stem cell may be another promising tool for human studies (Camp et al. 2015). Of course, cell atlas of healthy human brain could be also useful as a reference to calibrate results from patient-derived stem cell models.

Next-generation neuroimaging technologies are also developing our understanding about neural circuit as well as new noninvasive neuromodulation technologies, such as high-resolution fMRI for cortical column activation, molecular fMRI and magnetic resonance angiography (Bartelle et al. 2016; Feinberg et al. 2018; Lu et al. 2016). Other countries have also initiated their brain projects in succession, including the Chinese Brain Project, the Japanese Brain/MINDS project, the European Human Brain Project, and others. One of the Chinese Brain Project's main tasks is studying on the neuromechanism, early diagnosis, and treatment of developmental diseases (e.g., AD, depressive disorder, autistic disorder mental retardation) by using omics-based approaches, induced pluripotent stem cells and other advancing technologies. And through cooperation between neuroscientists and researchers of artificial intelligence, the Chinese Brain Project may get more knowledge about human brains in different profiles.

With these brain projects conducting, could we overcome the problems about depressive disorder or other mental diseases in the end? The answer is not quite sure. But we hold the promise that these projects may provide more clues for knowing this mood disorder better and deeper. These cues are still a far reach. More investment and patience are needed.

References

Bartelle BB, Barandov A, Jasanoff A (2016) Molecular fMRI. J Neurosci 36:4139–4148

Battle DE (2013) Diagnostic and statistical manual of mental disorders (DSM). Codas 25:191–192

Beck AT, Ward CH, Mendelson M, Mock J, Erbaugh J (1961) An inventory for measuring depression. Arch Gen Psychiatry 4:561–571

Bi Y, He Y (2014) Connectomics reveals faulty wiring patterns for depressed brain. Biol Psychiatry 76:515–516

Blamire AM (2008) The technology of MRI—the next 10 years? Br J Radiol 81:601–617

Bolton JM, Au W, Chateau D, Walld R, Leslie WD, Enns J, Martens PJ, Katz LY, Logsetty S, Sareen J (2016) Bereavement after sibling death: a population-based longitudinal case-control study. World Psychiatry 15:59–66

Bullmore E, Sporns O (2012) The economy of brain network organization. Nat Rev Neurosci 13:336–349

Burns MN, Begale M, Duffecy J, Gergle D, Karr CJ, Giangrande E, Mohr DC (2011) Harnessing context sensing to develop a mobile intervention for depression. J Med Internet Res 13:e55

Camp JG, Badsha F, Florio M, Kanton S, Gerber T, Wilsch-Brauninger M, Lewitus E, Sykes A, Hevers W, Lancaster M, Knoblich JA, Lachmann R, Paabo S, Huttner WB, Treutlein B (2015) Human cerebral organoids recapitulate gene expression programs of fetal neocortex development. Proc Natl Acad Sci USA 112:15672–15677

Capuron L, Pagnoni G, Drake DF, Woolwine BJ, Spivey JR, Crowe RJ, Votaw JR, Goodman MM, Miller AH (2012) Dopaminergic mechanisms of reduced basal ganglia responses to hedonic reward during interferon alfa administration. Arch Gen Psychiatry 69:1044–1053

Chibanda D, Verhey R, Gibson LJ, Munetsi E, Machando D, Rusakaniko S, Munjoma R, Araya R, Weiss HA, Abas M (2016) Validation of screening tools for depression and anxiety disorders in a primary care population with high HIV prevalence in Zimbabwe. J Affect Disord 198:50–55

Clarke H, Flint J, Attwood AS, Munafo MR (2010) Association of the 5-HTTLPR genotype and unipolar depression: a meta-analysis. Psychol Med 40:1767–1778

Conti B, Maier R, Barr AM, Morale MC, Lu X, Sanna PP, Bilbe G, Hoyer D, Bartfai T (2007) Region-specific transcriptional changes following the three antidepressant treatments electro convulsive therapy, sleep deprivation and fluoxetine. Mol Psychiatry 12:167–189

Copeland WE, Shanahan L, Worthman C, Angold A, Costello EJ (2012) Cumulative depression episodes predict later C-reactive protein levels: a prospective analysis. Biol Psychiatry 71:15–21

Crismon ML, Trivedi M, Pigott TA, Rush AJ, Hirschfeld RM, Kahn DA, DeBattista C, Nelson JC, Nierenberg AA, Sackeim HA, Thase ME (1999) The Texas medication algorithm project: report of the Texas consensus conference panel on medication treatment of major depressive disorder. J Clin Psychiatry 60:142–156

Cusin C, Ionescu DF, Pavone KJ, Akeju O, Cassano P, Taylor N, Eikermann M, Durham K, Swee MB, Chang T, Dording C, Soskin D, Kelley J, Mischoulon D, Brown EN, Fava M (2017) Ketamine augmentation for outpatients with treatment-resistant depression: preliminary evidence for two-step intravenous dose escalation. Aust N Z J Psychiatry 51:55–64

Czeh B, Fuchs E, Wiborg O, Simon M (2016) Animal models of major depression and their clinical implications. Prog Neuropsychopharmacol Biol Psychiatry 64:293–310

Dalton VS, Kolshus E, McLoughlin DM (2014) Epigenetics and depression: return of the repressed. J Affect Disord 155:1–12

Diamond PR, Farmery AD, Atkinson S, Haldar J, Williams N, Cowen PJ, Geddes JR, McShane R (2014) Ketamine infusions for treatment resistant depression: a series of 28 patients treated weekly or twice weekly in an ECT clinic. J Psychopharmacol 28:536–544

Ecker JR, Geschwind DH, Kriegstein AR, Ngai J, Osten P, Polioudakis D, Regev A, Sestan N, Wickersham IR, Zeng H (2017) The BRAIN initiative cell census consortium: lessons learned toward generating a comprehensive brain cell atlas. Neuron 96:542–557

Eisenberger NI, Berkman ET, Inagaki TK, Rameson LT, Mashal NM, Irwin MR (2010) Inflammation-induced anhedonia: endotoxin reduces ventral striatum responses to reward. Biol Psychiatry 68:748–754

Emerging Risk Factors Collaboration (2010) C-reactive protein concentration and risk of coronary heart disease, stroke, and mortality: an individual participant meta-analysis. Lancet 375:132–140.

Eyre H, Baune BT (2012) Neuroplastic changes in depression: a role for the immune system. Psychoneuroendocrinology 37:1397–1416

Feinberg DA, Vu AT, Beckett A (2018) Pushing the limits of ultra-high resolution human brain imaging with SMS-EPI demonstrated for columnar level fMRI. Neuroimage 164:155–163

Felger JC, Li L, Marvar PJ, Woolwine BJ, Harrison DG, Raison CL, Miller AH (2013) Tyrosine metabolism during interferon-alpha administration: association with fatigue and CSF dopamine concentrations. Brain Behav Immun 31:153–160

Gaska M, Kusmider M, Solich J, Faron-Gorecka A, Krawczyk MJ, Kulakowski K, Dziedzicka-Wasylewska M (2012) Analysis of region-specific changes in gene expression upon treatment with citalopram and desipramine reveals temporal dynamics in response to antidepressant drugs at the transcriptome level. Psychopharmacology 223:281–297

Gautier A, Gauron C, Volovitch M, Bensimon D, Jullien L, Vriz S (2014) How to control proteins with light in living systems. Nat Chem Biol 10:533–541

Golden RN, Gaynes BN, Ekstrom RD, Hamer RM, Jacobsen FM, Suppes T, Wisner KL, Nemeroff CB (2005) The efficacy of light therapy in the treatment of mood disorders: a review and meta-analysis of the evidence. Am J Psychiatry 162:656–662

Gong Q, He Y (2015) Depression, neuroimaging and connectomics: a selective overview. Biol Psychiatry 77:223–235

Grieve SM, Korgaonkar MS, Koslow SH, Gordon E, Williams LM (2013) Widespread reductions in gray matter volume in depression. Neuroimage Clin 3:332–339

Guo T, Xiang YT, Xiao L, Hu CQ, Chiu HF, Ungvari GS, Correll CU, Lai KY, Feng L, Geng Y, Feng Y, Wang G (2015) Measurement-based care versus standard care for major depression: a randomized controlled trial with blind raters. Am J Psychiatry 172:1004–1013

Hagan CC, Graham JM, Tait R, Widmer B, van Nieuwenhuizen AO, Ooi C, Whitaker KJ, Simas T, Bullmore ET, Lennox BR, Sahakian BJ, Goodyer IM, Suckling J (2015) Adolescents with current major depressive disorder show dissimilar patterns of age-related differences in ACC and thalamus. Neuroimage Clin 7:391–399

Hamilton M (1960) A rating scale for depression. J Neurol Neurosurg Psychiatry 23:56–62

Han KM, Choi S, Jung J, Na KS, Yoon HK, Lee MS, Ham BJ (2014) Cortical thickness, cortical and subcortical volume, and white matter integrity in patients with their first episode of major depression. J Affect Disord 155:42–48

Hollis F, Kabbaj M (2014) Social defeat as an animal model for depression. ILAR J 55:221–232

Hu Y, Hong W, Smith A, Yu S, Li Z, Wang D, Yuan C, Cao L, Wu Z, Huang J, Fralick D, Phillips MR, Fang Y (2017) Association analysis between mitogen-activated protein/extracellular signal-regulated kinase (MEK) gene polymorphisms and depressive disorder in the Han Chinese population. J Affect Disord 222:120–125

Januar V, Saffery R, Ryan J (2015) Epigenetics and depressive disorders: a review of current progress and future directions. Int J Epidemiol 44:1364–1387

Karpova NN, Pickenhagen A, Lindholm J, Tiraboschi E, Kulesskaya N, Agustsdottir A, Antila H, Popova D, Akamine Y, Bahi A, Sullivan R, Hen R, Drew LJ, Castren E (2011) Fear erasure in mice requires synergy between antidepressant drugs and extinction training. Science 334:1731–1734

Kiecolt-Glaser JK, Derry HM, Fagundes CP (2015) Inflammation: depression fans the flames and feasts on the heat. Am J Psychiatry 172:1075–1091

Kohler O, Benros ME, Nordentoft M, Farkouh ME, Iyengar RL, Mors O, Krogh J (2014) Effect of anti-inflammatory treatment on depression, depressive symptoms, and adverse effects: a systematic review and meta-analysis of randomized clinical trials. JAMA Psychiatry 71:1381–1391

Kohler CA, Freitas TH, Maes M, de Andrade NQ, Liu CS, Fernandes BS, Stubbs B, Solmi M, Veronese N, Herrmann N, Raison CL, Miller BJ, Lanctot KL, Carvalho AF (2017) Peripheral cytokine and chemokine alterations in depression: a meta-analysis of 82 studies. Acta Psychiatr Scand 135:373–387

Korgaonkar MS, Williams LM, Song YJ, Usherwood T, Grieve SM (2014) Diffusion tensor imaging predictors of treatment outcomes in major depressive disorder. Br J Psychiatry 205:321–328

Korte SM, Prins J, Krajnc AM, Hendriksen H, Oosting RS, Westphal KG, Korte-Bouws GA, Olivier B (2015) The many different faces of major depression: it is time for personalized medicine. Eur J Pharmacol 753:88–104

Kroenke K, Spitzer RL, Williams JB (2001) The PHQ-9: validity of a brief depression severity measure. J Gen Intern Med 16:606–613

Lake BB, Ai R, Kaeser GE, Salathia NS, Yung YC, Liu R, Wildberg A, Gao D, Fung HL, Chen S, Vijayaraghavan R, Wong J, Chen A, Sheng X, Kaper F, Shen R, Ronaghi M, Fan JB, Wang W, Chun J, Zhang K (2016) Neuronal subtypes and diversity revealed by single-nucleus RNA sequencing of the human brain. Science 352:1586–1590

Liang X, Zou Q, He Y, Yang Y (2013) Coupling of functional connectivity and regional cerebral blood flow reveals a physiological basis for network hubs of the human brain. Proc Natl Acad Sci U S A 110:1929–1934

Lu M, Zhu XH, Chen W (2016) In vivo (31) P MRS assessment of intracellular NAD metabolites and NAD(+)/NADH redox state in human brain at 4 T. NMR Biomed 29:1010–1017

Machado-Vieira R, Baumann J, Wheeler-Castillo C, Latov D, Henter ID, Salvadore G, Zarate CA (2010) The timing of antidepressant effects: a comparison of diverse pharmacological and somatic treatments. Pharmaceuticals (Basel) 3:19–41

Major Depressive Disorder Working Group of the Psychiatric (2013) A mega-analysis of genome-wide association studies for major depressive disorder. Mol Psychiatry 18:497–511

Malykhin NV, Coupland NJ (2015) Hippocampal neuroplasticity in major depressive disorder. Neuroscience 309:200–213

Mao R, Zhang C, Chen J, Zhao G, Zhou R, Wang F, Xu J, Yang T, Su Y, Huang J, Wu Z, Cao L, Wang Y, Hu Y, Yuan C, Yi Z, Hong W, Wang Z, Peng D, Fang Y (2018) Different levels of pro- and anti-inflammatory cytokines in patients with unipolar and bipolar depression. J Affect Disord 237:65–72

Mathews A, MacLeod C (2005) Cognitive vulnerability to emotional disorders. Annu Rev Clin Psychol 1:167–195

McGuffin P, Cohen S, Knight J (2007) Homing in on depression genes. Am J Psychiatry 164:195–197

Miller AH, Raison CL (2016) The role of inflammation in depression: from evolutionary imperative to modern treatment target. Nat Rev Immunol 16:22–34

Mohr DC, Burns MN, Schueller SM, Clarke G, Klinkman M (2013) Behavioral intervention technologies: evidence review and recommendations for future research in mental health. Gen Hosp Psychiatry 35:332–338

Montgomery SA, Asberg M (1979) A new depression scale designed to be sensitive to change. Br J Psychiatry 134:382–389

Nutt DJ (2008) Relationship of neurotransmitters to the symptoms of major depressive disorder. J Clin Psychiatry 69(Suppl E1):4–7

O'Connor E, Rossom RC, Henninger M, Groom HC, Burda BU, Henderson JT, Bigler KD, Whitlock EP (2016) Screening for depression in adults: an updated systematic evidence review for the U.S. preventive services task force, Rockville, MD

Orozco-Solis R, Sassone-Corsi P (2014) Circadian clock: linking epigenetics to aging. Curr Opin Genet Dev 26:66–72

Ota M, Noda T, Sato N, Hattori K, Hori H, Sasayama D, Teraishi T, Nagashima A, Obu S, Higuchi T, Kunugi H (2015) White matter abnormalities in major depressive disorder with melancholic and atypical features: a diffusion tensor imaging study. Psychiatry Clin Neurosci 69:360–368

Otte C, Gold SM, Penninx BW, Pariante CM, Etkin A, Fava M, Mohr DC, Schatzberg AF (2016) Major depressive disorder. Nat Rev Dis Primers 2:16065

Packer AM, Roska B, Hausser M (2013) Targeting neurons and photons for optogenetics. Nat Neurosci 16:805–815

Penninx BW, Milaneschi Y, Lamers F, Vogelzangs N (2013) Understanding the somatic consequences of depression: biological mechanisms and the role of depression symptom profile. BMC Med 11:129

Pradhan AD, Manson JE, Rifai N, Buring JE, Ridker PM (2001) C-reactive protein, interleukin 6, and risk of developing type 2 diabetes mellitus. JAMA 286:327–334

Raison CL, Rutherford RE, Woolwine BJ, Shuo C, Schettler P, Drake DF, Haroon E, Miller AH (2013) A randomized controlled trial of the tumor necrosis factor antagonist infliximab for treatment-resistant depression: the role of baseline inflammatory biomarkers. JAMA Psychiatry 70:31–41

Rudolf S, Greggersen W, Kahl KG, Huppe M, Schweiger U (2014) Elevated IL-6 levels in patients with atypical depression but not in patients with typical depression. Psychiatry Res 217:34–38

Rush AJ, Trivedi MH, Ibrahim HM, Carmody TJ, Arnow B, Klein DN, Markowitz JC, Ninan PT, Kornstein S, Manber R, Thase ME, Kocsis JH, Keller MB (2003) The 16-item quick inventory of depressive symptomatology (QIDS), clinician rating (QIDS-C), and self-report (QIDS-SR): a psychometric evaluation in patients with chronic major depression. Biol Psychiatry 54:573–583

Rush AJ, Trivedi MH, Wisniewski SR, Nierenberg AA, Stewart JW, Warden D, Niederehe G, Thase ME, Lavori PW, Lebowitz BD, McGrath PJ, Rosenbaum JF, Sackeim HA, Kupfer DJ, Luther J, Fava M (2006) Acute and longer-term outcomes in depressed outpatients requiring one or several treatment steps: a STAR*D report. Am J Psychiatry 163:1905–1917

Saeb S, Zhang M, Karr CJ, Schueller SM, Corden ME, Kording KP, Mohr DC (2015) Mobile phone sensor correlates of depressive symptom severity in daily-life behavior: an exploratory study. J Med Internet Res 17:e175

Salgado-Delgado R, Tapia Osorio A, Saderi N, Escobar C (2011) Disruption of circadian rhythms: a crucial factor in the etiology of depression. Depress Res Treat 2011:839743

Sampogna G (2017) ICD-11: draft diagnostic guidelines for mental disorders: a report for WPA membership. Psychiatr Pol 51:397–406

Schatzberg AF (2014) A word to the wise about ketamine. Am J Psychiat 171:262–264

Semenkovich K, Brown ME, Svrakic DM, Lustman PJ (2015) Depression in type 2 diabetes mellitus: prevalence, impact, and treatment. Drugs 75:577–587

Sexton CE, Mackay CE, Ebmeier KP (2009) A systematic review of diffusion tensor imaging studies in affective disorders. Biol Psychiatry 66:814–823

Shadrina M, Bondarenko EA, Slominsky PA (2018) Genetics factors in major depression disease. Front Psychiatry 9:334

Sharp T (2013) Molecular and cellular mechanisms of antidepressant action. Curr Top Behav Neurosci 14:309–325

Sillaber I, Panhuysen M, Henniger MS, Ohl F, Kuhne C, Putz B, Pohl T, Deussing JM, Paez-Pereda M, Holsboer F (2008) Profiling of behavioral changes and hippocampal gene expression in mice chronically treated with the SSRI paroxetine. Psychopharmacology 200:557–572

Singh JB, Fedgchin M, Daly EJ, De Boer P, Cooper K, Lim P, Pinter C, Murrough JW, Sanacora G, Shelton RC, Kurian B, Winokur A, Fava M, Manji H, Drevets WC, Van Nueten L (2016) A double-blind, randomized, placebo-controlled, dose-frequency study of intravenous ketamine in patients with treatment-resistant depression. Am J Psychiatry 173:816–826

Slattery DA, Cryan JF (2014) The ups and downs of modelling mood disorders in rodents. ILAR J 55:297–309

Soria V, Urretavizcaya M (2009) Circadian rhythms and depression. Actas Esp Psiquiatr 37:222–232

Srinivasan V, Pandi-Perumal SR, Trakht I, Spence DW, Hardeland R, Poeggeler B, Cardinali DP (2009) Pathophysiology of depression: role of sleep and the melatonergic system. Psychiatry Res 165:201–214

Stratmann M, Konrad C, Kugel H, Krug A, Schoning S, Ohrmann P, Uhlmann C, Postert C, Suslow T, Heindel W, Arolt V, Kircher T, Dannlowski U (2014) Insular and hippocampal gray matter volume reductions in patients with major depressive disorder. PLoS ONE 9:e102692

Sugimoto K, Kakeda S, Watanabe K, Katsuki A, Ueda I, Igata N, Igata R, Abe O, Yoshimura R, Korogi Y (2018) Relationship between white matter integrity and serum inflammatory cytokine levels in drug-naive patients with major depressive disorder: diffusion tensor imaging study using tract-based spatial statistics. Transl Psychiatry 8:141

Sullivan PF, Neale MC, Kendler KS (2000) Genetic epidemiology of major depression: review and meta-analysis. Am J Psychiatry 157:1552–1562

Tomasi D, Wang GJ, Volkow ND (2013) Energetic cost of brain functional connectivity. Proc Natl Acad Sci U S A 110:13642–13647

Trivedi MH, Rush AJ, Wisniewski SR, Nierenberg AA, Warden D, Ritz L, Norquist G, Howland RH, Lebowitz B, McGrath PJ, Shores-Wilson K, Biggs MM, Balasubramani GK, Fava M, Team SDS (2006) Evaluation of outcomes with citalopram for depression using measurement-based care in STAR*D: implications for clinical practice. Am J Psychiatry 163:28–40

Wang Z, Dai Z, Gong G, Zhou C, He Y (2015) Understanding structural-functional relationships in the human brain: a large-scale network perspective. Neuroscientist 21:290–305

Warner-Schmidt JL, Vanover KE, Chen EY, Marshall JJ, Greengard P (2011) Antidepressant effects of selective serotonin reuptake inhibitors (SSRIs) are attenuated by antiinflammatory drugs in mice and humans. Proc Natl Acad Sci U S A 108:9262–9267

Williams NR, Schatzberg AF (2016) NMDA antagonist treatment of depression. Curr Opin Neurobiol 36:112–117

World Health Organization (1992) International statistical classificationof diseases and related health problems. 10th Revision (ICD-10). World Health Organization, Geneva

Zhang C, Zhang DF, Wu ZG, Peng DH, Chen J, Ni J, Tang W, Xu L, Yao YG, Fang YR (2016) Complement factor H and susceptibility to major depressive disorder in Han Chinese. Br J Psychiatry 208:446–452

Zhao D, Wu Z, Zhang H, Mellor D, Ding L, Wu H, Wu C, Huang J, Hong W, Peng D, Fang Y (2018) Somatic symptoms vary in major depressive disorder in China. Compr Psychiatry 87:32–37

Zhou R, Wang F, Zhao G, Xia W, Peng D, Mao R, Xu J, Wang Z, Hong W, Zhang C, Wang Y, Su Y, Huang J, Yang T, Wang J, Chen J, Palaniyappan L, Fang Y (2018) Effects of tumor necrosis

factor-alpha polymorphism on the brain structural changes of the patients with major depressive disorder. Transl Psychiatry 8:217
Zuithoff NP, Vergouwe Y, King M, Nazareth I, van Wezep MJ, Moons KG, Geerlings MI (2010) The Patient Health Questionnaire-9 for detection of major depressive disorder in primary care: consequences of current thresholds in a crosssectional study. BMC Fam Pract 11:98
Zung WW, Richards CB, Short MJ (1965) Self-rating depression scale in an outpatient clinic: further validation of the SDS. Arch Gen Psychiatry 13:508–515

Chapter 2
Genetic Advance in Depressive Disorder

Chen Zhang and Han Rong

Abstract Major depressive disorder (MDD) and bipolar disorder (BPD) are both chronic, severe mood disorder with high misdiagnosis rate, leading to substantial health and economic burdens to patients around the world. There is a high misdiagnosis rate of bipolar depression (BD) just based on symptomology in depressed patients whose previous manic or mixed episodes have not been well recognized. Therefore, it is important for psychiatrists to identify these two major psychiatric disorders. Recently, with the accumulation of clinical sample sizes and the advances of methodology and technology, certain progress in the genetics of major depression and bipolar disorder has been made. This article reviews the candidate genes for MDD and BD, genetic variation loci, chromosome structural variation, new technologies, and new methods.

Keywords Major depression · Genetic progress · Bipolar depression

2.1 Introduction

Major depressive disorder (MDD) and bipolar disorder (BPD) are both chronic, severe mood disorder with high misdiagnosis rate, leading to substantial health and economic burdens to patients around the world (Smith 2014). There is a high misdiagnosis rate of bipolar depression (BD) just based on symptomology in depressed patients whose previous manic or mixed episodes have not been well recognized. In clinical practice, mood screening scales are commonly used as adjuvant tools to differentiate BD from MDD. However, misdiagnosis of MDD and BD is not inevitable. With the development of molecular biology, the advancement of genetic research in

---

C. Zhang (✉)
Division of Mood Disorders, Shanghai Mental Health Center, Shanghai Jiao Tong University School of Medicine, Shanghai, China
e-mail: zhangchen645@gmail.com

H. Rong
Department of Psychiatry, Shenzhen Kangning Hospital, Shenzhen, Guangdong, China
e-mail: szyhc@163.com

© Springer Nature Singapore Pte Ltd. 2019

Y. Fang (ed.), *Depressive Disorders: Mechanisms, Measurement and Management*, Advances in Experimental Medicine and Biology 1180,
https://doi.org/10.1007/978-981-32-9271-0_2

other complex diseases such as MDD and BD have made remarkable achievements, suggesting that the genetic research of depression has great potential and is expected to provide a new direction for the treatment of depression. Compared with a higher degree of autism, bipolar disorder, and schizophrenia, the heritability of MDD are between 31 and 42% (Sullivan et al. 2000), and the number of mutation loci found in genetic studies is far less than the above mental disorder. Revealing the genetic mechanism of depression requires a larger sample size, in recent years with Psychiatric Genomics Consortium (PGC), CONVERGE (China, Oxford, and Virginia Commonwealth University Experimental Research on Genetic Epidemiology), etc. The increase in the number of clinical cohort samples of depression in the organization, the genetic study of depression will usher in a new dawn. This article reviews the candidate genes for MDD and BD, genetic variation loci, chromosome structural variation, new technologies, and new methods.

2.2 MDD Candidate Gene Research

Candidate genes are mainly determined based on the hypothesis of the pathophysiological mechanism of the disease and the mechanism of action of the drug. According to the existing hypothesis, candidate genes for depression mainly focus on the neurotransmitter system, the hypothalamic–pituitary–adrenal axis (HPA), and the brain-derived neurotrophic factor (BDNF). As the understanding of gene–environment interactions deepens, current research on candidate genes is often associated with life events such as stress and trauma. A number of studies have found that dopamine receptor gene, dopamine β-hydroxylase (DBH) gene, dopamine transporter (DAT) gene, and brain-derived neurotrophic factor (BDNF). Upregulation of gene expression levels and 5-hydroxy tryptamine transporter (5-HTT) gene, monoamine oxidase A (MAOA) gene, catechol-O-methyltransferase (catechol- Downregulation of O-methyltransferase, COMT gene expression levels is associated with low ability to withstand stress and depression (Azadmarzabadi et al. 2018). Bustamante et al. (2018) found that the expression level of FKBP5 gene involved in HPA axis was related to the history of depression, however, it was not related to childhood abuse experience. In another meta-analysis, the FKBP5, 5-HTT-linked polymorphic region (5HTTLPR), gene, the tryptophan hyroxylase 2 (TPH2) gene, significant association between dopamine D2 receptor (DADR2) gene, BDNF gene and other candidate pathway genes and childhood trauma in depression patients, suggest that gene–environment interaction studies should further expand the sample size and patients' environmental exposure factors should be assessed more comprehensively and extensively (Van Der Auwera et al. 2018). The meta-analysis of the HPA axis also found that the corticotropin-releasing hormone binding protein (CRHBP) gene, the corticotropin-releasing hormone receptor 1 (CRHR1) gene, and the pro-opiomelanocortin (POMC) locus can predict the therapeutic effect of antidepressants

(Fischer et al. 2018). Another study further reported that the CRHBP gene polymorphism rs28365143 may predict the response to antidepressant therapy in patients with depression and the best antidepressant at the individual level (O'connell et al. 2018).

In recent years, the role of immunity in the pathogenesis of MDD has attracted the attention of scholars. For example, Zhang et al. (2016) found that IL6 gene is associated with depression in the Han population, while Labaka et al. (2017) confirmed such mechanism in mice. The decrease of IL10 gene expression in hippocampus under chronic stress is associated with monoaminergic neurotransmitter activity and anxiety and depressive symptoms. Sequencing studies have also found a correlation between immune-related variant loci and depression (Wong et al. 2017). In addition, scholars have also found that immune-related genes are involved in another research hotspot- the microbial–intestinal–brain axis. For example, Burokas et al. (2017) found that the simultaneous use of oligofructose and galactose to regulate intestinal microbiota in mice under chronic stress, antidepressant and antianxiety effects, and the simultaneous observation of human myeloid cell antigen (myeloid cell), the expression of nuclear differentiation antigen (MNDA) gene decreased, the expression of BDNF gene in hippocampus increased, and the expression of GABAB1 receptor gene and the GABAB2 receptor gene increased. With the application of genome-wide association study (GWAS) and gene sequencing technology, new candidate genes will continue to increase, which is conducive to further exploration of the biological mechanism of depression.

2.3 Common Variant Loci of MDD

The common single-nucleotide polymorphism (SNP) refers to the variation of the frequency of single nucleotides in the chromosome of the population is higher than 5%. The proportion of common mutations in the heritability of depression is about 4%, and therefore such genetic variations have attracted a large number of scholars to conduct related research. Compared with traditional candidate gene research, GWAS does not need to rely on prior knowledge, which can analyze common SNPs in case groups and control groups in a genome-wide scope, and find disease-related gene loci. Thus, GWAS has become a site for studying complex disease mutations. In the field of psychiatry, GWAS technology has made great achievements in finding significantly related loci such as schizophrenia (Biological insights from 108 schizophrenia-associated genetic loci 2014) and BDP (Ikeda et al. 2018), but the results of GWAS in major depression are far behind the above mental disorders. Levinson et al. (2014) argued that insufficient sample size and large heterogeneity may be responsible for this phenomenon. As such, subsequent large-scale and related studies that reduce heterogeneity are required to confirm this conjecture (Wray et al. 2018; Howard et al. 2018).

2.3.1 Large Sample Study of MDD

Throughout the large-scale analysis of MDD, when the sample size reached 18,759, 51,258, 161,460, 480,359, respectively, literature has documented 0 (Ripke et al. 2013), 1 (Hek et al. 2013), 2 (Okbay et al. 2016), and 44 (Wray et al. 2018) common variant loci associated with depression. The above results further indicate that the sample size needs to be several times as many as other mental disorders such as schizophrenia, in order to find a similar number of depression-related genetic variant loci. In addition, there are ongoing studies showing that when the sample size reaches hundreds of thousands, a new depression-susceptible site can be observed (Li et al. 2018; Xiao et al. 2017, 2018). Further analysis of the above loci revealed that depression-related variant genes are involved in obesity, chronic overactivation of HPA axis, presynaptic differentiation, neuroimmunity, neuronal calcium channels, dopaminergic neurotransmitters, and glutamatergic neurotransmitters. Pre-actoration vesicle trafficking, and some of them are highly correlated with schizophrenia; the analysis also found that these mutations are localized in the cerebral cortex, and the involved cells are neurons rather than oligodendrocytes or astrocytes, and are prominent Subsequence changes have no enrichment, suggesting that genetic variation involving changes in exon sequences may have little significance for depression (Wray et al. 2018). The above findings not only provide relevant evidence for the existing hypothesis of depression, but also provide clues for further understanding the biological mechanism of depression and subsequent research.

2.3.2 Heterogeneity of MDD and Related Research

The heterogeneity of MDD comes from many sources. First of all, depression has multi-gene genetic characteristics, and the number of genetic loci involved in depression is huge. The mutation loci of different depression individuals can be different or even overlap (Ostergaard et al. 2011). Second, there may be large differences in the symptoms of depressed patients who meet the current diagnostic criteria for depression. Others such as gender, ethnicity, onset time, recurrence, severity, and childhood traumatic history, medication history are all heterogeneous sources of depression. Therefore, research often increases statistical performance by reducing heterogeneity. For example, Howard et al. (2018) divided 322,580 British depression patients into three subgroups according to phenotype, and found that 17 loci were significantly associated with three subtypes, indicating that excitatory synapses may be involved in the pathology of depression. A study of patients with severe depression in Han women not only found common SNPs associated with depression, but annotated analysis suggested that these loci were located in the protein coding region (Peterson et al. 2017). Another study found that the 3p22.3 locus reached genotypic significance in male depression patients in Scotland and the United Kingdom (Hall et al. 2018).

It is worth mentioning that many of the above studies were analyzed using a polygenic risk score (PRS). PRS is a score calculated from the variation of multiple gene loci and its corresponding weights. It can predict the risk of individual disease, and can also measure the background of common genetic variation of a certain characteristic. It is easier to detect meaningful loci than GWAS alone. Therefore, it is often used for further analysis of GWAS. For example, related PRS studies (Andersen et al. 2017; Foo et al. 2018) have found a common genetic basis for depression and alcohol dependence. Wong et al. (2018) found that depression and cardiac metabolic diseases have a common genetic background, and the relevant loci were analyzed and found to be related to inflammation. Two other studies (Power et al. 2017; Rice et al. 2018) found that adult onset and early onset depression have different genetic susceptibility, and the former has greater genetic correlation with schizophrenia, bipolar disorder, attention deficit hyperactivity disorder, etc. Sexuality suggests that depression in early onset is closer to neurodevelopmental mechanisms. In addition, Zeng et al. (2017) found in the study that the NETRIN1 signaling pathway can be used as a candidate pathway for depression, which contributes to the formation of correct neural circuits during the development of the central nervous system, and that the pathway is in the stage of color formation of the thalamus. The amino acid pathway has a certain interaction, which complements the biological mechanism of depression to some extent.

2.4 Rare Variant Loci of MDD

The application of GWAS, PRS, etc. has made some progress in the study of common mutations in depression, but the common mutations that have been discovered still fail to fully reveal the genetic model of depression. Therefore, more and more scholars have turned their attention to rare mutation research. In recent years, the maturity of high-throughput sequencing technology and the decline in cost have made rare mutation research more convenient. Using high-throughput sequencing technology to sequence and analyze 1742 genes involved in synapses, Pirooznia et al. (2016) suggested that the etiology of depression may involve calcium channel and dendritic regulation. In addition, studies have found that non-synonymous mutations in the NKPD1 gene (Amin et al. 2017a) and missense mutations in the LIPG gene rs77960347 site (Amin et al. 2017b) are associated with depressive symptoms. Further analysis predicts that the NKPD1 gene may be involved in de novo synthesis of sphingomyelin, LIPG. Genes may be involved in carcass, cholesterol biosynthesis, and metabolism of thyroid hormones. Another study (Wong et al. 2017) performed genome-wide sequencing of American depression patients and healthy controls of Mexican and European origins, and found 44 cases involving immune responses, glutamate receptor signaling pathways, and olfactory organs perception of chemical stimuli. Common and rare variant loci in which the PPH21B gene mediates an individual's response to stress and are associated with depression. The study also suggests that rare variants may play different roles in people with depression in different races.

In addition, a family study (Amin et al. 2018) found that a rare missense mutation in the RCL1 gene (rs115482041) is associated with depression, which may be associated with cerebral cortical astrocytes, providing new mechanisms for the biological mechanisms of depression. These results not only help to reveal the genetic behavior of depression, but also provide evidence for further understanding of the pathological process of depression. In addition to the above research results, some scholars have put forward new insights and attempts in analytical methods. For example, some scholars have used multidimensional scaling (MDS) to show that mutations in low frequency and rare single nucleotide polymorphisms play an important role in the genetics of depression (Yu et al. 2018a). Rare variability, as a supplement to the study of common variability, maybe beyond the scientists' estimates and deserve further research and exploration.

2.5 Chromosome Structure Variation of MDD

Except for common and rare variant loci, which can explain the heritability of depression, some scholars have also regarded structural mutations such as short tandem repeats and copy number variants (CNV) as one of the sources of depression. The study found that short tandem repeats of 23 chromosomes in Mexican–American depression patients were significantly less than normal controls, although the same results were not found in Australian depression patients of European descent (Yu et al. 2018b), but also suggested that tandem repeats were in the role of depression, future studies may consider increasing the sample content to increase statistical power. The CNV study of the candidate gene CHRNA7 by Gillentine et al. (2018) found that the gene's copy number increased in patients with depression and anxiety, suggesting that the neuronal nicotinic acetylcholine receptor regulated by this gene may be used as a target for the treatment of depression. In addition, studies using whole-genome sequencing have found that CNV deletion mutations are associated with depression (Yu et al. 2017a), and studies have found that female children with depression and anxiety are associated with CNV (Martin et al. 2018), suggesting that the mechanism of depression may vary from gender to gender. The above suggests that further exploration of chromosome structural variation will contribute to the advancement of genetic research in depression.

2.6 Genetic Research Techniques and Analytical Method

The advancement of genetic research is inseparable from the advancement of technology and analytical methods. The rise and application of GWAS make genetic research of depression no longer limited to candidate genes, and sequencing technology, especially high-throughput sequencing technology, is conducive to rare mutation research. After discovering genetic mutation loci and genes related to diseases through gene chip and sequencing technology, gene annotation, pathway analysis, etc. are usually used to explore and analyze the cellular components, molecular functions and biological processes involved in related genes. As the discovery of genetic variation continues to increase, many scholars are gradually experimenting with new analytical methods to explore the genetic model of disease. For example, Yu et al. (2017b) suggested that the distribution of single-nucleotide variants (SNVs) in 12 different genomic regions may be related to depression, a new perspective on genetic mechanisms, and the use of Hamming. Distance and cluster analysis were subtyped in patients with depression based on genome-wide sequencing (Yu et al. 2017c). In addition, due to the emergence of some large-scale genome sequencing (WGS) research, Chen et al. (2018) proposed an analytical method that can be applied to complex samples and large-scale WGS research—variant-set hybrid model. Variant-set mixed-model association tests (SMMAT), although SMMAT has not been used in studies of depression, a similar approach will contribute to the genetic discovery of depression as the WGS study of depression increases.

2.7 Summary and Prospective

MDD is one of the most common mental illnesses, affecting not only the normal work and life of patients, but also the high risk of suicide in patients with depression (Fang et al. 2018; Chesney et al. 2014), which imposes a huge burden on society. Genetic research on depression is expected to bring new opportunities for the treatment of depression. However, due to the heterogeneity of depression and the large sample size required, the genetic research of depression faces enormous challenges. The genetic variation found in the current study can only explain part of the heritability, while the unresolved heritability of another part is often called the "lost heritability". The "missing heritability" includes common genetic loci mutations, rare genetic locus variations, and chromosomal structural variations that are not currently discovered. In addition, gene–gene interactions and gene–environment interactions also constitute inheritance. Further revealing the genetic mechanism of depression needs to be improved in many ways on the existing basis. First, to expand the sample size, when the sample size exceeds 100,000, the site variable energy that achieves the significance of genome-wide association research increases with the increase of sample size (Nishino et al. 2018), and the large sample can overcome the heterogeneity interference of some depression. The number of PGC depressive samples has increased year by year. The latest one has included 480,359 samples, including 135,458 patients with depression (Wray et al. 2018), and found 44 depression-related mutation loci. Second, to reduce heterogeneity, people with different characteristics of depression or different depression may have different genetic structures, and reducing heterogeneity will help to better reveal the genetic characteristics of depression. For example, the heterogeneity of depression patients included in the CONVERGE organization is low, that is, mainly for Chinese women with major depression, which helps to discover the genetic mechanism of depression.

In addition, new technologies and analytical methods need to be sought. The research of rare mutations relies heavily on the advancement of sequencing technology, but the price of whole-genome sequencing is still relatively expensive, which is also an important factor limiting the sample size of the study, so it is technically Seeking a breakthrough will also benefit genetic research in depression. In terms of analytical methods, traditional post-research studies commonly use gene annotation and pathway analysis to study the biological mechanisms involved in the relevant genetic loci, while new analytical methods such as the distribution of variant loci in different regions and new computational methods are possible and bring new clues to genetic research in depression. Finally, in the genetic study of depression, we should use existing rules to reflect on the biological mechanism and classification of depression, such as PRS research suggesting depression and neurosis, alcohol dependence, bipolar disorder, and schizophrenia. Other mental disorders such as other mental disorders and cardiac metabolic diseases have a partial genetic background, suggesting that the diagnosis and classification of existing mental illnesses can be further improved. In summary, the genetic research of depression has not yet reached the maturity level, but with the continuous accumulation of sample size and technological advancement, the genetic research of depression is expected to make breakthrough progress, in order to further understand the molecular mechanism of depression and seek new treatments to provide clues.

2.8 BD Genetic Research

Recent molecular studies, using both traditional approaches and new procedures such as Whole-Genome Sequencing (WGS), have suggested that genetic factors could significantly contribute to the development of BD, with heritability estimates of up to 85% (Smith 2014). However, it is assumed that BD is a multigenic and multifactorial illness with environmental factors that strongly contribute to disease development/progression, which means that progress in genetic knowledge of BD might be difficult to interpret in clinical practice.

2.9 Neurotransmitters Signaling Related Genes

Genes involved in neurotransmitter metabolism pathways have been a primary focus in candidate studies of BP. Genes of current interest include those encoding the serotonin, dopamine, glutamate and GABA receptor signaling-related protein. Association studies are summarized, which support a possible role for numerous candidate genes in BPD including COMT, DAT, MAO, DRD1-4, HTR4, HTR2A, and 5-HTT.

Dopamine (DA) is one of the predominant catecholamine neurotransmitters that is involved in bipolar disorder and many other psychiatry diseases. It's well accepted that excessive dopamine neurotransmission is involved in the development of manic symptoms of BD (Sullivan et al. 2000). Up to now, compelling data have shown that drugs control the symptoms of mania were mediated by inhibiting the dopamine receptor signaling. The function of dopamine depends on multiple elements such as dopamine transporter (DAT), dopamine receptors, the enzymes metabolizing dopamine such as monoamine oxidase (MAO) and catechol-Omethyltransferase (COMT) (Azadmarzabadi et al. 2018). Among them, dopamine receptors have been found to play an important role in the pathogenesis of BD. The role of dopamine receptors in the development of BD has been intensively investigated. Several linkage studies show that the dopamine receptor D1 (DRD1) gene is associated with BD (Bustamante et al. 2018; Van Der Auwera et al. 2018). The dopamine receptor D2 (DRD2) polymorphism has also been associated with mood disorder (Azadmarzabadi et al. 2018; Fischer et al. 2018). One study reported that dopamine receptor D3 (DRD3) gene may be a susceptibility gene for BD (O'connell et al. 2018). Dopamine receptor D4 (DRD4) gene polymorphisms being associated with BD has been revealed by several studies (Zhang et al. 2016). Monoamine oxidase A (MAOA), coded by the *MAOA* gene, is a mitochondrial enzyme involved in the metabolism of several biological amines such as dopamine, norepinephrine, and serotonin, which are important neurochemicals in the pathogenesis of BP. Several studies confirmed that *MAOA* variants confer risk of BP and other psychiatric disorders (Labaka et al. 2017). The dopamine transporter (DAT), a presynaptic plasma membrane protein and responsible for reuptake of the dopamine, the level of which is correlated with dopamine activity. It has been shown that polymorphisms of *DAT1* play a role in predisposition to BD disorder (Sullivan et al. 2000). The catechol-O-methyltransferase (COMT) gene plays an important role in the clearance and metabolic inactivation of DA and norepinephrine in regions with a paucity of DAT expression. Studies data indicated relationships between variant of COMT gene and rapid cycling or cognitive dysfunction in BD (Wong et al. 2017). G-protein receptor kinase 3 (GRK3) appears to regulate the brains response to DA is also an excellent candidate risk gene for bipolar disorder (Burokas et al. 2017).

The serotonin neurotransmitter has also been implicated for several reasons as having a major role in the pathophysiology of psychiatry disease, including bipolar disorder. In particular, common polymorphisms in serotonin transporter (5-HTT), receptors (5HTR1A, 1B, 5HTR4, and 5HTR2A), and enzymes involved in serotonin metabolism: synthesis (Tryptophan Hydroxylase, TPH1, and TPH2) genes have been linked to bipolar disorder (Biological insights from 108 schizophrenia-associated genetic loci 2014).

Dysfunction in gamma amino butyric acid (GABA) system activity and glutamatergic system have been hypothesized to play a role in BP vulnerability. GABA receptor (GABRA3) genotype was reported to be associated with bipolar disorder (Ikeda et al. 2018). Glutamate is the principal excitatory neurotransmitter in the central nervous system. In mature brains, it is critically involved in neuroplasticity and, at high levels, neurotoxicity. GRIN1 codes for the zeta-1 subunit of NMDA receptor, GRM3 code for metabotropic glutamate receptor 3 (GRM3). These gene variants were associated with bipolar disorder too (Levinson et al. 2014; Wray et al. 2018).

There is another robust gene named G72, located at 13q34, the common linkage locus in bipolar disorder, and it encodes d-amino acid oxidase activator (DAOA) (Howard et al. 2018).

2.10 Mitochondrial Function

Over the past 20 years, there has been increasing evidence that psychotic, mood, anxiety, and personality disorders may be part of the clinical manifestations of mitochondrial diseases, as well as that psychiatric disorders could represent a complication of mitochondrial function (Ripke et al. 2013; Hek et al. 2013).

The mitochondrion is an indispensable organelle of eukaryotic cells that plays many important roles required for cell survival and well-being. It is thought to descend from prokaryotic bacteria through endosymbiotic evolution (Okbay et al. 2016; Li et al. 2018).

Mitochondria have two membranes (outer and inner), an intermembrane space, and an internal matrix. The inner mitochondrial membrane contains the electron transport chain (ETC), the molecular machinery for energy production (Hek et al. 2013; Xiao et al. 2018). The ETC contains five protein complexes, of these, three (I, III, and IV) pump protons (H1) through the inner membrane, generating an H1 gradient required for the synthesis of ATP at complex V (ATP synthase). The mitochondrial genome codes for 13 of the ETC proteins (Hek et al. 2013; Xiao et al. 2017, 2018).

Mitochondria produce the most of energy-rich molecules of adenosine triphosphate (ATP) by oxidative phosphorylation, apart from energy production they are involved in other functions: regulation of free radicals, antioxidant defenses, lipid peroxidation, calcium metabolism and participate in the intrinsic pathway of apoptosis (Hek et al. 2013; Ostergaard et al. 2011). Also, mitochondria are crucial for neurogenesis and neuronal functions, especially in energy production, the generation of reactive oxygen species, and calcium signaling (Hek et al. 2013; Peterson et al. 2017).

Neurons are especially dependent on mitochondria, partly because of their high energy demands. As a result, mitochondrial dysfunction leads to multiple types of brain disorders. These changes include impaired energy metabolism in the brain, comorbidity with mitochondrial diseases, the effects of psychotropics on mitochondrial function, increased mitochondrial DNA (mtDNA) deletion in the brain, and association with mtDNA polymorphisms. Additionally, psychotropic drug treatments can alter energy metabolism and may affect mitochondrial processes (Hek et al. 2013; Okbay et al. 2016; Hall et al. 2018; Andersen et al. 2017; Foo et al. 2018). Multiple lines of evidence have suggested the possible involvement of mitochondrial deficits in the pathophysiology of CNS diseases including bipolar disorder (BD), major depressive disorder, and schizophrenia (Hek et al. 2013; Okbay et al. 2016).

Bipolar disorder (BD) is a common mood disorder associated with chronic course alternating between mania and depression, with symptoms characterized by alternating increases and decreases in energy and activity. It is currently one of the main focuses of psychiatry research effort (Wong et al. 2018; Power et al. 2017). The World Health Organization (WHO) identifies BD as among the top ten leading cause of lifelong disability worldwide accounting for approximately 1–2% of the population affected (Rice et al. 2018). The underlying etiology of BD remains elusive, rendering translational strategies for therapeutic development very challenging (Zeng et al. 2017; Pirooznia et al. 2016). Although the underlying pathophysiological mechanisms of BD remain largely unknown, as a phasic symptomatic disorder of energy, accumulating evidence is beginning to link BD to perturbations in mitochondrial functioning (Zeng et al. 2017; Amin et al. 2017a, b).

Symptomatically, BD is a biphasic disorder of energy availability; increased in mania and decreased in depression. Indices of increased mitochondrial respiration and ATP production in bipolar mania stand in contrast with decreased mitochondrial function in patients in the euthymic or depressive phase of the illness (Amin et al. 2018; Yu et al. 2018a, b; Gillentine et al. 2018; Yu et al. 2017a). BD is a very complex illness with genetics being a significant contributing factor. However, no single gene has been consistently identified, making it challenging to come into a general consensus. Exploring mitochondrial gene expression may help uncover genes that are up- or downregulated in BD and thereby provide molecular insight on its genetic foundation that can help explain molecular and regulatory mechanisms in a pathway involved in the pathophysiology of BD. Gene expression studies of the prefrontal cortex from BD subjects have been widely conducted owing to the fact that most meaningful findings were detected in this brain region (Martin et al. 2018).

According to increasing evidence dysfunctions of mitochondria are associated with affective disorders, a hypothesis of impaired mitochondrial functions has been proposed in bipolar disorder pathogenesis. Mitochondrial DNA mutations and/or polymorphisms, impaired phospholipid metabolism and glycolytic shift, decrease in ATP production, increased oxidative stress and changes of intracellular calcium are concerned in mood disorders and effects of mood stabilizers. Recent studies have also provided data about the positive effects of chronic treatment by mood stabilizers on mitochondrial functions (Ostergaard et al. 2011) (Figs. 2.1, 2.2, 2.3, and 2.4).

Mitochondrial diseases have been extensively investigated over the last three decades, but many questions regarding their underlying aetiologias remain unanswered. Mitochondrial dysfunction is not only responsible for a range of neurological and myopathy diseases but also considered pivotal in a broader spectrum of common diseases such as epilepsy, autism, and bipolar disorder. These disorders are a challenge to diagnose and treat, as their etiology might be multifactorial (Chen et al. 2018).

Exact pathophysiological mechanisms of bipolar disorder have not been sufficiently clarified. Mitochondrial dysfunctions are ones of the characteristic features, which can be manifested in BD symptoms. Disturbances, such as mitochondrial genetic variations and mtDNA mutations, impaired energy metabolism, OXPHOS and respiratory chain activities, diminished mitochondrial biogenesis and dynamics, impaired calcium homeostasis, oxidative stress, neuronal survival, and apoptotic processes, could be reflected in mitochondrial pathology. In spite of the increasing

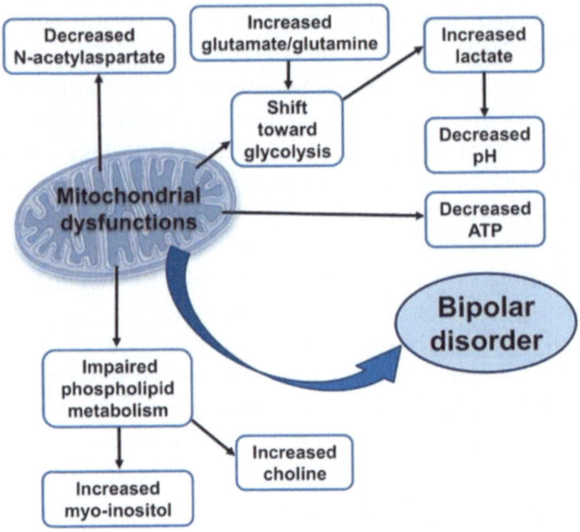

Fig. 2.1 Mitochondrial dysfunctions in bipolar disorder (Ostergaard et al. 2011). There are many potential mediators of mitochondrial function which collectively are implicated in bipolar disorder. Levels of oxidative stress, pro-inflammatory cytokines nand intracellular calcium ions are all higher in bipolar mania than in the euthymic and depressive phases of the illness. Increased levels of calcium ions can partly account for increased oxidative phosphorylation via well-documented pathways such as the modulation of the F1-FO elements of ATP synthase. Likewise, increased levels of oxidative stress and pro-inflammatory cytokines lead to the upregulation of AMPK, SIRT-1, SIRT-3, and NAD+ which directly stimulate oxidative phosphorylation. Uric acid and melatonin are also differentially elevated in bipolar mania and both molecules stimulate the production of ATP. The pro-apoptotic, neurotoxic and mitotoxic effects of elevated glutamate, dopamine and GSK-3 in bipolar mania may be counterbalanced by higher basal levels and activity of p53, Bcl-2, PI3 K, and Akt in an environment of elevated uric acid and decreased BDNF (Ostergaard et al. 2011; Amin et al. 2018)

evidence, that these dysfunctions play a central role in the pathophysiology of BD, it remains unclear, whether they are a primary or secondary cause of this mental disease. Mood stabilizers cause changes in signaling intracellular pathways and should improve energy metabolism (Ostergaard et al. 2011; Fang et al. 2018).

Lithium and VPA attenuate formation of mitochondrial permeability transition pores, and consequently release of cytochrome c and other proapoptotic factors from the intermembrane space of mitochondria; they participate in the prevention of excitotoxicity and inhibit enzymes with pathological impact on mitochondrial metabolism. Other antiepileptic drugs very often interfere with mitochondrial bioenergetics. Therefore, the effects of mood stabilizers on mitochondria should be investigated in the context to clinical symptoms of BD. In conclusion, a complex view of the cellular pathology in BD is crucially important for therapeutic strategies for BD, and the development of new diagnostic tools (Ostergaard et al. 2011).

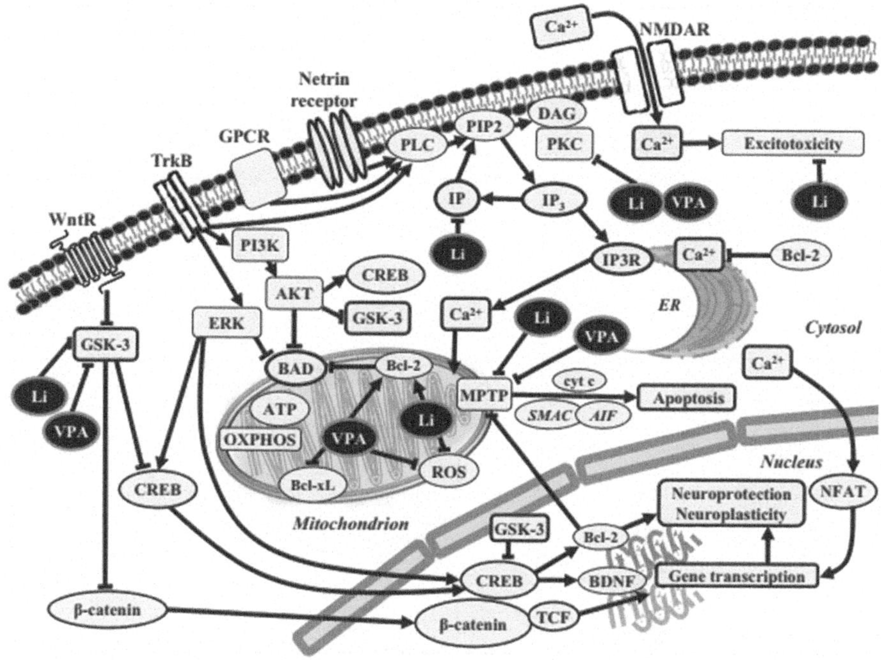

Fig. 2.2 Effects of mood stabilizers on energy metabolism. Both lithium (Li) and valproate (VPA) have neuroprotective properties at the level of apoptosis reduction. Mood stabilizers attenuate formation of mitochondrial permeability transition pores (MPTP), release of proapoptotic factors and cytochrome c (cyt c), and other proapoptotic factors from mitochondria. Lithium and VPA inhibit glycogen synthase kinase 3 (GSK-3). Activation of Wnt receptor (WntR) leads to inhibition of β-catenin phosphorylation; unphosphorylated β-catenin enters the nucleus and promotes T cell factor (TCF)-mediated gene transcription. Phosphatidylinositol-3-kinase (PI3 K)/protein kinase B (Akt) pathway is a signaling pathway promoting the survival in response to extracellular signals. ERK pathway regulates cytosolic targets and many cellular transcription factors, including cAMP response element-binding protein (CREB). Phospholipase C (PLC), which may be activated via neurotransmitter G-protein-coupled receptors (GPCR), neurotrophin receptors (TrkB) or netrin receptors, catalyses the hydrolysis of phoshatidylinositol 4, 5-bisphosphate (PIP2) and give arise to inositol triphosphate (IP3) and diacylglycerol (DAG) as the second messengers. Inhibition of recovery of PIP2 due to inhibition of inositol monophosphatase (IP) may be a key step in the action of Li in treating BD. DAG activates protein kinase C (PKC); Li and VPA indirectly inhibit PKC. IP3 is binding to intracellular receptors (IP3R), and is causing the trigger of Ca^{2+} from the endoplasmic reticulum (ER). Excessive entry of Ca^{2+} , e.g., through activated N-methyl-D-aspartate receptor (NMDAR), may affect neuroplasticity and induce excitotoxicity. Intracellular Ca^{2+} increase activates nuclear factor of activated T cells (NFAT) proteins, involved in axon growth. Release of cyt c, the second mitochondria-derived activator of caspases (SMACs) and apoptosis-inducing factor (AIF) from mitochondria induce the apoptosis (Ostergaard et al. 2011)

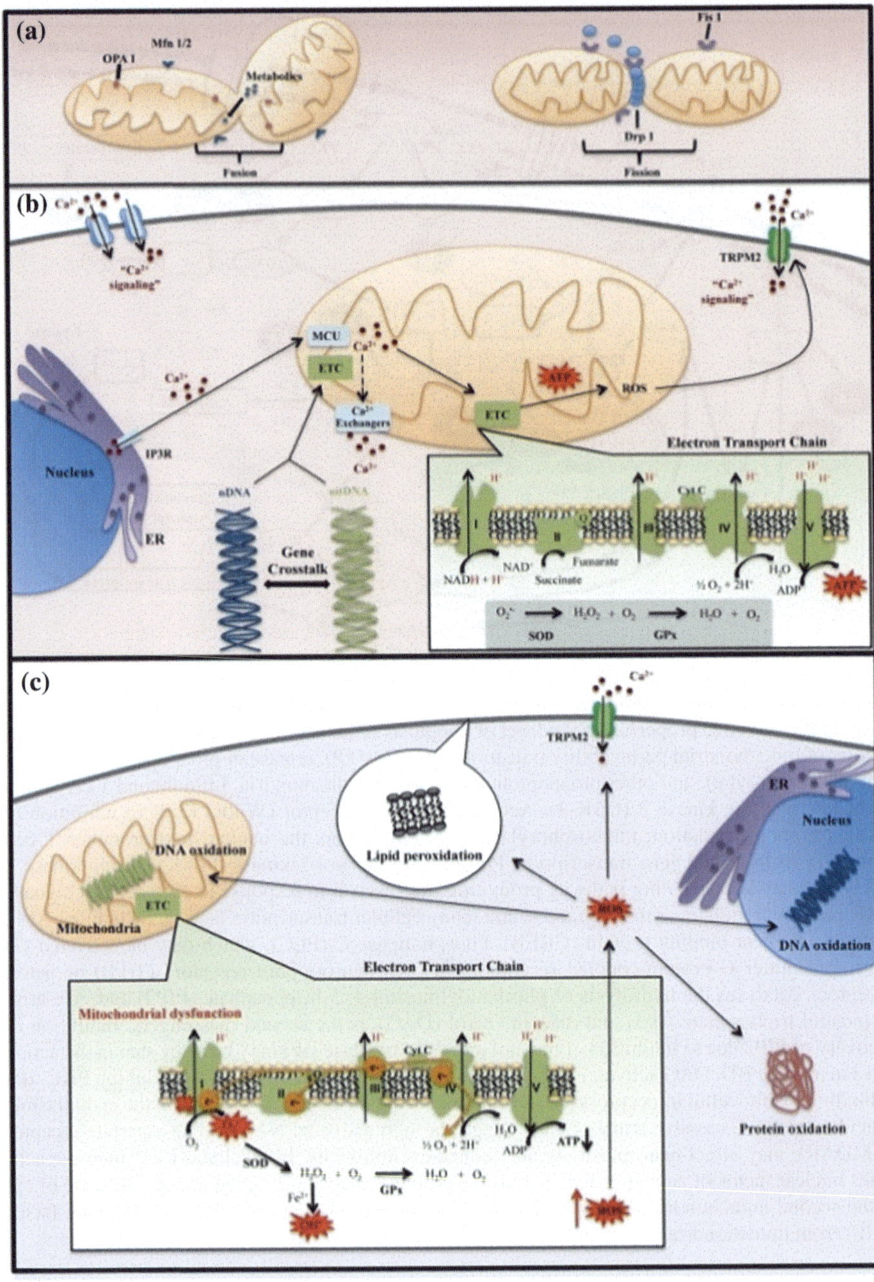
(a)
OPA 1
Mfn 1/2
Metabolites
Fusion
Fis 1
Drp 1
Fission
(b)
"Ca2+ signaling"
TRPM2
MCU
ETC
Ca2+ Exchangers
ATP
ROS
Nucleus
IP3R
ER
nDNA
mtDNA
Gene Crosstalk
Electron Transport Chain
CytC
NADH + H+
NAD+
Succinate
Fumarate
½ O2 + 2H+
H2O
ADP
SOD
GPx
(c)
TRPM2
ER
Nucleus
Lipid peroxidation
DNA oxidation
Mitochondria
ROS
Electron Transport Chain
Mitochondrial dysfunction
Protein oxidation

◀**Fig. 2.3** Mitochondrial dynamic. **a** Fusion process is important for mitochondrial function by diffusion of metabolites and enzymes between mitochondria, as well as dilution of damaged proteins and DNA. The fusion mediators are Mfn1 and Mfn2, which is present on the outer mitochondrial membrane, and Opa1, which is located in the inner mitochondria membrane. Fission process can isolate injured mitochondria, contributing to mitochondrial quality control. The fission mediators are Fis1 and Drp1. Fis1 recruits Drp1 to mitochondria, and it permits the development of fission process. **b** Normal mitochondrial function: mitochondrial and electron transport chain (ETC) assembly and function are dependent on nuclear DNA (nDNA) and mitochondrial DNA DNA (mtDNA)-encoded proteins. nDNA-encoded proteins regulate mitochondrial replication, transcription, and repair, allowing for crosstalk between nDNA and mtDNA. Mitochondria take up Ca^{2+} primarily through mitochondrial Ca^{2+} uniporter (MCU). Ca^{2+} is then extruded from mitochondria through ion exchangers that are coupled to adenosine triphosphate (ATP) production. Mitochondria are localized close to sites of Ca^{2+} entry, such as the endoplasmic reticulum (ER) and membrane channels, allowing them to buffer cytosolic Ca^{2+} concentrations. **c** Mitochondrial dysfunction: ETC impairment increases the amount of electrons leakage, resulting in increased reactive oxygen species (ROS) production. High levels of ROS coupled with low antioxidant defenses disrupt redox homeostasis, leading to cellular oxidative stress. Antioxidant defenses include superoxide dismutase (SOD) and glutathione peroxidase (GPx). If these high levels of ROS are not sufficiently detoxified by these antioxidant enzymes, it can cause oxidative damage to proteins, lipids, and nucleic acids (Yu et al. 2017b)

Normal mitochondrial physiology is integral to healthy well-being. After decades of research in interpreting mitochondrion function, there is currently no treatment against mitochondrial diseases, which reflects the complexity of dysfunction when it occurs. Environmental factors are now thought to be a potential etiology to some mitochondrial diseases. Understanding the extent of genotoxic and/or epigenetic influences will enable us to move toward novel research techniques, develop diagnostic tests and perhaps influence lifestyle changes. Undoubtedly, more research is required, as current therapeutic approaches mostly employ palliative therapies rather than targeting primary mechanisms or prophylactic approaches. Much effort is needed into further unraveling the relationship between xenobiotics and mitochondria, as the extent of influence in mitochondrial pathogenesis is increasingly recognized (Chen et al. 2018).

2.11 Epigenetics and microRNA

During the last decade and a half, there has been an explosion of data regarding epigenetic changes in psychotic diseases. The examination of potential roles of epigenetic alterations in the pathogenesis of psychotic diseases has become an essential alternative in recent years as genetic studies alone are yet to uncover major gene(s) for psychosis (Chesney et al. 2014). Epigenetic regulation plays an important role in the development of the embryo and nervous system and plays an important role in the pathophysiology of psychiatric disorders, including depression, bipolar disorder, and schizophrenia (Nishino et al. 2018). So far, genetic studies of treatment response

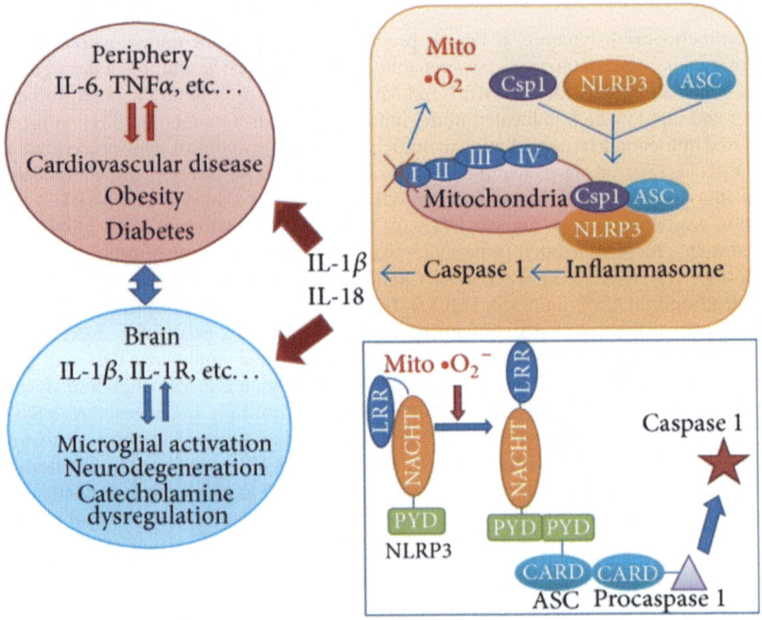

Fig. 2.4 Mitochondrial complex I dysfunction in patients with BD could lead to increased release of superoxide anions, resulting in greater reactive oxygen species (ROS) production. This release of ROS causes a conformational change in NLRP3 such that the pyrin domain (PYD) becomes available recruit ASC. The combining of NLRP3 and ASC that allows for the recruitment of caspase 1 (csp1) through ASC's CARD domain, causing the formation of the NLRP3 inflammasome. The inflammasome then migrates to the mitochondria, allowing it to be close to the site of damage. Activated NLRP3 inflammasome releases caspase 1 into the cytosol, which then cleaves and activates two downstream cytokines, Il-1beta and Il-18, causing them to be released into the extracellular space. These two cytokines cause the activation of downstream pathways, which may differ depending on the type of immune cell. Indeed, NLRP3 inflammasome activation may underlie the different patterns of cytokine activation observed in the brain and peripheral samples of patients with BD, where alterations in cytokines pertaining to the IL-1 pathway have been reported for the brain, while a more general pattern of cytokine activation involving IL-6 and TNF-alpha has been reported in the periphery. Cytokine activation in the periphery can lead to various immune disorders, including cardiovascular disease and diabetes, while, in the brain, it could lead to alterations in neurotransmitters and neurodegeneration (Yu et al. 2017c)

in schizophrenia, bipolar disorder, and major depression have returned results with limited clinical utility. A gene × environment interplay has been proposed as a factor influencing not only pathophysiology but also the treatment response. Therefore, epigenetics has emerged as a major field of research to study the treatment of these three disorders (Orru and Carta 2018). In addition, neural basis of cognition, learning and memory, neurodevelopmental plasticity, and neurotransmission are influenced by epigenetic regulations (Ashok et al. 2017; Kaalund et al. 2014; Zhao et al. 2015; Rybakowski et al. 2009). Since epigenetic modifications are involved in the pathophysiology of psychiatric disorders including depression, bipolar disorder, and schizophrenia, there is considerable interest in targeting these for drug discovery (Nishino et al. 2018; Hu et al. 2013; Chang et al. 2013). Advances in our understanding of the epigenetic mechanisms that control gene expression in the central nervous system (CNS) and their role in neuropsychiatric disorders are paving the way for a potential new therapeutic approach that is focused on reversing the epigenetic underpinnings of neuropsychiatric conditions (Chang et al. 2013). Epigenetic drugs, including histone deacetylation and DNA methylation inhibitors have received increased attention for the management of psychiatric diseases (Nishino et al. 2018) (Fig. 2.5).

Epigenetic mechanisms are heritable changes in gene expression that control transcriptional regulation via modifications of DNA and histone. Epigenetic profiles, unlike DNA mutations, are reversible. In addition, epigenetic modifications are affected by environmental factors. Methylation of DNA sequences at key nucleotides and modification of histone and non-histone-binding proteins, such as scaffold and polycomb proteins, represent the main epigenetic mechanisms (Nishino et al. 2018). The gene-specific and genome-wide studies of postmortem brain and blood cells indicating that aberrant DNA methylation, histone modifications and dysregulation of micro-RNAs are linked to the pathogenesis of mental diseases. There is also strong evidence supporting that all classes of psychiatric drugs modulate diverse features of the epigenome. While comprehensive environmental and genetic/epigenetic studies are uncovering the origins, and the key genes/pathways affected in psychotic diseases, characterizing the epigenetic effects of psychiatric drugs may help to design novel therapies in psychiatry (Chesney et al. 2014; Lopez Leon et al. 2005).

Among the epigenetic marks that can modify gene expression, DNA methylation is the best studied. DNA methylation in higher eukaryotes is the addition of a methyl (CH3) group to the carbon-5 of cytosines in the DNA sequence, generating a modified nucleotide called 5-methylcytosine (5mC). Formation of 5mC has been classically involved in gene silencing and repressive chromatin (heterochromatin), but recent evidence suggests additional functions for DNA methylation. In vertebrates, DNA methylation occurs throughout the entire genome at cytosines of CpG dinucleotides. Even though these dinucleotides are underrepresented in the whole genome, they are frequently enriched around promoter regions, in so-called CpG islands (CGI) (Lopez Leon et al. 2005; Reif et al. 2014; Taylor 2018; Zhou et al. 2008; Ottenhof et al. 2018).

Bipolar disorder (BD) is a multifactorial illness thought to result from an interaction between genetic susceptibility and environmental stimuli. Epigenetic mechanisms, including DNA methylation, can modulate gene expression in response to the environment, and therefore might account for part of the heritability reported for BD. This paper aims to review evidence of the potential role of DNA methylation in the pathophysiology and treatment of BD. In summary, several studies suggest that alterations in DNA methylation may play an important role in the dysregulation of gene expression in BD, and some actually suggest their potential use as biomarkers to improve diagnosis, prognosis, and assessment of response to treatment. This is also supported by reports of alterations in the levels of DNA methyltransferases in patients and in the mechanism of action of classical mood stabilizers. In this sense,

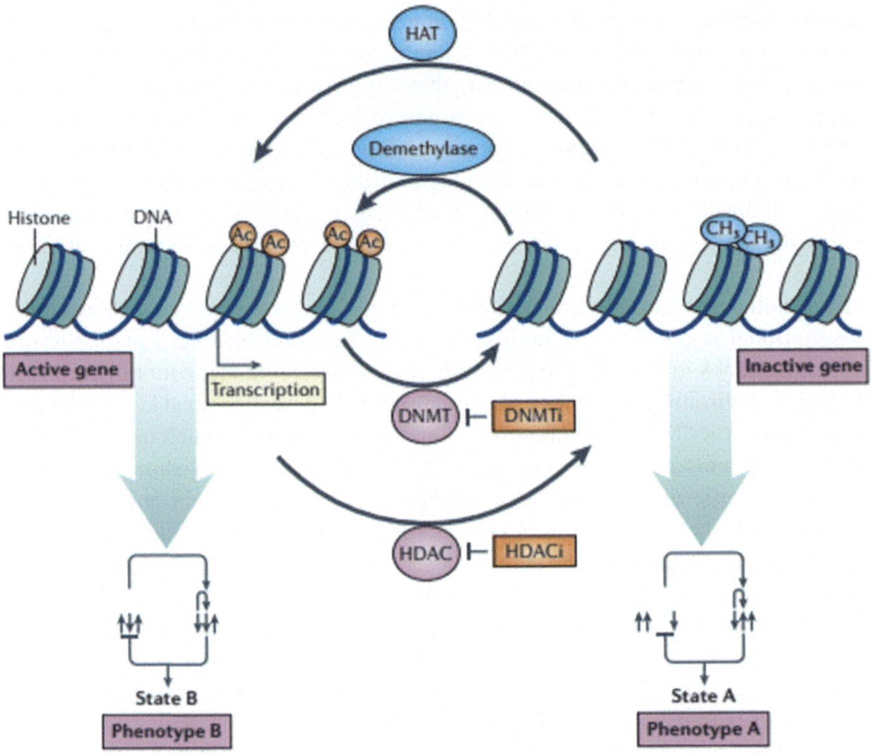

Fig. 2.5 Rationale for epigenetic drugs, reversibility of epigenetic reactions, and epigenetic states and sites of pharmacological modulators. Epigenetic states are defined by a balance of biochemical reactions that modify either the DNA or the histones associated with the DNA. Active genes are characterized by unmethylated promoters and regulatory regions, and histones associated with active genes are usually acetylated (for example, on the tail of the H3 histone at lysine 9). The promoters of inactive genes are often methylated (CH3) and the histones associated with these regions are deacetylated. A balance of DNA methyltransferases (DNMTs) and demethylases defines the state of methylation of a gene, and the balance of histone acetyltransferases (HATs) and histone deacetylases (HDACs) defines the state of histone acetylation. Pharmacological inhibitors of these enzymes (epigenetic drugs) can "tilt" the epigenetic balance and thereby alter gene expression. The lower part of the figure illustrates how brain phenotypes are maintained by a balance of activities of epigenetic enzymes that act on groups of genes (nodes). Each arrow indicates a gene; an upward arrow represents an active gene and a downward arrow represents a silenced gene. An inhibitory interaction between the nodes or genes is indicated by a blunted arrow. The balance of the gene expression state is defined by a balance of activities of several isoforms of DNMTs, HATs, DNA demethylases, HDACs, and numerous other epigenetic processes that are not listed here. Inhibition of one or several epigenetic enzymes will tilt the balance and could reverse the gene expression state from A to B and, as a consequence, change the phenotype from A to B. Ac, acetyl moiety; DNMTi, DNMT inhibitor; HDACi, HDAC inhibitor (Szyf 2015)

targeting specific alterations in DNA methylation represents exciting new treatment possibilities for BD, and the "plastic" characteristic of DNA methylation accounts for a promising possibility of restoring environment-induced modifications in patients (Lopez Leon et al. 2005) (Table 2.1).

Epigenetic mechanisms have been suggested to play a key role in the pathophysiology of bipolar disorder (BD), among which microRNAs (miRNAs) may be of particular significance according to recent studies. Recent studies investigating an association between BD and miRNAs. They report miRNA alterations in postmortem brain tissues and in the periphery, cell culture and preclinical findings, genetic associations, and the effects of medications. Several studies report changes in miRNA expression levels in postmortem brain and in the periphery of patients, although most of the results so far have not been replicated and are not concordant between different populations. Genetic studies also suggest that miRNA genes are located within susceptibility loci of BD, and also a putative role of miRNAs in modulating genes previously shown to confer risk of BD. miRNA findings in BD significantly vary between studies but are consistent to suggest a key role for these molecules in BD's pathophysiology and treatment, particularly miR-34a and miR-137. Accordingly, miRNA might represent important biomarkers of illness to be used in the clinical settings, and potentially also for the development of novel therapeutics for BD in the near future (Massat et al. 2002) (Tables 2.2 and 2.3).

2.12 Circadian Rhythm Disturbances in BD

The circadian system plays a fundamental role in overall health and longevity. This is also true for mental disorders since misalignment between the endogenous circadian system and the sleep/wake cycle is a critical factor in the clinical status of many psychiatric disorders. This chapter examines the evidence for circadian disturbances in severe psychiatric disorders such as chronic schizophrenia and bipolar affective disorder (BPD), describes circadian-related interventions that have been used successfully to treat these disorders, and discusses current research on the role of clock genes in mental illness.

Circadian rhythms, sleep, and bipolar affective disorder

The introduction of sleep deprivation therapy (SDT) in the early 1970s (Kandaswamy et al. 2013) was a major breakthrough in view of its rapid (within 24 h) reversal of depressive symptoms. Coupled with new technological advances to study the complexities of the circadian clock gene machinery, clues have emerged as to the role of clock genes and their contributions to depressive illness. Additionally, the evolutionary conservation of the clock gene machinery has helped to provide the template for circadian regulation from Drosophila to humans. The mechanism of action of sleep deprivation therapy has been investigated over a number of decades. Although there are persuasive hypotheses, no proven mechanism has yet been identified. The focus

Table 2.1 Alterations in DNA methylation in bipolar disorder

Gene	Cell/tissue	Method	Main findings
5-HTR1A	Leukocytes	HRM	Hypermethylation of *5HTR1A* gene
5-HTR2A	Saliva	MSP and bisulfite sequencing	Hypomethylation of the 5-*HTR2A* gene at the T102C polymorphic region
5-HTR2A	Postmortem frontal lobe	MSP	Hypermethylation of the 5-*HTR2A* promoter region around—1438A/G polymorphic region. Hypomethylation of the T102C region in patients
5-HT3AR	Peripheral blood	Bisulfite pyrosequencing	Methylation of two CpGs mediated the effects of childhood maltreatment on the severity of the disorder in adulthood
5-HTT	Postmortem frontal lobe	MSP	Tendency of a hypermethylation of *5HTT* in antipsychotic-free BD patients
BDNF	Peripheral blood mononuclear cells	Bisulfite sequencing	Hypermethylation of *BDNF* in BD II compared to BD I. Hypermethylation in depressed patients compared with manic/mixed patients. Lower methylation levels in patients on lithium and valproate
BDNF	Peripheral blood mononuclear cells	MSP	*BDNF* exon I methylation was increased in MDD subjects compared to BD patients and controls. Increased methylation associated with antidepressant use
BDNF	Venous blood	MassARRAY platform	Different methylation degree between BD and controls for 11 of 36 CpG units
BDNF	Peripheral blood mononuclear cells	MSP	Increased *BDNF* methylation in BD II (but not type I). Negative correlation between methylation and gene expression
Chromosome X	Buccal mucosa and peripheral leukocytes	Restriction enzymes analysis	Discordant twins for BD are more discordant for the methylation of chromosome X, especially compared to concordant twins for BD

(continued)

Table 2.1 (continued)

Gene	Cell/tissue	Method	Main findings
DTNBP1	Saliva	Bisulfite sequencing	Higher *DTNBP1* promoter methylation in BD patients with psychotic depression compared to other BD patients
FAM63B	Whole peripheral blood	iPLEX assay	Hypomethylation of two CpG sites in the exon 9 of the *FAM63B* gene
FKBP5	Peripheral blood mononuclear cells	Bisulfite pyrosequencing	Increased *FKBP5* methylation
GAD1 regulatory network genes	Postmortem hippocampus	Bisulfite sequencing	Diagnosis- and circuit-specific methylation changes at a subset of *GAD1* regulatory network genes
Genome-wide methylation	Postmortem brain region BA9	MeDIP-seq	16,599 differentially methylated regions in BD compared to controls (6,836 hypermethylated and 9,763 hypomethylated)
Genome-wide methylation; *CYP11A1*	Blood	Illumina Human-Methylation450 BeadChip and pyrosequencing	Hypomethylation at the *CYP11A1* gene in manic patients. Association of this methylation with inflammatory markers
Genome-wide methylation	Blood	Illumina Human-Methylation450 BeadChip	Altered methylation patterns in BD patients in use of quetiapine and valproic acid after adjusting for drug-related changes on cell-type composition
Genome-wide methylation	Blood	Illumina Human-Methylation450 BeadChip	Altered methylation in carriers of a haplotype linked to BD and MDD, including genes related to neurodevelopment and ion channel activity (such as *FANCI*)
Genome-wide methylation	Blood	MeDIP-Seq	Thousands of differentially methylation regions located preferentially in promoters 3′-UTRs and 5′-UTRs of genes
Genome-wide methylation	Frontal cortex and anterior cingulate	MeDIP-seq	Several differentially methylated regions in BD compared to controls. Different distributions of methylation across the genome between the two brain regions

(continued)

Table 2.1 (continued)

Gene	Cell/tissue	Method	Main findings
Genome-wide methylation	Cerebellum	Illumina Infinium HumanMethylation27	CpG gene pairs were found to significantly correlate with the differential expression and methylation of *PIK3R1*, *BTN3A3*, *NHLH1*, and *SLC16A7* genes
Genome-wide methylation	Peripheral blood and postmortem brain	Illumina Infinium HumanMethylation27	Hypomethylation of the promoter region of the *ST6GALNAC1* gene in affected twins with psychosis and in brains from patients
Genome-wide methylation	Postmortem frontal cortex and germline	CpG-island microarrays	Specific differences in methylation between males and females. No difference between controls and patients in the methylome of germlines
Global methylation	Whole blood	ELISA-based assay	Lower 5hmC levels in BD patients, but no difference in global 5mC levels compared to controls
Global methylation	Leukocyte	Luminometric Methylation Assay	Methylation in BD patients stable on medication was significantly influenced by insulin resistance, second-generation antipsychotic use, and smoking
Global methylation	Transformed lymphoblasts	ELISA	Decreased methylation in BD subjects and their relatives compared to controls
Global methylation	Leukocytes	Cytosine-extension assay	No differences between patients and controls
Global methylation, *COX-2*, *BDNF*, debrin-like protein	Postmortem frontal cortex	MSP and ELISA (for global methylation)	Global hypermethylation; *COX-2* hypomethylation; *BDNF* hypermethylation; debrin-like protein gene hypermethylation
HCG9	Postmortem prefrontal cortex	Bisulfite pyrosequencing	Significant differences between patients and controls detected in CpG modifications
HCG9	Postmortem brain tissues, peripheral blood cells, and sperm	Bisulfite pyrosequencing	Low *HCG9* methylation in patients

(continued)

Table 2.1 (continued)

Gene	Cell/tissue	Method	Main findings
KCNQ3	Postmortem prefrontal cortex	Bisulfite pyrosequencing	Lower methylation of the *KCNQ3* exon 11 in BD
MB-COMT	Saliva	MSP	Hypomethylation of the *MB-COMT* promoter
MB-COMT	Postmortem frontal lobe	MSP and bisulfite sequencing	Hypomethylation of the *MB-COMT* promoter
RELN	Postmortem forebrain	Restriction enzymes analysis and PCR	No correlation between *RELN* methylation and age (which was shown in controls)
RGS4	Dorsolateral prefrontal cortex	Bisulfite sequencing	No alteration in the methylation status of CpG islands of the *RGS4* gene
SLC6A4	Lymphoblastoid cell lines and postmortem prefrontal cortex	Bisulfite sequencing	Hypermethylation of the *SLC6A4* gene
Upstream regions of *SMS* and *PPIEL*	Transformed lymphoblastoid cells	MS-RDA	Aberrant methylation in upstream regions of the *SMS* and *PPIEL* genes

5mC—5-methylcytosine; 5hmC—5-hydroxymethylcytosine; 5-HT3AR—serotonin receptor 3A; 5HTR1A—serotonin receptor 1A; 5HTR2A—serotonin receptor 2A; 5-HTT—serotonin transporter; BD—bipolar disorder; BDNF—brain-derived neurotrophic factor; BTN3A3—butyrophilin, subfamily 3, member A3; COX-2—cyclooxygenase-2; CYP11A1—Cytochrome P450, Family 11, Subfamily A, Polypeptide 1; DTNBP1—dystrobrevin binding protein 1; ELISA—enzyme-linked immunosorbent assay; FAM63B—family with sequence similarity 63, member B; FANCI—Fanconi anemia, complementation group I; FKBP5—FK506-binding protein 5; GAD1—glutamate decarboxylase 1; HCG9—HLA Complex Group 9; HRM—high-resolution melting; KCNQ3—potassium channel, voltage gated KQT-like subfamily Q, member 3; MB-COMT—membrane-bound catechol-O-methyltransferase; MDD—major depressive disorder; MeDIP—Methylated DNA immunoprecipitation; MSP—methylation-specific PCR; MS-RDA—methylation-sensitive representational difference analysis; NHLH1—nescient helix loop helix 1; PIK3R1—phosphoinositide-3-kinase, regulatory subunit 1; PPIEL—peptidylprolyl isomerase E-like; RGS4—regulator of G protein signaling 4; RELN—reelin; SLC16A7—solute carrier family 1 (neurotransmitter transporter, serotonin) member 7; SLC6A4—solute carrier family 6 (neurotransmitter transporter, serotonin) member 4; SMS—spermine synthase

of this review is on clues to circadian-related mechanisms of action underlying the rapid treatment of depressive symptoms with SDT.

Bipolar disorder (BD) and major depressive disorder (MDD) are heritable neuropsychiatric disorders associated with disrupted circadian rhythms. The hypothesis that circadian clock dysfunction plays a causal role in these disorders has endured for decades but has been difficult to test and remains controversial. In the meantime, the discovery of clock genes and cellular clocks has revolutionized our understanding of circadian timing. Cellular circadian clocks are located in the suprachiasmatic nucleus (SCN), the brain's primary circadian pacemaker, but also throughout the

Table 2.2 miRNA findings in postmortem tissue from patients with bipolar disorder

Tissue	Sample	Main findings
ACC (BA 24)	Patients with BD ($n = 5$) and controls ($n = 6$)	Higher levels of miR-149 in exosomes from patients compared to controls, likely due to increased expression of miR-149 in glial cells (not neurons)
ACC (BA 24)	Patients with BD ($n = 8$), major depression ($n = 15$), and controls ($n = 14$)	Four miRNAs (miR-132, miR-133a, miR-212, and miR-34a) showed nominal p-values <0.05 (downregulation in BD compared to controls), but not after multiple testing correction. A trend was found for miR-195, as well
DLPFC (BA 46)	Patients with BD ($n = 32$), schizophrenia ($n = 35$), and controls ($n = 34$)	Nonsignificant trend for reduced expression of miR-346 and significant lower expression of GRID1 (which encodes for this miRNA) in BD patients compared to controls. Strong correlation between miRNA-346 and GRID1 expression levels
DLPFC (BA 46)	Patients with BD ($n = 31$), schizophrenia ($n = 35$), and controls ($n = 34$)	Ten significantly different miRNAs between BD patients and controls (miR-383, miR-32, miR-490-5p, miR-196b, miR-513-5p, miR-876-3p, miR-449b, miR-297, miR-188-5p, miR-187), all upregulated
PFC (BA 9)	Patients with BD ($n = 35$), schizophrenia ($n = 35$) and controls ($n = 35$)	19% of miRNAs were altered in BD and schizophrenia patients. Both diagnoses were associated with reduced miRNA expression levels, especially BD
PFC (BA 9)	Patients with BD ($n = 9$), schizophrenia ($n = 8$), and controls ($n = 13$)	miR-29c was significantly increased in extracellular vesicles extracted from prefrontal cortex from BD samples compared to controls
PFC (BA9), ACC, and cerebellum	Patients with BD (frontal cortex, $n = 31$; ACC, $n = 29$; cerebellum, $n = 29$) and controls (frontal cortex, $n = 33$; ACC, $n = 29$; cerebellum, $n = 34$)	Increased miR-34a expression in the cerebellum of BD patients, with no differences in the other brain regions. The increase was only significant in patients without treatment. Significant correlations between miR-34a and age of onset and psychotic symptoms. Increased miR-34a levels were also detected in BD-derived hNPCs and neurons

(continued)

Table 2.2 (continued)

Tissue	Sample	Main findings
PFC (BA 10)	Patients with BD ($n = 15$), major depression ($n = 15$), schizophrenia ($n = 15$), and controls ($n = 15$)	Nine miRNAs were significantly altered in BD (fourupregulated—miR-17-5p, miR-579, miR-106b-5p, miR-29c-3p; and 5 downregulated—miR-145-5p, miR-485-5p, miR-370, miR-500a-5p, miR-34a-5p)
PFC (BA 46)	Patients with BD ($n = 35$), schizophrenia ($n = 35$), and controls ($n = 34$)	Fifteen miRNAs were differentially expressed between BD patients and controls (upregulated: miR-504, miR-145, miR-22, miR-145*, miR-133b, miR-154*, and miR-889; downregulated: miR-454*, miR-29a, miR-520c-3p, miR-140-3p, miR-767-5p, miR-874, miR-32, and miR-573)
Several brain regions	Cohort 1: Patients with BD ($n = 9$), schizophrenia ($n = 7$) and controls ($n = 10$); Cohort 2: Patients with BD ($n = 31$), schizophrenia ($n = 35$), and controls ($n = 33$)	No significant differences in miR-137 expression between groups. Lower miR-137 expression in homozygous TT subjects for the SNP rs1625579 in controls
Hippocampal dentate gyrus granule cells	Patients with BD ($n = 16$), schizophrenia ($n = 17$), major depression ($n = 17$), and controls ($n = 29$)	Cumulative comparisons identified 141 genes as likely to be involved in shaping dentate gyrus expression in mental illness. Twenty miRNAs or miRNA families targeted at least 25% of the most heavily weighted genes (of which miR-182 remained significant after statistical correction)

Legend: ACC, anterior cingulate cortex; BA, Brodmann area; BD, bipolar disorder; DLPFC, dorsolateral prefrontal cortex; hNPC, human neuroprogenitor cells; miRNA, microRNA; PFC, prefrontal cortex. The * symbol denotes that this miRNA comes from the same hairpin miRNA as the "main" miRNA, but is the less predominant form

brain and peripheral tissues. In BD and MDD patients, defects have been found in SCN-dependent rhythms of body temperature and melatonin release. However, these are imperfect and indirect indicators of SCN function. Moreover, the SCN may not be particularly relevant to mood regulation, whereas the lateral habenula, ventral tegmentum, and hippocampus, which also contain cellular clocks, have established roles in this regard. Dysfunction in these non-SCN clocks could contribute directly to the pathophysiology of BD/MDD. We hypothesize that circadian clock dysfunction in non-SCN clocks is a trait marker of mood disorders, encoded by pathological

Table 2.3 Circulating miRNA findings in patients with bipolar disorder

Tissue	Sample	Main findings
Lymphoblastoid cell lines	Patients with BD (n = 10) and discordant unaffected siblings (n = 10)	Lithium treatment changed the expression of 7 miRNAs at day 4 (upregulated: miR-221, miR-152, miR-15a, miR-155, miR-34a, miR-181c; downregulated: miR-494), and 4 miRNAs at day 16 of treatment (upregulated: miR-34a, miR-221, miR-152, miR-155)
Peripheral blood monocytes	Patients with postpartum psychosis (n = 28), among which patients with a previous diagnosis of BD (n = 11), and controls (n = 28)	Reduced expression of miR-146a in patients with a previous history of BD compared to controls. Reduced expression of miR-212 and miR-92a compared to controls and first-onset postpartum psychosis patients
Plasma	Patients with BD (n = 21) and controls (n = 21)	Decrease of miR-134 plasma levels in drug-free, 2-week medicated, and 4-week medicated bipolar mania patients compared to controls
Whole blood	Patients with BD (n = 20), major depression (n = 20), and controls (n = 20)	Thirteen miRNAs were altered in BD patients as compared to controls: miR-720, miR-140-3p, miR-1973, miR-30d-5p, miR-3158-3p, miR-330-5p, miR-378a-5p, miR-4521, miR-21-3p (upregulated) and miR-1915-5p, miR-1972, miR-4793-3p, miR-4440 (downregulated)
Whole blood	Patients with BD (n = 10), before and after treatment with antipsychotics	Sixteen miRNAs were differentially expressed after treatment with asenapine (14 upregulated: miR-18a-5p, miR-19b-3p, miR-145-5p, miR-27a-3p, miR-148b-3p, miR-210-3p, miR-17-3p, miR-30b-5p, miR-106b-5p, miR-339-5p, miR-106a-5p, miR-20a-5p, miR-17-5p, miR-15a-5p; and 2 downregulated: miR-92b-5p, 1343-5p). Three miRNAs were downregulated after risperidone treatment (miR-664b-5p, miR-6778-5p, miR-146b-5p)
Whole blood	Relatives of patients with BD (n = 34, high-risk individuals) and controls (n = 46)	Three miRNAs, miR-15b, miR-132, and miR-652) were upregulated in the high-risk individuals

Legend: BD, bipolar disorder

genetic variants. Because network features of the SCN render it uniquely resistant to perturbation, previous studies of SCN outputs in mood disorders patients may have failed to detect genetic defects affecting non-SCN clocks, which include not only mood-regulating neurons in the brain but also peripheral cells accessible in human subjects. Therefore, reporters of rhythmic clock gene expression in cells from patients or mouse models could provide a direct assay of the molecular gears of the clock, in cellular clocks that are likely to be more representative than the SCN of mood-regulating neurons in patients. This approach, informed by the new insights and tools of modern chronobiology, will allow a more definitive test of the role of cellular circadian clocks in mood disorders.

Mood

Diurnal variation in mood may offer some clues to underlying clock gene function. A subset of depressed patients shows marked diurnal patterns in mood that can persist throughout the course of a depressive episode (Mundo et al. 2003). These patients may experience a severe psychotic depression in the early morning hours and improve to a relatively euthymic state by evening (Detera-Wadleigh and McMahon 2006; Iwamoto et al. 2005). Diurnality in mood is associated with higher rates of response to SDT (Adzic et al. 2016; Srivastava et al. 2018).

Sleep

Sleep disturbances are one of the most prevalent symptoms of depression and affects an estimated 70–80% of patients. Disruptions in sleep include early morning awakening and difficulties in initiating and maintaining sleep, as well as a decrease in latency from the time of sleep onset to the onset of rapid eye movement (REM) sleep (Gray 2017; Mattson et al. 2008). Severe insomnia is associated with a more severe form of depression (Herrero-Mendez et al. 2009) and may include an increased risk for suicidal ideation (Cikankova et al. 2017) and a higher probability for poor treatment outcome. Insomnia may also be a biomarker for the onset of a depressive episode, while normalization of sleep patterns can be an early predictor of antidepressant response (Uguz et al. 2016). In nondepressed individuals, subjects with insomnia have a twofold risk for developing depression (Goh et al. 2014). Although alterations in sleep architecture in BPD and MDD are severe and probably involve phase changes, they are not unique to mood disorders.

Cortisol

A literature review of data, including 1080 plasma cortisol assays in depressed patients and control subjects, reported elevated cortisol levels in depressed patients versus control subjects at all time points around a 24 h period. Cortisol levels in depressed patients peaked earlier (3:00–4:00 am) than control subjects (7:00–8:00 am), suggesting a 3–4 h phase advance (Bhat et al. 2015), although early morning awakening could account for the earlier rise in cortisol. Additional factors implicating the hypothalamic–pituitary–adrenal function include hypersecretion of corticotropin-releasing factor, decreased corticosteroid receptor function, and blunted dexamethasone suppression tests (Sorrentino et al. 2018; Malhi et al. 2015).

Hypothalamic–pituitary–adrenal function may normalize as depressive symptoms remit (Anderson et al. 2012).

Temperature

Elevated nocturnal core body temperature is associated with depressive symptoms and may also normalize as patients improve (Woods 2000; Andreazza et al. 2010). Higher minimal core body temperature may predict response to SDT (Berk and Nierenberg 2015; Kato and Kato 2000). Clinical data suggest a relationship between core body temperature, severity of depression, and the timing of sleep (Scola et al. 2013).

Clock genes and bipolar disorder

Of all major mental disorders, the evidence for genetic abnormalities associated with clock genes is strongest in BPD. An analysis of 46 SNPs of 8 clock genes (BMAL1, CLOCK, PER 1, 2, 3, CRY 1, 2, and TIM) revealed association of BMAL1 and TIM with BPD using family-based samples with BPD or schizophrenia. The intracellular clock gene machinery involves complex multiple interlocking transcriptional—translational loops containing positive and negative transcription factors that adjust rhythms to an approximate 24 h cycle. Even though these were modest associations found using very liberal analyses, the association with BMAL1 has been independently confirmed using haplotype analysis. The same study also demonstrated an association with PER3. Studies examining other genes have found negative results: screening for human PER2 mutations at the CKIδ/ε binding site showed no difference in frequency between BPD patients and controls (103), nor was there any evidence for linkage or association of CRY1 with BPD. Interestingly, a recent study pointed to a general reduction in the amplitude of clock gene mRNAs infibroblasts taken from BPD patients. The core loop includes a BMAL1/CLOCK (or the paralog to CLOCK, NPAS2) heterodimer that binds to Enhancer Box (E-box) containing elements on the promoters of the core clock genes, period (PER1, PER2, PER3) and cryptochrome (CRY1, CRY2), in the nucleus to drive transcription of the period and cryptochrome genes. Period and cryptochrome messenger RNAs move into the cytoplasm where their proteins are transcribed. Following translation, the period and cryptochrome proteins form heterodimers and translocate back into the nucleus to inhibit their own transcription. The rhythmic expression of BMAL1 is modulated by additional loops, which consist of retinoid-related orphan receptors, D-box-binding protein (activator), and Rev-erbα (repressor) of BMAL1 gene transcription. Rev-erbα is also activated by BMAL1/CLOCK and represents a link between the positive and negative loops of the circadian clock (Morris et al. 2017). Relay projections from the SCN to regions, including the hypothalamus (e.g., dorsomedial and ventrolateral preoptic areas), midbrain, and brain arousal systems, regulate the sleep/wake cycle (Cataldo et al. 2010). A family of nuclear receptors control circadian rhythms of energy, glucose, and lipid metabolism. These nuclear receptors include Rev-Erbα, RORα, and peroxisome proliferator-activated receptor-alpha. PER2 interacts with PPAR-α and Rev-erbα to modulate the BMAL1 gene (Clay et al. 2011).

Sleep deprivation differentially affects neurotransmitter systems, including serotoninergic, cholinergic, noradrenergic, and dopaminergic function (de Sousa et al. 2014). One of the most consistent findings comes from data showing that it enhances serotonergic function, similar to that of the actions of many antidepressant medications (Morris and Berk 2015) in humans (Duong et al. 2016) and animals (Machado et al. 2016). Functional polymorphisms within the promoter of the serotonin transport gene, serotonin transporter-linked polymorphic region, may influence SDT response as seen in BPD patients who were homozygotic for the long variant of serotonin transporter-linked polymorphic region. Those patients with the long variant showed significantly greater improvement with SDT than patients with the short allele (Kim et al. 2015), although this was not replicated in an independent study (Paraskevaidi et al. 2017).

Accumulating evidence implicating abnormal clock gene machinery in mood disorders comes from studies documenting an association between single nucleotide polymorphisms (SNPs) in circadian genes and depression. Variants associated with MDD include RORB (rs2028122) and CRY1 (rs2287161), while variants in BPD include RORB (rs7022435, rs3750420, rs1157358, rs3903529, and rs10869435) and NR1D1 (Rev-erbα) (rs2314339), DEC1 (rs1537720, rs10982664), and BMAL1 (ARNTL) (rs747601) (Hroudova and Fisar 2011; Abdolmaleky et al. 2015). In addition to SDT, only one other treatment approach, low-dose ketamine, has been shown to robustly and rapidly decrease depressive symptoms within 24 h in 50–70% of the patient (Karsli-Ceppioglu 2016; Goud Alladi 2018). We hypothesized that the rapid antidepressant effects of ketamine could also involve the modulation of clock genes and circadian rhythms similar to that proposed for SDT. Compatible with this, our recent study (Reichenberg et al. 2009) in neuronal cell culture showed that ketamine influences BMAL1/CLOCK (NPAS2) function leading to altered circadian gene expression. Ketamine blocked the BMAL1 transcription of PER1. This effect was abolished when the E-boxes to which BMAL1/CLOCK (NPAS2) binds were mutated to impair binding. Furthermore, GSK3B inhibition blocked ketamine's modulation of BMAL1/CLOCK (NPAS2) driven transcription. Thus, a potential mechanism of the rapid action of both ketamine and SDT could involve resetting clock gene machinery to normalize dysregulated circadian rhythms and rapidly treat mood disorders.

Clinical research into the mechanisms of action of SDT has been, theoretically, limited by the relative inability to blind subjects to treatment. Efforts to blind subjects have included comparisons in efficacy between the more effective late partial sleep deprivation with early partial sleep deprivation. It can be argued, however, that many mood disorder patients who have experienced severe depressive symptoms for many months rapidly and dramatically improve within 24 h following SDT. The timing of the rapid antidepressant response significantly increases confidence that SDT is associated with a marked decrease in the symptoms in depression. Thus far, research has not identified the specific components of clock gene machinery that regulate sleep processes, although it has been shown that BMAL1/CLOCK are key players in this process. Another limitation is that it has not been possible to measure clock gene machinery in living human brain or to correlate it with the peripheral expression of clock genes.

2.13 Brain-Derived Neurotrophic Factor in Bipolar Disorder

Candidate Gene Analysis

Early studies of imaging genetics included association analyses between brain morphology and a single nucleotide polymorphism whose function was well known. Numerous researchers tested their hypotheses using imaging genetics according to the function of the Val158Met single-nucleotide polymorphism (SNP) of catechol-O-methyltransferase (COMT) (Jiang et al. 2008), the Val66Met SNP of brain-derived neurotrophic factor (BDNF) (Borrelli et al. 2008), and the Ser704Cys SNP of disrupted-in-schizophrenia 1 (DISC1) (Peedicayil 2011), among other genes.

The COMT gene is one of the most frequently investigated gene targets of functional imaging genetics and structural imaging genetics. COMT is a major mammalian enzyme involved in the metabolic degradation of catecholamines; this activity is higher in the Val158 protein than in the Met158 protein. The Val158 genotype was associated with lower executive performance, less efficient physiological response in the prefrontal cortex and a risk for schizophrenia (Millan 2013; Szyf 2015). Many subsequent neuroimaging studies using structural MRI have been reported using the functional imaging genetics method. Four of six studies in healthy controls or the general population in adults and children showed decreased hippocampal gray matter volume in the Val158 allele, although two studies failed to replicate this finding (Fries et al. 2016; Leon et al. 1983). Two studies in patoemts schizophrenia and individuals at high genetic risk for schizophrenia had smaller gray matter density in the anterior cingulate cortex (Schubeler 2015; Reddington et al. 2013).

BDNF modulates hippocampal plasticity and hippocampal-dependent memory in cell models and animals. BDNF is a very important molecule in the neuroscience field, and it has also been implicated in several neuropsychiatric disorders (Lo and Weksberg 2014; Fries et al. 2018; Pflug and Tolle 1971). The functional Val66Met polymorphism is well studied in imaging genetics. At the cellular level, the Met66-type BDNF protein showed a lower depolarization-induced secretion of BDNF and failed to localize to secretory granules or synapses compared to the Val66-type BDNF protein (Martiny 2018; Takaesu 2018; Wehr 2018). That study also indicated that Met66 was associated with poorer episodic memory, abnormal hippocampal activation and lower hippocampal n-acetyl aspartate (NAA) in human subjects. There are more than 20 association studies between the BDNF Val66Met SNP and hippocampal volume. A recent systematic review and meta-analysis analyzed 25 studies including healthy controls and patients with schizophrenia, bipolar disorder, and major depressive disorder and showed that carrying a met allele at the BDNF val66met locus is associated with lower hippocampal volumes (Haug 1992). Four studies examined the gene–environment interaction between BDNF Val66Met and childhood abuse in healthy controls and in patients with psychiatric disease. Three studies showed reduced hippocampal volume in Met carriers with childhood abuse or early life stress (Riemann et al. 1991; Manber et al. 2008; Mendlewicz 2009), and one study demonstrated a smaller volume of the anterior cingulate cortex in Met carriers with a history of childhood adversity (O'Brien et al. 2011). Few studies reported an association between gray matter volume reduction related to disease progression and BDNF Val66Met, which would be suspected because BDNF is an essential neurotrophic factor in the brain. Met allele carriers had significantly greater reductions in frontal gray matter volume over an average of 3 years with schizophrenia (Krakow et al. 2011).

Chromosome translocation analysis first identified DISC1 as a candidate gene for psychiatric illness, and this gene segregated with schizophrenia, bipolar disorder and major depressive disorder in a single pedigree (Gupta and Lahan 2011). A substantial body of evidence demonstrated the role of this gene in multiple pathways, such as cell migration, neurite organization, the cytoskeleton, mitochondrial function, and glutamate signaling, implicated in psychiatric disorders (Baglioni et al. 2011). The Ser704Cys SNP in the DISC1 gene has been studied extensively in imaging genetics because this SNP is a functional SNP. The Cys704-type DISC1 protein showed less ERK activation compared to the Ser704-type DISC1 protein in primary neuron cultures (Carpenter and Bunney 1971; Nemeroff 1996; Bao et al. 2008; Scharnholz et al. 2010; Souetre et al. 1988). ERK is a key molecule in neuronal plasticity, development, and death. A recent systematic review summarized the effects of the Ser704Cys SNP on volumes in frontal and hippocampal regions (Power et al. 2017). The results of four studies in frontal regions have not been replicated, and four reports focusing on hippocampal structures were inconsistent. Six studies examined the volumetric effects of the Leu607Phe SNP, but their results have not been replicated.

Association studies between other genes that are implicated in psychiatric disorders, especially schizophrenia, such as dystrobrevin binding protein 1 (DTNBP1), neuregulin-1 (NRG1), and v-akt murine thymoma viral oncogene homolog 1 (AKT1), and brain regional volumes have been reported. Four studies examined the association between an SNP or schizophrenia-risk haplotype in DTNBP1 and hippocampal, prefrontal, and total brain volumes. Three studies found an influence of genetic variants on brain volumes in different regions, but one study failed to find this association (Avery et al. 1999; Duncan et al. 1980; Elsenga and Van den Hoofdakker 1988; Hasler et al. 2010; Zhang and Kay 2010). Two NRG1 studies showed an association of an SNP or schizophrenia-risk haplotype and temporal region volumes (Zhang and Kay 2010; Robillard et al. 2018; Wright et al. 2012; Basheer et al. 2007), but another study failed to replicate those findings (Ressler and Nemeroff 2000).

Acknowledgements Parts of the contents in this chapter were reused from the articles (Maturana et al. 2015; Zant et al. 2011) with permissions.

References

Abdolmaleky HM, Zhou JR, Thiagalingam S (2015) An update on the epigenetics of psychotic diseases and autism. Epigenomics 7(3):427–449

Adzic M et al (2016) Antidepressant action on mitochondrial dysfunction in psychiatric disorders. Drug Dev Res 77(7):400–406

Amin N, Belonogova NM, Jovanova O et al (2017a) Nonsynonymous variation in NKPD1 increases depressive symptoms in european populations. Biol Psychiat 81(8):702–707
Amin N, Jovanova O, Adams HH et al (2017b) Exome-sequencing in a large population-based study reveals a rare Asn396ser variant in the LIPG gene associated with depressive symptoms. Mol Psychiatry 22(4):537–543
Amin N, De Vrij FMS, Baghdadi M et al (2018) A rare missense variant in RCL1 segregates with depression in extended families. Mol Psychiatry 23(5):1120–1126
Anastasia A et al (2013) Val66Met polymorphism of BDNF alters prodomain structure to induce neuronal growth cone retraction. Nat Commun 4:2490
Ancin I et al (2011) Sensory gating deficit is associated with catechol-O-methyltransferase polymorphisms in bipolar disorder. World J Biol Psychiatry 12(5):376–384
Anderson IM, Haddad PM, Scott J (2012) Bipolar disorder. BMJ 345:e8508
Andersen AM, Pietrzak RH, Kranzler HR et al (2017) Polygenic scores for major depressive disorder and risk of alcohol dependence. JAMA Psychiatry 74(11):1153–1160
Andreazza AC et al (2010) Mitochondrial complex I activity and oxidative damage to mitochondrial proteins in the prefrontal cortex of patients with bipolar disorder. Arch Gen Psychiatry 67(4):360–368
Ashok AH et al (2017) The dopamine hypothesis of bipolar affective disorder: the state of the art and implications for treatment. Mol Psychiatry 22(5):666–679
Avery DH et al (1999) Nocturnal sweating and temperature in depression. Acta Psychiatr Scand 100(4):295–301
Azadmarzabadi E, Haghighatfard A, Mohammadi A (2018) Low resilience to stress is associated with candidate gene expression alterations in the dopaminergic signalling pathway. Psychogeriatr: Off J Jpn Psychogeriatr Soc 18(3):190–201
Baghai TC et al (2003) No influence of a functional polymorphism within the serotonin transporter gene on partial sleep deprivation in major depression. World J Biol Psychiatry 4(3):111–114
Baglioni C et al (2011) Insomnia as a predictor of depression: a meta-analytic evaluation of longitudinal epidemiological studies. J Affect Disord 135(1–3):10–19
Bao AM, Meynen G, Swaab DF (2008) The stress system in depression and neurodegeneration: focus on the human hypothalamus. Brain Res Rev 57(2):531–553
Basheer R et al (2007) Sleep deprivation upregulates A1 adenosine receptors in the rat basal forebrain. NeuroReport 18(18):1895–1899
Bellet MM et al (2011) Ketamine influences CLOCK:BMAL1 function leading to altered circadian gene expression. PLoS ONE 6(8):e23982
Benedetti F et al (1999) Influence of a functional polymorphism within the promoter of the serotonin transporter gene on the effects of total sleep deprivation in bipolar depression. Am J Psychiatry 156(9):1450–1452
Berk M, Nierenberg AA (2015) Three paths to drug discovery in psychiatry. Am J Psychiatry 172(5):412–414
Berman RM et al (2000) Antidepressant effects of ketamine in depressed patients. Biol Psychiatry 47(4):351–354
Bhat AH et al (2015) Oxidative stress, mitochondrial dysfunction and neurodegenerative diseases; a mechanistic insight. Biomed Pharmacother 74:101–110
Borrelli E et al (2008) Decoding the epigenetic language of neuronal plasticity. Neuron 60(6):961–974
Bunney B, Bunney WE (2013) Mechanisms of rapid antidepressant effects of sleep deprivation therapy: clock genes and circadian rhythms. Biol Psychiat 73(12):1164–1171
Burokas A, Arboleya S, Moloney RD et al (2017) Targeting the microbiota-gut-brain axis: prebiotics have anxiolytic and antidepressant-like effects and reverse the impact of chronic stress in mice. Biol Psychiat 82(7):472–487
Bustamante AC, Aiello AE, Guffanti G et al (2018) Fkbp5 DNA methylation does not mediate the association between childhood maltreatment and depression symptom severity in the detroit neighborhood health study. J Psychiatr Res 96:39–48

Calabro M et al (2018) Neuroplasticity, neurotransmission and brain-related genes in major depression and bipolar disorder: focus on treatment outcomes in an asiatic sample. Adv Ther 35(10):1656–1670

Callicott JH et al (2005) Variation in DISC1 affects hippocampal structure and function and increases risk for schizophrenia. Proc Natl Acad Sci U S A 102(24):8627–8632

Cardoner N et al (2013) Val66Met BDNF genotypes in melancholic depression: effects on brain structure and treatment outcome. Depress Anxiety 30(3):225–233

Carpenter WT Jr, Bunney WE Jr (1971) Adrenal cortical activity in depressive illness. Am J Psychiatry 128(1):31–40

Carter CJ (2007) Multiple genes and factors associated with bipolar disorder converge on growth factor and stress activated kinase pathways controlling translation initiation: implications for oligodendrocyte viability. Neurochem Int 50(3):461–490

Cataldo AM et al (2010) Abnormalities in mitochondrial structure in cells from patients with bipolar disorder. Am J Pathol 177(2):575–585

Chang YH et al (2013) Genetic variants of the BDNF and DRD3 genes in bipolar disorder comorbid with anxiety disorder. J Affect Disord 151(3):967–972

Cheah SY et al (2015) Dysbindin (DTNBP1) variants are associated with hallucinations in schizophrenia. Eur Psychiatry 30(4):486–491

Chen H, Huffman JE, Brody JA et al (2018) Efficient variant set mixed model association tests for continuous and binary traits in large-scale whole-genome sequencing studies. Am J Hum Genet 104(2):260–274

Chesney E, Goodwin GM, Fazel S (2014) Risks of all-cause and suicide mortality in mental disorders: a meta-review. World Psychiatry: Off J World Psychiatr Assoc (WPA) 13(2):153–160

Cikankova T et al (2017) Mitochondrial dysfunctions in bipolar disorder: effect of the disease and pharmacotherapy. CNS Neurol Disord: Drug Targets 16(2):176–186

Clay HB, Sillivan S, Konradi C (2011) Mitochondrial dysfunction and pathology in bipolar disorder and schizophrenia. Int J Dev Neurosci 29(3):311–324

Craddock N, O'Donovan MC, Owen MJ (2005) The genetics of schizophrenia and bipolar disorder: dissecting psychosis. J Med Genet 42(3):193–204

de Sousa RT et al (2014) Targeting mitochondrially mediated plasticity to develop improved therapeutics for bipolar disorder. Expert Opin Ther Targets 18(10):1131–1147

Detera-Wadleigh SD, McMahon FJ (2006) G72/G30 in schizophrenia and bipolar disorder: review and meta-analysis. Biol Psychiatry 60(2):106–114

Dong Z et al (2016) Association of DISC1, BDNF, and COMT polymorphisms with exploratory eye movement of schizophrenia in a Chinese Han population. Psychiatr Genet 26(6):258–265

Duff BJ et al (2013) Human brain imaging studies of DISC1 in schizophrenia, bipolar disorder and depression: a systematic review. Schizophr Res 147(1):1–13

Duncan WC Jr et al (1980) Relationship between EEG sleep patterns and clinical improvement in depressed patients treated with sleep deprivation. Biol Psychiatry 15(6):879–889

Duong A et al (2016) Regulators of mitochondrial complex I activity: a review of literature and evaluation in postmortem prefrontal cortex from patients with bipolar disorder. Psychiatry Res 236:148–157

Elsenga S, Van den Hoofdakker RH (1988) Body core temperature and depression during total sleep deprivation in depressives. Biol Psychiatry 24(5):531–540

Fallin MD et al (2005) Bipolar I disorder and schizophrenia: a 440-single-nucleotide polymorphism screen of 64 candidate genes among Ashkenazi Jewish case-parent trios. Am J Hum Genet 77(6):918–936

Fang X, Zhang C, Wu Z et al (2018) Prevalence, risk factors and clinical characteristics of suicidal ideation in Chinese patients with depression. J Affect Disord 235:135–141

Fischer S, Gardini ES, Haas F et al (2018) Polymorphisms in genes related to the hypothalamic-pituitary-adrenal axis and antidepressant response—systematic review. Neurosci Biobehav Rev 96:182–196

Foo JC, Streit F, Treutlein J et al (2018) Shared genetic etiology between alcohol dependence and major depressive disorder. Psychiatr Genet 28(4):66–70

Fries GR et al (2016) The role of DNA methylation in the pathophysiology and treatment of bipolar disorder. Neurosci Biobehav Rev 68:474–488

Fries GR, Carvalho AF, Quevedo J (2018) The miRNome of bipolar disorder. J Affect Disord 233:110–116

Fukuda S et al (2010) A functional polymorphism in the disrupted-in schizophrenia 1 gene is associated with chronic fatigue syndrome. Life Sci 86(19–20):722–725

Gillentine MA, Lozoya R, Yin J et al (2018) CHRNA7 copy number gains are enriched in adolescents with major depressive and anxiety disorders. J Affect Disord 239:247–252

Goh S et al (2014) Mitochondrial dysfunction as a neurobiological subtype of autism spectrum disorder: evidence from brain imaging. JAMA Psychiatry 71(6):665–671

Goud Alladi C et al (2018) DNA methylation as a biomarker of treatment response variability in serious mental illnesses: a systematic review focused on bipolar disorder, schizophrenia, and major depressive disorder. Int J Mol Sci 19(10)

Gray MW (2017) Lynn Margulis and the endosymbiont hypothesis: 50 years later. Mol Biol Cell 28(10):1285–1287

Gupta R, Lahan V (2011) Insomnia associated with depressive disorder: primary, secondary, or mixed? Indian J Psychol Med 33(2):123–128

Hall LS, Adams MJ, Arnau-Soler A et al (2018) Genome-wide meta-analyses of stratified depression in generation scotland and UK biobank. Transl Psychiatry 8(1):9

Hashimoto R et al (2006) Impact of the DISC1 Ser704Cys polymorphism on risk for major depression, brain morphology and ERK signaling. Hum Mol Genet 15(20):3024–3033

Hashimoto R et al (2015) Imaging genetics and psychiatric disorders. Curr Mol Med 15(2):168–175

Hasler BP et al (2010) Phase relationships between core body temperature, melatonin, and sleep are associated with depression severity: further evidence for circadian misalignment in non-seasonal depression. Psychiatry Res 178(1):205–207

Haug HJ (1992) Prediction of sleep deprivation outcome by diurnal variation of mood. Biol Psychiatry 31(3):271–278

Havelka Mestrovic A et al (2018) Significant association between catechol-O-methyltransferase (COMT) Val(158/108) Met polymorphism and cognitive function in veterans with PTSD. Neurosci Lett 666:38–43

Hek K, Demirkan A, Lahti J et al (2013) A genome-wide association study of depressive symptoms. Biol Psychiat 73(7):667–678

Herrero-Mendez A et al (2009) The bioenergetic and antioxidant status of neurons is controlled by continuous degradation of a key glycolytic enzyme by APC/C-Cdh1. Nat Cell Biol 11(6):747–752

Howard DM, Adams MJ, Shirali M et al (2018) Genome-wide association study of depression phenotypes in UK biobank identifies variants in excitatory synaptic pathways. Nat Commun 9(1):1470

Hroudova J, Fisar Z (2011) Connectivity between mitochondrial functions and psychiatric disorders. Psychiatry Clin Neurosci 65(2):130–141

Hu MC et al (2013) Association study of DRD2 and MAOA genes with subtyped alcoholism comorbid with bipolar disorder in Han Chinese. Prog Neuropsychopharmacol Biol Psychiatry 40:144–148

Ikeda M, Saito T, Kondo K et al (2018) Genome-wide association studies of bipolar disorder: a systematic review of recent findings and their clinical implications. Psychiatry Clin Neurosci 72(2):52–63

Iwamoto K, Bundo M, Kato T (2005) Altered expression of mitochondria-related genes in postmortem brains of patients with bipolar disorder or schizophrenia, as revealed by large-scale DNA microarray analysis. Hum Mol Genet 14(2):241–253

Jiang Y et al (2008) Epigenetics in the nervous system. J Neurosci 28(46):11753–11759

Jiang X et al (2009) Human BDNF isoforms are differentially expressed in cocaine addicts and are sorted to the regulated secretory pathway independent of the Met66 substitution. Neuromolecular Med 11(1):1–12

Joo EJ et al (2007) Dysbindin gene variants are associated with bipolar I disorder in a Korean population. Neurosci Lett 418(3):272–275

Kaalund SS et al (2014) Contrasting changes in DRD1 and DRD2 splice variant expression in schizophrenia and affective disorders, and associations with SNPs in postmortem brain. Mol Psychiatry 19(12):1258–1266

Kailainathan S et al (2016) Activation of a synapse weakening pathway by human Val66 but not Met66 pro-brain-derived neurotrophic factor (proBDNF). Pharmacol Res 104:97–107

Kandaswamy R et al (2013) Genetic association, mutation screening, and functional analysis of a Kozak sequence variant in the metabotropic glutamate receptor 3 gene in bipolar disorder. JAMA Psychiatry 70(6):591–598

Kang Y et al (2015) Crystal structure of rhodopsin bound to arrestin by femtosecond X-ray laser. Nature 523(7562):561–567

Karsli-Ceppioglu S (2016) Epigenetic mechanisms in psychiatric diseases and epigenetic therapy. Drug Dev Res 77(7):407–413

Kato T, Kato N (2000) Mitochondrial dysfunction in bipolar disorder. Bipolar Disord 2(3 Pt 1):180–190

Kim HK, Chen W, Andreazza AC (2015) The potential role of the NLRP3 Inflammasome as a link between mitochondrial complex I dysfunction and inflammation in bipolar disorder. Neural Plast 2015:408136

Kirenskaya AV et al (2018) COMT and GAD1 gene polymorphisms are associated with impaired antisaccade task performance in schizophrenic patients. Eur Arch Psychiatry Clin Neurosci 268(6):571–584

Kishimoto M et al (2008) The dysbindin gene (DTNBP1) is associated with methamphetamine psychosis. Biol Psychiatry 63(2):191–196

Krakow B et al (2011) Sleep disturbances and suicidal ideation in sleep medical center patients. J Affect Disord 131(1–3):422–427

Labaka A, Gomez-Lazaro E, Vegas O et al (2017) Reduced hippocampal IL-10 expression, altered monoaminergic activity and anxiety and depressive-like behavior in female mice subjected to chronic social instability stress. Behav Brain Res 335:8–18

Lelli-Chiesa G et al (2011) The impact of the Val158Met catechol-O-methyltransferase genotype on neural correlates of sad facial affect processing in patients with bipolar disorder and their relatives. Psychol Med 41(4):779–788

Leon M, Morelis R, Louisot P (1983) Demonstration of a double subcellular localization of cerebral fucosyltransferase. Arch Int Physiol Biochim 91(2):121–125

Levinson DF, Mostafavi S, Milaneschi Y et al (2014) Genetic studies of major depressive disorder: why are there no genome-wide association study findings and what can we do about it? Biol Psychiat 76(7):510–512

Li M, Chang H, Xiao X (2016) BDNF Val66Met polymorphism and bipolar disorder in European populations: a risk association in case-control, family-based and GWAS studies. Neurosci Biobehav Rev 68:218–233

Li X, Luo Z, Gu C et al (2018) Common variants on 6q16.2, 12q24.31 and 16p13.3 are associated with major depressive disorder. Neuropsychopharmacol: Off Publ Am Coll Neuropsychopharmacol 43(10):2146–2153

Liu RT (2010) Early life stressors and genetic influences on the development of bipolar disorder: the roles of childhood abuse and brain-derived neurotrophic factor. Child Abuse Negl 34(7):516–522

Liu B et al (2015) DISC1 Ser704Cys impacts thalamic-prefrontal connectivity. Brain Struct Funct 220(1):91–100

Lo R, Weksberg R (2014) Biological and biochemical modulation of DNA methylation. Epigenomics 6(6):593–602

Lopez Leon S et al (2005) The dopamine D4 receptor gene 48-base-pair-repeat polymorphism and mood disorders: a meta-analysis. Biol Psychiatry 57(9):999–1003

Machado AK et al (2016) Upstream pathways controlling mitochondrial function in major psychosis: a focus on bipolar disorder. Can J Psychiatry 61(8):446–456

Malhi GS et al (2015) Agitation for recognition by DSM-5 mixed features specifier signals fatigue? Aust NZ J Psychiatry 49(6):499–501

Manber R et al (2008) Cognitive behavioral therapy for insomnia enhances depression outcome in patients with comorbid major depressive disorder and insomnia. Sleep 31(4):489–495

Marrocco J, Einhorn NR, Petty GH et al (2018) Epigenetic intersection of BDNF Val66Met genotype with premenstrual dysphoric disorder transcriptome in a cross-species model of estradiol add-back. Mol Psychiatry. https://doi.org/10.1038/s41380-018-0274-3

Martin J, Tammimies K, Karlsson R et al (2018) Copy number variation and neuropsychiatric problems in females and males in the general population. Am J Med Genet B Neuropsychiatr Genet 180(6):341–350

Martiny K et al (2018) Psychiatry and circadian rhythms. Ugeskr Laeger 180(36)

Massat I et al (2002) Excess of allele1 for alpha3 subunit GABA receptor gene (GABRA3) in bipolar patients: a multicentric association study. Mol Psychiatry 7(2):201–207

Mattson MP, Gleichmann M, Cheng A (2008) Mitochondria in neuroplasticity and neurological disorders. Neuron 60(5):748–766

Maturana MJ et al (2015) REM sleep deprivation reverses neurochemical and other depressive-like alterations induced by olfactory bulbectomy. Mol Neurobiol 51(1):349–360

McCarthy MJ et al (2012) A survey of genomic studies supports association of circadian clock genes with bipolar disorder spectrum illnesses and lithium response. PLoS ONE 7(2):e32091

Mendlewicz J (2009) Sleep disturbances: core symptoms of major depressive disorder rather than associated or comorbid disorders. World J Biol Psychiatry 10(4):269–275

Millan MJ (2013) An epigenetic framework for neurodevelopmental disorders: from pathogenesis to potential therapy. Neuropharmacology 68:2–82

Montag C et al (2010) The BDNF Val66Met polymorphism and anxiety: support for animal knock-in studies from a genetic association study in humans. Psychiatry Res 179(1):86–90

Morris G, Berk M (2015) The many roads to mitochondrial dysfunction in neuroimmune and neuropsychiatric disorders. BMC Med 13:68

Morris G et al (2017) A model of the mitochondrial basis of bipolar disorder. Neurosci Biobehav Rev 74(Pt A):1–20

Mundo E et al (2003) Evidence that the N-methyl-D-aspartate subunit 1 receptor gene (GRIN1) confers susceptibility to bipolar disorder. Mol Psychiatry 8(2):241–245

Nemeroff CB (1996) The corticotropin-releasing factor (CRF) hypothesis of depression: new findings and new directions. Mol Psychiatry 1(4):336–342

Nishino J, Ochi H, Kochi Y et al (2018) Sample size for successful genome-wide association study of major depressive disorder. Front Genet 9:227

Norkett R et al (2016) DISC1-dependent regulation of mitochondrial dynamics controls the morphogenesis of complex neuronal dendrites. J Biol Chem 291(2):613–629

O'Brien EM et al (2011) Severe insomnia is associated with more severe presentation and greater functional deficits in depression. J Psychiatr Res 45(8):1101–1105

O'connell CP, Goldstein-Piekarski AN, Nemeroff CB et al (2018) Antidepressant outcomes predicted by genetic variation in corticotropin-releasing hormone binding protein. Am J Psychiatry 175(3):251–261

Okbay A, Baselmans BM, De Neve JE et al (2016) Genetic variants associated with subjective well-being, depressive symptoms, and neuroticism identified through genome-wide analyses. Nat Genet 48(6):624–633

Orru G, Carta MG (2018) Genetic variants involved in bipolar disorder, a rough road ahead. Clin Pract Epidemiol Ment Health 14:37–45

Ostergaard SD, Jensen SO, Bech P (2011) The heterogeneity of the depressive syndrome: when numbers get serious. Acta Psychiatr Scand 124(6):495–496

Ottenhof KW et al (2018) TPH2 polymorphisms across the spectrum of psychiatric morbidity: a systematic review and meta-analysis. Neurosci Biobehav Rev 92:29–42

Paraskevaidi M et al (2017) Underlying role of mitochondrial mutagenesis in the pathogenesis of a disease and current approaches for translational research. Mutagenesis 32(3):335–342

Peedicayil J (2011) Epigenetic management of major psychosis. Clin Epigenetics 2(2):249–256

Peterson RE, Cai N, Bigdeli TB et al (2017) The genetic architecture of major depressive disorder in han chinese women. JAMA Psychiatry 74(2):162–168

Pflug B, Tolle R (1971) Disturbance of the 24-hour rhythm in endogenous depression and the treatment of endogenous depression by sleep deprivation. Int Pharmacopsychiatry 6(3):187–196

Pirooznia M, Wang T, Avramopoulos D et al (2016) High-throughput sequencing of the synaptome in major depressive disorder. Mol Psychiatry 21(5):650–655

Poletti S et al (2017) Brain-derived Neurotrophic Factor (BDNF) and gray matter volume in bipolar disorder. Eur Psychiatry 40:33–37

Power RA, Tansey KE, Buttenschon HN et al (2017) Genome-wide association for major depression through age at onset stratification: Major depressive disorder working group of the psychiatric genomics consortium. Biol Psychiat 81(4):325–335

Ratta-Apha W et al (2013) Association analysis of the DISC1 gene with schizophrenia in the Japanese population and DISC1 immunoreactivity in the postmortem brain. Neurosci Res 77(4):222–227

Raybould R et al (2005) Bipolar disorder and polymorphisms in the dysbindin gene (DTNBP1). Biol Psychiatry 57(7):696–701

Raznahan A et al (2011) Catechol-o-methyl transferase (COMT) val158met polymorphism and adolescent cortical development in patients with childhood-onset schizophrenia, their non-psychotic siblings, and healthy controls. Neuroimage 57(4):1517–1523

Reddington JP, Pennings S, Meehan RR (2013) Non-canonical functions of the DNA methylome in gene regulation. Biochem J 451(1):13–23

Reichenberg A, Mill J, MacCabe JH (2009) Epigenetics, genomic mutations and cognitive function. Cogn Neuropsychiatry 14(4–5):377–390

Reif A et al (2014) MAOA and mechanisms of panic disorder revisited: from bench to molecular psychotherapy. Mol Psychiatry 19(1):122–128

Ressler KJ, Nemeroff CB (2000) Role of serotonergic and noradrenergic systems in the pathophysiology of depression and anxiety disorders. Depress Anxiety 12(Suppl 1):2–19

Rice F, Riglin L, Thapar AK et al (2019) Characterizing developmental trajectories and the role of neuropsychiatric genetic risk variants in early-onset depression. JAMA Psychiatry 76(3):306–313

Richter-Schmidinger T et al (2011) Influence of brain-derived neurotrophic-factor and apolipoprotein E genetic variants on hippocampal volume and memory performance in healthy young adults. J Neural Transm (Vienna) 118(2):249–257

Riemann D, Wiegand M, Berger M (1991) Are there predictors for sleep deprivation response in depressed patients? Biol Psychiatry 29(7):707–710

Ripke S, Wray NR, Lewis CM et al (2013) A mega-analysis of genome-wide association studies for major depressive disorder. Mol Psychiatry 18(4):497–511

Robillard R et al (2018) Circadian rhythms and psychiatric profiles in young adults with unipolar depressive disorders. Transl Psychiatry 8(1):213

Rybakowski JK et al (2009) Dopamine D1 receptor gene polymorphism is associated with prophylactic lithium response in bipolar disorder. Pharmacopsychiatry 42(1):20–22

Scharnholz B et al (2010) Does night-time cortisol excretion normalize in the long-term course of depression? Pharmacopsychiatry 43(5):161–165

Schizophrenia Working Group of the Psychiatric Genomics Consortium (2014) Biological insights from 108 schizophrenia-associated genetic loci. Nature 511(7510):421–427

Schubeler D (2015) Function and information content of DNA methylation. Nature 517(7534):321–326

Scola G et al (2013) A fresh look at complex I in microarray data: clues to understanding disease-specific mitochondrial alterations in bipolar disorder. Biol Psychiatry 73(2):e4–e5

Smith K (2014) Mental health: a world of depression. Nature 515(7526):181

Soria V et al (2010) Differential association of circadian genes with mood disorders: CRY1 and NPAS2 are associated with unipolar major depression and CLOCK and VIP with bipolar disorder. Neuropsychopharmacology 35(6):1279–1289

Sorrentino V, Menzies KJ, Auwerx J (2018) Repairing mitochondrial dysfunction in disease. Annu Rev Pharmacol Toxicol 58:353–389

Souetre E et al (1988) Twenty-four-hour profiles of body temperature and plasma TSH in bipolar patients during depression and during remission and in normal control subjects. Am J Psychiatry 145(9):1133–1137

Srivastava R et al (2018) Dynamic changes of the mitochondria in psychiatric illnesses: new mechanistic insights from human neuronal models. Biol Psychiatry 83(9):751–760

Sullivan PF, Neale MC, Kendler KS (2000) Genetic epidemiology of major depression: review and meta-analysis. Am J Psychiatry 157(10):1552–1562

Szczepankiewicz A (2013) Evidence for single nucleotide polymorphisms and their association with bipolar disorder. Neuropsychiatr Dis Treat 9:1573–1582

Szyf M (2015) Prospects for the development of epigenetic drugs for CNS conditions. Nat Rev Drug Discov 14(7):461–474

Takaesu Y (2018) Circadian rhythm in bipolar disorder: a review of the literature. Psychiatry Clin Neurosci 72(9):673–682

Taylor S (2018) Association between COMT Val158Met and psychiatric disorders: a comprehensive meta-analysis. Am J Med Genet B Neuropsychiatr Genet 177(2):199–210

Teh CA et al (2012) Bipolar disorder, brain-derived neurotrophic factor (BDNF) Val66Met polymorphism and brain morphology. PLoS ONE 7(7):e38469

Tripp A et al (2012) Brain-derived neurotrophic factor signaling and subgenual anterior cingulate cortex dysfunction in major depressive disorder. Am J Psychiatry 169(11):1194–1202

Trost S et al (2013) The DTNBP1 (dysbindin-1) gene variant rs2619522 is associated with variation of hippocampal and prefrontal grey matter volumes in humans. Eur Arch Psychiatry Clin Neurosci 263(1):53–63

Uguz AC, Demirci K, Espino J (2016) The importance of melatonin and mitochondria interaction in mood disorders and schizophrenia: a current assessment. Curr Med Chem 23(20):2146–2158

Van Der Auwera S, Peyrot WJ, Milaneschi Y et al (2018) Genome-wide gene-environment interaction in depression: a systematic evaluation of candidate genes: the childhood trauma working-group of PGC-MDD. Am J Med Genet Part B, Neuropsychiatr Genet: Off Publ Int Soc Psychiatr Genet 177(1):40–49

Wehr TA (2018) Bipolar mood cycles associated with lunar entrainment of a circadian rhythm. Transl Psychiatry 8(1):151

Wong ML, Arcos-Burgos M, Liu S et al (2017) The PHF21B gene is associated with major depression and modulates the stress response. Mol Psychiatry 22(7):1015–1025

Wong BC, Chau CK, Ao FK et al (2019) Differential associations of depression-related phenotypes with cardiometabolic risks: polygenic analyses and exploring shared genetic variants and pathways. Depress Anxiety 36(4):330–344

Woods SW (2000) The economic burden of bipolar disease. J Clin Psychiatry 61(Supp 13):38–41

Wray NR, Ripke S, Mattheisen M et al (2018) Genome-wide association analyses identify 44 risk variants and refine the genetic architecture of major depression. Nat Genet 50(5):668–681

Wright KP, Lowry CA, Lebourgeois MK (2012) Circadian and wakefulness-sleep modulation of cognition in humans. Front Mol Neurosci 5:50

Xiao X, Wang L, Wang C et al (2017) Common variants at 2q11.2, 8q21.3, and 11q13.2 are associated with major mood disorders. Transl Psychiatry 7(12):1273

Xiao X, Zheng F, Chang H et al (2018) The gene encoding protocadherin 9 (PCDH9), a novel risk factor for major depressive disorder. Neuropsychopharmacol: Off Publ Am Coll Neuropsychopharmacol 43(5):1128–1137

Yu C, Baune BT, Wong ML et al (2017a) Investigation of copy number variation in subjects with major depression based on whole-genome sequencing data. J Affect Disord 220:38–42

Yu C, Baune BT, Licinio J et al (2017b) Whole-genome single nucleotide variant distribution on genomic regions and its relationship to major depression. Psychiatry Res 252:75–79
Yu C, Arcos-Burgos M, Licinio J et al (2017c) A latent genetic subtype of major depression identified by whole-exome genotyping data in a mexican-american cohort. Transl Psychiatry 7(5):e1134
Yu C, Arcos-Burgos M, Baune BT et al (2018a) Low-frequency and rare variants may contribute to elucidate the genetics of major depressive disorder. Transl Psychiatry 8(1):70
Yu C, Baune BT, Wong ML et al (2018b) Investigation of short tandem repeats in major depression using whole-genome sequencing data. J Affect Disord 232:305–309
Zant JC et al (2011) Increases in extracellular serotonin and dopamine metabolite levels in the basal forebrain during sleep deprivation. Brain Res 1399:40–48
Zarate CA Jr et al (2006) A randomized trial of an N-methyl-D-aspartate antagonist in treatment-resistant major depression. Arch Gen Psychiatry 63(8):856–864
Zeng Y, Navarro P, Fernandez-Pujals AM et al (2017) A combined pathway and regional heritability analysis indicates NETRIN1 pathway is associated with major depressive disorder. Biol Psychiat 81(4):336–346
Zeni CP et al (2016) BDNF Val66Met polymorphism and peripheral protein levels in pediatric bipolar disorder and attention-deficit/hyperactivity disorder. Acta Psychiatr Scand 134(3):268–274
Zhang EE, Kay SA (2010) Clocks not winding down: unravelling circadian networks. Nat Rev Mol Cell Biol 11(11):764–776
Zhang C, Wu Z, Zhao G et al (2016) Identification of IL6 as a susceptibility gene for major depressive disorder. Sci Rep 6:31264
Zhao L et al (2015) Association study of dopamine receptor genes polymorphism with cognitive functions in bipolar I disorder patients. J Affect Disord 170:85–90
Zhou X, Barrett TB, Kelsoe JR (2008) Promoter variant in the GRK3 gene associated with bipolar disorder alters gene expression. Biol Psychiatry 64(2):104–110

Chapter 3
Neuroimaging Advance in Depressive Disorder

Daihui Peng and Zhijian Yao

Abstract Neuroimaging shed light on the understanding of psychopathological mechanisms underlying major depressive disorder, despite its inconsistent findings. Noninvasive neuroimaging studies have indicated that various behavioral deficits in major depressive disorder are implicated with structural and functional abnormalities in specific brain regions. Moreover, disrupted brain morphological and functional properties may map out the underlying pathways from genetic and environmental factors to the prognosis of depression. Molecular neuroimaging studies have also provided novel method to probe transmitters and metabolites in brain regions rather than simply measuring brain morphological changes. Recent advanced neuroimaging approaches (e.g., pattern recognition) provides great opportunity to probe neuroimaging biomarkers that may contributes to improving diagnostic accuracy and predicting treatment outcomes. In this chapter, we conclude neuroimaging studies in the research field of depression from psychopathological, molecular, genetic/environmental, diagnostic, and therapeutic perspectives.

Keywords Neuroimaging · Major depressive disorder · Genetic · Molecular · Treatment · Classification

3.1 Neuroimaging Techniques Involved in MDD

Noninvasive neuroimaging techniques may provide the greatest hope for exploring biological substrates of depression, identifying depression subtypes, predicting treatment outcomes (Masdeu 2011; Linden 2012). Many forms of neuroimaging

D. Peng (✉)
Division of Mood Disorder, Shanghai Mental Health Center, School of Medicine, Shanghai Jiao Tong University, Shanghai, China
e-mail: pdhsh@126.com

Z. Yao (✉)
Department of Psychiatry, The Affiliated Brain Hospital of Nanjing Medical University, Nanjing, China
e-mail: zjyao@njmu.edu.cn

© Springer Nature Singapore Pte Ltd. 2019

Y. Fang (ed.), *Depressive Disorders: Mechanisms, Measurement and Management*, Advances in Experimental Medicine and Biology 1180,
https://doi.org/10.1007/978-981-32-9271-0_3

techniques have been applied to the research filed of depression, including structural magnetic resonance imaging (sMRI), functional MRI (fMRI), magnetic resonance spectroscopy (MRS), diffusion tensor imaging (DTI), positron emission tomography (PET), and near-infrared spectroscopy (NIRS) (Masdeu 2011; Bunge and Kahn 2009).

Numerous sMRI studies measures the structure of the brain, particularly the size, volume, or area of specific brain regions, which provide high-resolution gray matter abnormality in MDD patients (Symms et al. 2004). On the other hand, DTI techniques enable to specifically visualize the integrity of white matter fiber tracts (Alexander et al. 2007). fMRI is currently the most widely used brain imaging technique for measuring brain activity due to its good spatial resolution, widespread availability of MRI scanner and low costs. fMRI enable to detect the changes in blood oxygenation and flow responding to neural activity in specific brain regions either in the "resting state" (e.g., when patients are not involving in any particular stimulus) or in the "task state" that induces a cognitive or emotional response (Glover 2011). What's more, state-of-art techniques (e.g., machine learning and deep neural learning) have been applied to fMRI data to identify brain networks and connectivity (Librenza-Garcia et al. 2017).

NIRS is an optical technique that uses light transmission and absorption to measure blood oxygenation in the brain. More specifically, shine light in the near-infrared part of the spectrum (700–900 nm) through the skull, then detect how much the light is attenuated, thus provide an indirect measure of neural activity (Satomura et al. 2019). Another indirect measure of neural activity is PET through tracing amounts of short-lived radioactive material to observe metabolic processes in the brain. Furthermore, MRS also provides a noninvasive, ionizing-radiation-free analytical technique to detect metabolic changes in the brain (Su et al. 2014). Other types of neuroimaging for depression include electroencephalography (EEG) and magnetoencephalography (MEG). Noninvasive EEG can be able to detect changes in the electrical activity in the brain with high temporal resolution. Likewise, MEG is an imaging technique to measure the magnetic fields elicited by electrical activity in the brain. Detailed information of neuroimaging studying in MDD will be explained in the following section.

3.2 Abnormal Neuroimaging in MDD

Although substantial studies focused on identifying neurobiological underpinning of depression during the past decades, the pathophysiological mechanisms underlying depression still need to be elucidated. Considerable neuroimaging studies have provided robust evidence that various clinical symptoms in MDD are implicated with structural and functional abnormalities in specific brain regions, especially those regions supporting emotion regulation, cognitive control, and rewards. Therefore, we summarize the following neuroimaging studies to further understanding of the development of MDD through detecting structural and functional alteration in brain.

3.2.1 Abnormal Structural Neuroimaging in MDD

Although numerous neuroimaging studies aimed at identifying neurobiological substrates of depression, the neuro-anatomical underpinnings of MDD are still not clear. Previous structural neuroimaging studies found that depressed subjects are accompanied by structural abnormalities in many brain regions, including dorsomedial prefrontal cortex, orbitofrontal cortex, anterior cingulate cortex, hippocampus, striatum, amygdala, and caudate nucleus (Zhang et al. 2018a, 2019; Han et al. 2014; Pirnia et al. 2016). However, neuroimaging studies always failed to replicate previous findings. Inconclusive findings are owing to the limited sample size, differences in scanning and analysis methods, disease heterogeneity and clinical variations (e.g. medication status, illness duration, and other comorbidities).

In here, we present some largest coordinated worldwide meta-analyzed studies to identify brain cortical and subcortical morphological abnormality in MDD relative to healthy controls. Suh et al. (2019) performed a comprehensive and well-established literature search for neuroimaging studies measuring cortical thickness. This systematic review and meta-analysis indicated that depressed subjects had significant decreased cortical thickness in the left calcarine fissure/lingual gyrus, left pars opercularis, and bilateral orbitofrontal gyrus, as well as significant increased cortical thickness in the supramarginal gyrus. What's more, the largest ever worldwide study by the ENIGMA (Enhancing Neuro Imaging Genetics through Meta-Analysis) Major Depressive Disorder Working Group revealed some robust and rich findings on the cortical and subcortical abnormalities in patients with MDD (Renteria et al. 2017; Schmaal et al. 2016, 2017). As for cortical structural abnormalities, significant decreased cortical gray matter was observed in the anterior and posterior cingulate, orbitofrontal cortex, insula and temporal lobe in adults with MDD, especially in the first episode and adult-onset patients. Moreover, significant decreased total surface area, but no differences in cortical thickness, were observed in adolescents with MDD, especially in recurrent adolescent patients. Meanwhile, adolescents with MDD had significant regional reductions in frontal region and primary and higher order somatosensory, visual, and motor areas. As for the subcortical structural abnormalities, depressed patients with recurrent episode status and age of onset ≤ 21 showed significantly smaller hippocampal volumes relative to healthy controls. Therefore, depression may affect brain morphology in a highly dynamic way, which has different patterns of abnormalities at different stages of life.

Diffusion MRI studies using the diffusion tensor imaging (DTI) technique have used to identify microstructural brain abnormalities, particularly in the integrity of white matter. Whole-brain neuroimaging studies have indicated that MDD patients showed widely white matter abnormalities, including frontal cortex, thalamus, right parietal lobe, corpus callosum, cerebellum, and superior longitudinal fasciculus. Several recent meta-analysis studies revealed that depressed subjects showed fractional anisotropy alterations in the right inferior longitudinal fasciculus, right inferior fronto-occipital, and the right posterior thalamic radiation (Jiang et al. 2017; Wise et al. 2016). However, neuroimaging studies of DTI are hardly to replicate previous findings.

3.2.2 *Abnormal Functional Neuroimaging in MDD*

Recent advances in neuroimaging techniques (e.g., graph theory, functional connectivity, and effective connectivity) highlight disrupted topological characteristics of functional neuroimaging in depression. Rich and robust evidence has supported that MDD is implicated with widespread local alterations in many brain regions (Qiu et al. 2018; Zhang et al. 2016a), disrupted neural circuits (Zhang et al. 2018a), and abnormal large-scale brain connectomics (Gong and He 2015).

Considerable neuroimaging studies revealed that significant brain abnormalities were observed in prefrontal cortex, temporal lobe, and limbic system (Sheline et al. 2010; Arnone 2019; Tang et al. 2018). Consistently, disrupted neural circuits in depression, contributing to the neural substrates of various clinical symptoms, included prefrontal-subcortical circuit, dorsolateral prefrontal circuit, orbitofrontal prefrontal circuit, anterior cingulate-prefrontal circuit, prefrontal-hippocampal circuit, and fronto thalamic circuit (Zhang et al. 2018a; Marchand 2010; Tekin 2002; Wegener et al. 2015). Moreover, disrupted brain connectivity networks have been found in the default mode network (DMN), the central executive network (CEN), and the salience network (SN) (Zhang et al. 2016a). More specifically, DMN, predominantly consisting of medial prefrontal cortex, posterior cingulate cortex, precuneus and temporo-parietal cortices, involved in the psychological process of introspection. Although previous findings showed inconsistent, most neuroimaging studies indicated hyperactivity of DMN in MDD (Nejad et al. 2013). CEN consists of dorsolateral prefrontal cortex and pregenual anterior cingulate cortex, which supports the top–down modulation of attention and the regulation of emotional responses (Phillips et al. 2003). As for SN, this network, predominantly consisting of anterior insular cortex and dorsal anterior cingulate cortex, is thought to measure the correlation of external and internal stimuli to produce direct behaviors and appropriate responses (Liu et al. 2015a). Enough evidence indicated that MDD was implicated with several neural networks in numerous brain regions (Zhang et al. 2016a; Gong and He 2015; Sheline et al. 2010). In terms of the topological architecture of whole-brain network, MDD is also implicated with abnormal topological characteristics of the brain networks, including disrupted regional connectivity and global integrity (Wang et al. 2014a; Zhang et al. 2011). For instance, Zhang et al. (2011) revealed that depressed patients showed decreased nodal centralities in the frontal (orbital part), occipital, and temporal regions, and increased nodal centralities in the caudate nucleus and DMN regions (e.g., hippocampus, medial frontal, inferior parietal, and parietal regions).

As mentioned above, neuroimaging techniques provide a noninvasive method to explore the neural substrates underlying clinical symptoms. Previous studies have showed that affective, cognitive, and psychomotor deficits, as well as various combinations were implicated with specific brain regions, especially in prefrontal cortex, and limbic system (Maglanoc et al. 2019). For example, Yang et al. (2017a) found that anhedonia severity was associated with functional connectivity between ventral caudate and cuneus/middle temporal gyrus. Gong et al. (2018) further revealed that imbalance between reward network and cognitive control network activity also was implicated with anhedonia severity in patients with MDD. A meta-analysis from the ENIGMA-MDD working group revealed that significant smaller total intracranial volume was observed in patients reporting suicidal attempts or plans relative to healthy controls (Renteria et al. 2017). Interestingly, some pioneering neuroimaging studies have found neuroimaging biomarkers define neurophysiological subtypes (biotypes) of MDD instead of clinical symptoms (phenotype), despite inconsistent findings (Beijers et al. 2019; Feder et al. 2017; Price et al. 2017a, b). Drysdale et al. 2017 categorized individuals with MDD into four neurophysiological subtypes: low anhedonia-related connectivity, high anxiety-related connectivity; low anhedonia-related connectivity, low anxiety-related connectivity; high anhedonia-related connectivity, low anxiety-related connectivity; high anhedonia-related and high anxiety-related connectivity (Drysdale et al. 2017). Feder et al. (2017) subdivided depressed subjects into two biotypes: higher connectivity (longer duration, more sever); lower connectivity (shorter duration, less severe). Taken together, neural connectivity-based categorization revealed atypical patterns of brain connectivity in MDD patients, which would been overlooked in group comparisons (Price et al. 2017a, b).

In summary, these neuroimaging studies provide new perspectives to identify the psychopathology of depression, classify depression subtypes, and even enhance understanding of the biological substrates of therapeutic effects to optimize the treatment response in depression. Details will be described in the following paragraph at this chapter.

3.2.3 Other Neuroimaging Modalities in MDD

Emerging evidence has indicated that individuals with MDD suffered metabolic and neurotransmitters changes in the specific brain. Using 18F-fluoradeoxyglucose (18F-FDG) PET, significantly reduced regional cerebral glucose metabolism were observed in the bilateral insula, left lentiform nucleus, as well as right caudate and cingulate cortex, whereas higher metabolism was measured in the thalamus and cerebellum (Su et al. 2014). As for neurotransmitters, serotonergic system plays vital roles in the psychopathological mechanisms underlying MDD. Neuroimaging studies found that MDD patients showed higher serotonin 1A (5-HT1A) receptor binding relative to controls in anterior cingulate cortex, amygdala, hippocampus, and dorsolateral prefrontal cortex (Miller et al. 2013). On the other hand, serotonin transporter availability in the striatum, thalamus, and midbrain was significant reduction, which was associated with suicidal behaviors in depressed subjects

(Yeh et al. 2014). Moreover, midbrain serotonin transporter availability also was related with depression severity. Taken together, MDD was implicated with abnormalities of the serotonergic system.

In contrast, proton magnetic resonance spectroscopy (^{1}H-MRS) is a non-radioactive and noninvasive method measuring metabolites levels in specific brain regions, which further enhance the understanding of neuronal abnormalities at the metabolic and cellular levels. Abnormal amino acid neurotransmitter (e.g., glutamate and GABA) levels detected by ^{1}H-MRS were observed in the dorsomedial prefrontal cortex, dorsolateral prefrontal cortex, dorsoanterolateral prefrontal cortex, anterior cingulate cortex (Lener et al. 2017; Rosa et al. 2017; Shirayama et al. 2017). What's more, using combined fMRI-MRS methods, Zhang et al. (2016b) found that high spontaneous brain activity was associated with Glu concentration in medial prefrontal cortex. Taken together, MRS and PET shed light on the quantitative levels of metabolites and neurotransmitters underlying depression rather than simple quality changes of brain morphology.

3.3 Genetic and Environmental Influences on Neuroimaging in MDD

Considerable studies of genetic epidemiology of MDD revealed that both environment and genes contribute to MDD susceptibility (Flint and Kendler 2014; Zhang et al. 2018b). For instance, exposure to stress or family history boosts the risk of depression (Richter-Levin and Xu 2018). However, the underlying mechanism remains to be elucidated. Recently, genetic-neuroimaging studies have provided new insights into this field. Substantial studies suggested that risk genotypes of candidate genes or psychological stress do not directly modulate clinical symptoms, but regulate brain structure/function through cellular and molecular mechanisms (Pezawas and Meyer-Lindenberg 2010).

3.3.1 Genetic Influences on Neuroimaging in MDD

Genetic-neuroimaging studies suggested that candidate genes are associated with some common hypotheses of the pathogenesis of MDD, such as monoamine-deficiency hypothesis, altered glutamatergic neurotransmission, hypothalamic–pituitary–cortisol hypothesis, and neurotrophic factors hypothesis (Zhang et al. 2018b). Significant genetic-neuroimaging associations were found in brain-derived neurotrophic factor (BDNF) gene, monoaminergic genes, glutamatergic genes, hypothalamic-pituitary-adrenal (HPA) axis genes, renin–angiotensin–aldosterone (RAA) system-related genes, and the other common genes.

The effect of BDNF genetic variations on brain morphology and function is mainly found on the hippocampal and frontal regions, such as caudal middle frontal, rostral middle frontal, and anterior cingulate (Carballedo et al. 2013; Cardoner et al. 2013). BDNF involves in neurogenesis and synaptic plasticity in the brain, which may contribute to the pathophysiology of depression. However, some studies found that there was no significant BDNF genotype-diagnosis association on above-mentioned brain regions (Jaworska et al. 2016). Furthermore, significant interaction between stress and BDNF polymorphisms was observed (Hosang et al. 2014b). Therefore, BDNF polymorphisms-neuroimaging association remains inconclusive, and still needs to be clarified in future research.

Monoaminergic genes associated with brain structure/function alterations were those regulating the function of monoaminergic neurotransmitters, including tryptophan hydroxylase-1 (TPH1) gene, tryptophan hydroxylase-2 (TPH-2) gene, serotonin receptor 1A (HTR1A) gene, serotonin receptor 2A (HTR2A) gene, serotonin transporter polymorphism (5-HTTLPR/SLCA4) gene, norepinephrine transporter (NET) gene, catechol-O-methyltransferase (COMT) gene, and monoamine oxidase A (MAOA) gene (Zhang et al. 2018b). Genetic-neuroimaging studies suggested that genetic variants of serotonin genes were significant associated with abnormal morphology and activation of emotion-related brain regions, such as prefrontal cortex, amygdala, hippocampus, and anterior cingulate cortex (Frodl et al. 2010; Han et al. 2017; Lee et al. 2009; Guo et al. 2014; Shelton et al. 2009). What's more, significant NET gene-neuroimaging associations were observed with dorsolateral prefrontal cortex volume and hippocampal volume (Phillips et al. 2015a). These findings further provide robust evidence for the monoamine-deficiency hypothesis that serotonin and norepinephrine are implicated in the pathophysiology of MDD. Especially, the depletes of serotonin in brain may enhance emotional dysregulation in depression, such as depressive mood, fear, anxiety, self-accusation and criticism, and loneliness (Liu et al. 2018).

Aberrant brain morphology and function were associated with two glutamatergic genes: SLC6A15 and D-amino acid oxidase activator (DAOA). While DAOA genetic variants–diagnosis interaction effects were found on the abnormality of regional homogeneity (Chen et al. 2012), there was significant association between SLC6A15 polymorphisms and specific brain region alterations (Li et al. 2013a). These genes may regulate glutamate synthesis, and influences the expression of glutamate receptors, which in turn cause glutamate excitotoxicity for brain. Furthermore, evidence indicates that HPA axis and RAA system play an important role in the pathogenesis of MDD. Numerous genes in the HPA axis and RAA system were significantly implicated in neuroimaging alterations in depression. Both FK506-binding protein 51 (FKBP5) and corticotropin-releasing hormone receptor 1 (CRHR1) genes were significant associated with neuroimaging alteration via regulating HPA-negative feedback loop (Hsu et al. 2012; Tozzi et al. 2016). Wang et al. (2012) found that the effect of angiotensin-converting enzyme insertion/deletion (ACE I/D) genetic variants on neuroimaging alterations were observed in anterior cingulate gyrus, middle temporal gyrus, default-model network in remitted geriatric depression. Significant genetic-brain alteration association was observed in type 1 angiotensin II receptor (AGTR1) polymorphisms (Taylor et al. 2012). The other most frequently studied candidate genes associated with neuroimaging alterations were the apolipoprotein Eε4 (ApoE-ε4) and glycogen synthase kinase-3 beta (GSK3β) (Zhang et al. 2018b).

Although meta-analysis estimated heritability of MDD is up to 37%, genetic studies have failed to provide robust evidence for particular genes (Singleton and Hardy 2011). The generalized explanation might rely on that susceptibility genes alone generate minor biological effects on brain morphology in MDD; rather, the joint effects of numerous candidate genes may exert relatively large effects (Bick et al. 2019). Previous studies found that a significant synergistic effect on neuroimaging alteration was observed in 5-HTR2A and MAOA, 5-HTTLPR and 5-HTR1A, COMT and MTHFR polymorphisms, as well as 5-HTTLPR, BDNF, and COMT polymorphisms (Zhang et al. 2018b). Although precise mechanisms of genetic-neuroimaging alterations associations need to be clarified, genetic variation may provide new endophenotypes of MDD via affect morphological and functional brain alterations.

3.3.2 Environment Influences on Neuroimaging in MDD

Psychological stress (e.g., childhood maltreatment and chronic life stress) has a crucial role in the pathogenesis of depression. Exposure to childhood early life stress boosts risk of MDD in adulthood, which further causes recurrent and persistent depression, worsens outcomes, and leads to heavy financial burdens for individuals and society (Saleh et al. 2017). Main mediations of the influence of psychological stress on depression are increased levels of inflammatory markers, elevated neuroendocrine stress responses, and brain morphology and function alterations (Opel et al. 2019; Kelly et al. 2013). Enough evidence has suggested that psychological stress is implicated with reductions in cortical thickness, subcortical volumes, and cortical surface area in specific brain regions supporting emotion regulation, cognition control, and rewards (Zhai et al. 2019; Schoenfeld et al. 2017). More specifically, preclinical studies revealed that chronic stress causes hippocampal volume loss and suppression of adult neurogenesis and atrophy of CA3 pyramidal cells in rodents (Dinarello 1996). Possible mechanisms for tissue loss include repeated episodes of hypercortisolemia, increased glutamate neurotoxicity, reduction in neurotrophic factors. Therefore, early parenting interventions and adverse family environments may support more normative and optimal patterns of neural function in middle childhood, which might relieve the risk of depression in adulthood (Saleh et al. 2017).

3.4 Neuroinflammatory Influence of Cytokines and Other Metabolites in MDD

MDD is regarded as a psychoneuroimmunological disorder due to the close relationship between inflammation and depression. To date, there are numerous evidences which support the hypothesis that inflammatory cytokines play a prominent role in the etiopathological mechanisms underlying depression.

Cytokines, which are cell-signaling molecules that mediate and control immune responses, are classified into two subtypes: pro-inflammatory such as interleukin (IL) IL-1β, IL-6, IL-8 and tumor necrosis factor alpha (TNF-α); and anti-inflammatory including IL-1Ra, IL-4, IL-10 and transforming growth factor beta-1 (TGF-β1) (Sharpley and Agnew 2011; Jehn et al. 2010) These immunomodulating cytokines are believed to induce hypothalamic–pituitary–adrenal (HPA) axis dysregulations (Mastorakos et al. 1993; Himmerich et al. 2006; Myint and Kim 2003), alterations in cortical 5-HT systems (Catenadell'Osso et al. 2013; Pemberton et al. 1997; Zalcman et al. 1994), abnormalities in noradrenergic systems (Ando and Dunn 1999; Morón et al. 2003), dysfunctional dopaminergic pathways (Wu et al. 2007; Viviani et al. 2003), glutamatergic system malfunctions (Kaneko et al. 2006) and impaired neuroplasticity (Hayley et al. 2005; Dantzer and Capuron 2017; Huang and Lin 2015) in depression. In addition to bestowing increased concentrations of neuroinflammatory cytokines and cell-mediated immune (CMI) activation, MDD has been linked with aberrant pathways involving platelet-activating factors (PAF) as well as oxidative and nitrosative stress (O&NS) (Raison et al. 2006). MDD has also been associated with elevated C-reactive protein levels, increased microglial cell numbers, mitochondrial dysfunction, increased expressions of acute phase proteins, chemokines, and cellular adhesion molecules (CAMs) (Zunszain et al. 2012; Mayer et al. 2014).

Furthermore, the microbiome–gut–brain axis is of particular importance in the etiology and pathogenesis of MDD because it has a bidirectional gut–brain communication which takes place via multiple, complex neural, immune, endocrine, and metabolic pathways (Hestad et al. 2017). A putative mechanism for inflammation-associated depression is the tryptophan/kynurenine pathway (Dantzer et al. 2008). Tryptophan is an essential amino acid that is transported into the brain across the blood–brain barrier for the synthesis of serotonin via the kynurenine pathway (Myint et al. 2007). While an increase in tryptophan breakdown and kynurenine formation was found in MDD (Peng et al. 2018; Christmas et al. 2011), the metabolic enzyme indoleamine 2,3-dioxygenase (IDO) acts as a biological mediator linking immune functioning with altered neurotransmission (Frodl and Amico 2014). However, the neural inflammatory changes associated with the pathogenesis of MDD remains to be further explored.

3.4.1 Structural Neuroimaging Findings of Inflammation in MDD

There is a multitude of neuropsychiatric diseases including MDD, schizophrenia, and Alzheimer's disease which observed the consequences of inflammatory processes on brain structure (Delany et al. 2016; Ac et al. 2009). In the past, preclinical and clinical MRI research have indicated that neuroinflammation in MDD could be associated with structural and functional anomalies in various regions of the CNS such as the hippocampus, amygdala, anterior cingulate cortex, basal ganglia, and prefrontal cortex (Wilkin et al. 2009; Macqueen and Frodl 2011; Savitz et al. 2013). Savitz et al. reported several inflammatory and neurological disease-related protein-coding genes which were significantly correlated with the thickness of the left subgenual anterior cingulate cortex, hippocampus, and caudate in MDD patients (Marsland et al. 2008). The subgenual anterior cingulate cortex, hippocampus and caudate have also been previously implicated with MDD (Delany et al. 2016; Frodl et al. 2012; Wohleb et al. 2011). Wohleb et al. proposed that MDD patients could benefit from therapeutic interventions, such as β-adrenergic receptor antagonism, which target stress-related neuronal changes in the hippocampus (Jefferson et al. 2007). Moreover, a smaller whole brain volume (Rajeev et al. 2015) and thinner cortical volume (Eisenberger et al. 2009) were also associated with inflammation in depression.

A recent study demonstrated that inflammation was associated with higher reward-related neural activity in the dorsal anterior cingulate cortex and anterior insula (Eisenberger et al. 2010). A reduction of ventral striatum activation to reward predicting cues was associated with increase in self-reported depressed mood (Felger et al. 2016). Felger et al. performed resting-state fMRI in medically stable and unmedicated MDD patients and reported an increase in CRP and a lower functional connectivity between the ventral striatum and ventromedial prefrontal cortex as well as the dorsal striatum to the ventromedial prefrontal cortex and presupplementary motor area. These findings correlated with self-reported symptoms of anhedonia and psychomotor retardation (Harrison et al. 2011). Harrison et al. showed that an increased IL-6 serum was responsible for the hypoconnectivity of the subgenual anterior cingulate cortex to amygdala, medial prefrontal cortex, nucleus accumbens, and superior temporal sulcus (Kullmann et al. 2013). Furthermore, a higher TNF-α plasma concentration was associated with an increased activation of the right inferior orbitofrontal cortexin response to emotional visual stimuli (Mayer et al. 2015).

In addition, the connections between the gut and the brain affect emotional and cognitive processes (Tillisch et al. 2013). In an fMRI study, there was a significantly decreased neural emotional reactivity when fermented milk was given to healthy women during emotional faces recognition tasks (Hannestad et al. 2012). (PET) has the potential to understand the role of increased inflammation in cerebral structures pertaining to psychiatric disorders, and could determine the role of anti-inflammatory therapies in reducing inflammation in the brain. In a study using [11C] PBR28, Hannestad et al. found no differences in the binding potential (BP_{ND}) between mild to moderate MDD patients and controls (Setiawan et al. 2015). In contrast,

Setiawan et al. used the 18F analogue of [11C]PBR28, [18F]FEPPA, to examine the microglia activity of severely depressed subjects and observed significant increase in BP_{ND} in the striatum, hippocampus, insula and prefrontal areas (Takizawa et al. 2014).

3.4.2 Other Neuroimaging Findings of Inflammation in MDD

fNIRSis a promising neuroimaging tool that is based on the principle of brain oxygenation. fNIRS has the capability to detect and measure cortical hemodynamic changes noninvasively using light in the near-infrared spectrum (between 700 and 1000 nm) (Eschweiler et al. 2000). Although fNIRS could aid in the differential diagnosis of MDD, there are few studies that investigated inflammatory markers of MDD. One study by Eschweiler et al. examined the hemodynamic changes in the prefrontal cortex during mental work pre and post rTMS in MDD and found an absence of a task-related increase of total hemoglobin concentrations at the stimulation site (Luborzewski et al. 2007). Future research should make use of fNIRS to explore the neuroinflammatory markers of MDD.

MRS is a research technique that exploits biochemical changes in the brain. MRS makes use of the magnetic properties of certain atomic nuclei to assess the chemical composition of different metabolites. Luborzewski et al. applied proton MRS to MDD patients after rTMS treatment and reported an increase in glutamate concentration in the left DLPFC (Haroon et al. 2016). With the use of MRS, Haroon et al. showed that an increase in CRP is associated with increased basal ganglia glutamate in MDD patients. In addition, log left basal ganglia glutamate was correlated with anhedonia and psychomotor retardation (Fung et al. 2015).

3.4.3 Conclusions and Future Direction

There are rich and robust evidences which indicate that inflammatory processes affect the progression and development of MDD. These findings, based primarily on clinical and translational neuroimaging results, have indicated that inflammatory mechanisms are strongly associated with distinct symptoms of MDD such as anhedonia, low mood, lethargy, cognitive impairment, and suicidality. In the future, there is the necessity to exploit the various inflammatory stimuli in the brain so as to develop preventive and treatment strategies to reduce the vulnerability of depression and mood related symptoms.

3.5 Potential Diagnostic Neuroimaging Biomarkers of MDD

The current diagnostic approaches to MDD are mainly based on subjective and clinical examinations of depressive symptoms. Since there was an absence of noninvasive and quantifiable assessments in the past, the risk of possible misdiagnosis was elevated. At present, the neurobiological underpinnings of MDD are increasingly being investigated with the assistance of structural and functional neuroimaging techniques. While neuroimaging studies have the advantage of being noninvasive and repeatable, it is imperative to identify the neuroimaging biomarkers that will effectively guide diagnostic assessment of MDD. This section focuses on the diagnostic neuroimaging findings of MDD which provide valuable insights for differential diagnosis.

3.5.1 Structural Neuroimaging Advances of Diagnostic Biomarkers for MDD

In recent years, numerous achievements have been made in the study of MDD using MRI techniques. Prior studies have revealed that alterations in various cerebral structures contributed to the etiology and pathophysiology of depression. MDD patients possessed morphological abnormalities such as decreased cortical thickness and decreased gray matter volume in several brain regions. Although there are some inconsistencies in the existing literature, the aforementioned findings indicated that the differences in gray matter volumes could potentially serve as a differential diagnostic marker for MDD (Kambeitz et al. 2017). For the diagnostic accuracy in sMRI studies, a recent meta-analysis literature revealed that there was a specificity of 61–89%, and a sensitivity of 60–92% for identifying MDD patients using multivariant pattern recognition approach (Wu et al. 2015). Wu et al. (2015) indicated that morphological abnormalities in the right temporal pole and the right thalamus were the most hubs in identifying patients with MDD from healthy controls with a 78.4% accuracy (Redlich et al. 2014). A VBM morphometry-pattern recognition approach provided up to 79.3% accuracy by differentiating bipolar depression from unipolar depression (Yang et al. 2017b).

DTI is an innovative diagnostic method with the ability to demonstrate changes in white matter volumes. There exist some imaging parameters which are currently used as biomarkers for MDD diagnostic studies. The most commonly employed DTI parameters in neurological studies are fractional anisotropy and mean diffusivity. Previous DTI studies have shown a decreased fractional anisotropy in the cingulum (Chen et al. 2017), hippocampus, parietal regions, inferior temporal gyrus, and superior frontal gyrus, which were validated in a recent meta-analysis (Deng et al. 2017). There were alterations in white matter tracts including the right uncinate fasciculus and the left anterior thalamic radiation, which are linked to brain areas involved in emotion regulation and processing in MDD patients (Sacchet et al. 2015).

As for multivariant pattern recognition approach identifying MDD patients in DTI studies, previous studies showed that there was specificity of 71–98% and sensitivity of 70–98% (Wu et al. 2015). Using support vector machine, diffusion-weighted neuroimaging and graph theory was able to classify depressed subject from healthy controls with a diagnostic accuracy of 71.88% (Sambataro et al. 2014).

3.5.2 *Functional Neuroimaging Advances of Diagnostic Biomarkers for MDD*

Functional imaging techniques have assisted researchers to uncover important findings in MDD. fMRI studies revealed abnormal functional connectivity in the DMN (Chen et al. 2015; Alexopoulos et al. 2012; Zeng et al. 2012; Shen et al. 2015), CCN (Zeng et al. 2012; Liu et al. 2015), SN (Andrei et al. 2013; Zhang et al. 2014), and affective network (AN) in MDD patients (Shen et al. 2015; Phillips et al. 2015b). Apart from the above-mentioned networks, the brain regions exhibiting functional activity changes, including the orbitofrontal cortex, dorsolateral prefrontal lobe, ventromedial prefrontal cortex, insula, amygdala, and anterior cingulate cortex in MDD patients. These abnormal functional topologies have been associated with cognitive functions such as reward processing, cognitive control, and regulation of emotions and could therefore, potentially serve as biomarkers of MDD (Jie et al. 2015).

Compared with sMRIstudies, fMRI studies showed higher specificity and sensitivity for identifying individuals with MDD (Wu et al. 2015). More specifically, there was a specificity of 82.59% (95% CI: 74.64–88.43%), and a sensitivity of 85.39% (95% CI: 74.75–92.02%) in resting-state fMRI studies (Lois and Wessa 2016; Yang et al. 2016). As for the task-related fMRI studies, there was a specificity of 77.20% (95%CI: 69.92–83.15%), and a sensitivity of 74.06% (95% CI: 67.17–79.94%) (Rosa et al. 2015; Gao et al. 2018). Interestingly, the most discriminating functional biomarkers were observed in DMN, AN, visual cortical areas, and cerebellum, which was consistent with previous neuroimaging (Smitha 2009).

3.5.3 *Other Neuroimaging Technologies*

Notwithstanding the progresses made by PET and SPECT neuroimaging techniques, the idea that neurochemical processes related to MDD could be assessed by the current radio-ligands (Amen et al. 2017). One quantitative perfusion SPECT neuroimaging study found that there was diagnostic accuracy of 90.8% and cross-validated diagnostic accuracy of 88.6% distinguishing depression from dementia and those with both comorbidities (Almeida et al. 2013). Meanwhile, one study using arterial spin labeling method measured subdivisions of anterior cingulate cortex blood flow at rest to distinguish unipolar depression from bipolar depression with an 81% diagnostic accuracy (Olbrich and Arns 2013).

The application of EEG, which measures wavelengths reflecting different brain activities, demonstrated an alteration in α wave activity in MDD patients relative to healthy controls (Hosang 2014a). In addition, a low α wave activity mediated the effects of val66met polymorphism in BDNFgene, which was associated with elevated risk of developing anxiety and MDD (Tekin et al. 2015). Consistently, Erguzel et al. (2015) found promising quantitative EEG biomarkers to classify unipolar depression with 80.19% overall classification accuracy (Domschke et al. 2016). Moreover, the use of MEG to investigate functional abnormalities is becoming increasingly common. Domschke et al. described a lower response in the right parietal areas and regions around the right temporo-parietal junction, in addition to aberrant emotional valence-specific effects in the right parietal and bilateral dorsolateral prefrontal cortex areas (Hirakawa et al. 2017). MDD patients had significantly lower magnetic global field power (GFP) of the MEG equivalent of mismatch negativity (MMNm) in the right hemisphere of MDD patients as well as early bilateral MMNm peak latencies (Lu et al. 2013). As for the disease classification for depression, the classification accuracy could reach up to 87.5% using a windowed granger causality analysis of MEG recordings (Ball et al. 2014).

3.5.4 Conclusion and Future Direction

The recent advances of in vivo imaging techniques have considerably enriched our knowledge on the pathophysiology of MDD. Robust neuroimaging diagnostic biomarkers which could be used in clinical practice are yet to be found and thus, further studies are needed to elucidate the inconsistencies in the existing literature. One of the causes of such discrepancies might be the result of the heterogeneity nature of patients under investigation. A particular challenge to researchers for the development of a true MDD diagnostic biomarker is the variances in individual patients' symptomatology and severity, as well as the overlapping of symptoms from other psychiatric disorders. In order to overcome these problems, homogenous patients should be selected while taking the demographic variables, such as age, gender, medication, disease duration, and symptom dimensions into careful consideration. It is also noteworthy that most of the aforementioned studies had small sample sizes, which increased the false-positive and/or false-negative rates of their results. Therefore, multicenters collaboration would not solve the problem of sample size but would additionally allow for more in-depth researches. The relationship between the structural and functional changes seen in MDD remains largely unclear. Future multimodal neuroimaging studies should be conducted to ascertain the associations between structural and functional abnormalities in MDD. Lastly, if the findings of neuroimaging studies described in this section are combined with other objective biomarkers (e.g., genomics, transcriptomic, proteomics, metabolomics) and clinical assessments, they have the potential to provide additional support for a diagnosis.

3.6 Potential Therapeutic Neuroimaging Biomarkers of MDD

3.6.1 Functional Neuroimaging Biomarkers

A disruption in the activations of a number of cortical-limbic structures is consistently found in the literature, notably in the amygdala, anterior cingulatecortex, and prefrontal cortex (Gong and He 2015; Phillips et al. 2015c; Lener and Iosifescu 2015). The hyperactivity of anterior cingulate cortex is considered as a potential biomarker of pharmacotherapy outcome across a range of antidepressant, including selective serotonin reuptake inhibitors, serotonin and norepinephrine reuptake inhibitors, norepinephrine reuptake inhibitors, and tricyclic antidepressants, and methodologies (Phillips et al. 2015c; Dichter et al. 2015; Breitenstein et al. 2014; Pizzagalli 2011; Alexopoulos et al. 2012), while anterior cingulate cortex hypoactivity was demonstrated as a biomarker of treatment nonresponsiveness in MDD. An increase in functional connectivity between the frontal and limbic brain regions is also a good predictor of treatment outcomes (Pizzagalli 2011; Alexopoulos et al. 2012).

Dysfunction in the resting state networks of the brain, particularly a reduced activity of the DMN and CCN, predicted poorer antidepressant outcome in adults (Li et al. 2013b). In general, the higher response in DMN is regulated with the use of antidepressant therapy. However, persisting hyper connectivity of the DMN after treatment could be used to distinguish treatment-sensitive from treatment-resistant MDD patients (Breitenstein et al. 2014). Furthermore, medication therapy selectively normalized the posterior DMN but not the anterior DMN (Milak et al. 2009).

Other functional modality studies revealed some potential therapeutic biomarkers which could be used as predictors of antidepressant treatment. A baseline hypermetabolism or hypometabolism at the pregenual, subgenual anterior cingulate cortex, anterior insula, parahippocampal gyrus, putamen, lateral globus pallidus, and thalamus was associated with antidepressant response (McGrath et al. 2013; Konarski et al. 2009; Nugent et al. 2016). MEG studies demonstrated an attenuation of hyper connectivity between the temporal areas and amygdala in post treatment patients (Straub et al. 2017).

Functional neuroimaging studies performed on patients receiving cognitive behavior treatment uncovered the anterior cingulate cortex, prefrontal cortex, amygdala/temporal lobe, and insula as response biomarkers. A previous study showed that a higher functional connectivity at baseline in the amygdala, left dorsolateral prefrontal cortex, and leftanterior insula predicted an improvement of depressive symptoms in the resting-state patients (Dunlop et al. 2017) while another investigation demonstrated a greater functional connectivity of the subcallosal cingulate cortex with the dorsal midbrain, ventrolateral prefrontal cortex/insula, and ventromedial prefrontal cortex was associated with remission of MDD symptoms (Fu et al. 2008). Taken together, these findings support the idea that a higher fronto-limbic connectivity during resting-state at baseline is predictive of treatment response to a psychotherapeutic intervention.

In task-based studies, the anterior, amygdala and prefrontal cortex also predicted the response of MDD patients to cognitive behavior treatment (Costafreda et al. 2009a; Ritchey et al. 2011; Siegle et al. 2006, 2012; McGrath et al. 2013). Responders to cognitive behavior treatment had dorsal anterior cingulate cortex hypoactivity and subgenual anterior cingulate cortex hyporeactivity in response to sad faces and negative words, respectively (Costafreda et al. 2009a; Siegle et al. 2006). On the other hand, patients with higher sustained reactivity in the amygdala during an emotional information processing task performed before cognitive behavior treatment presented the strongest improvement with cognitive behavior treatment (McGrath et al. 2013). Therefore, it can be inferred that responders to psychological interventions are characterized by a pattern of higher amygdala reactivity and lower anterior cingulate cortex activity in response to emotional stimuli.

Other imaging modalities were similarly employed to predict response to psychotherapy. Studies pertaining to cerebral glucose metabolism revealed that nonresponders to cognitive behavior treatment exhibited hypermetabolism in the insula and at the interface of the pregenual and subgenual cingulate cortices, while predicting response to escitalopram pharmacotherapy (Konarski et al. 2009; Kennedy et al. 2007). Another study demonstrated a hypermetabolism in the hippocampus and dorsal mid-cingulate, as well as a hypometabolism in the dorsolateral prefrontal cortex, ventrolateral prefrontal cortex, superior and inferior medial frontal regions, posterior cingulate cortex, inferior parietal, and inferior temporal cortex in MDD patients undergoing cognitive behavior treatment (Chi et al. 2015). In conclusion, a higher metabolism in the limbic system and a lower metabolism in the prefrontal cortex could predict better response with psychotherapy.

3.6.2 Structural Neuroimaging Biomarkers

Structural MRI studies revealed that altered brain structures associated with AD treatment outcomes were observed in the prefrontal cortex and hippocampus. Several investigations demonstrated that smaller hippocampal volume was also present in MDD patients and therefore, an increase in hippocampal volumes could be used to differentiate between treatment responders and nonresponders (Pizzagalli 2011; Leuchter et al. 2010; El-Hage et al. 2013; Frodl et al. 2004, 2008; Hsieh et al. 2002) whereas lower volumes after treatment predicted a relapse or lack of remission (Sheline et al. 2012; MacQueen et al. 2008; Dusi et al. 2015). Additionally, cortical thickness, gray matter volumes, and white matter integrity of the prefrontal cortex have all been associated with treatment outcome. A remission after antidepressant medication was accompanied by preserved volumes of anterior cingulate, dorsolateral prefrontal cortex, and dorsomedial prefrontal cortex in MDD patients (Taylor et al. 2011).

Numerous studies based on DTI technique investigated the integrity of the white matter in predicting treatment efficacy, and its implication in the cortico–striato–limbic systems in distinguishing treatment responders from nonresponders. Remitted MDD patients exhibited a significantly greater change in fractional anisotropy of the anterior cingulate cortex compared to non-remitters (Korgaonkar et al. 2014). In addition, fractional anisotropy values of the cingulum part of the cingulate and stria terminalis tracts were useful predictors of antidepressant treatment outcome (Goldapple et al. 2004). It was previously reported that successful treatment with escitalopram or cognitive behavior treatment was associated with modulation of resting-state brain activity or metabolic changes within the fronto-limbic circuit (Wang et al. 2014b; Qin et al. 2017). The modular reconfiguration of hubs within the fronto-limbic circuit is believed to be the result of antidepressant medication treatment (Fujino et al. 2015).

So far, there is a scarcity of studies investigating underlying brain structures relationship with response to cognitive behavior treatment in MDD patients. Fujino et al. discovered that the regional gray matter volume in the caudal portion of the anterior cingulate cortex was positively correlated with improvements of depressive symptoms after cognitive behavior treatment (Costafreda et al. 2009b). In contrast, no positive response to cognitive behavior treatment was found with whole brain structural neuroanatomy in MDD patients (Widge et al. 2018).

3.7 Conclusion

Functional and structural neuroimaging modalities have identified potential neural biomarkers of antidepressant treatment outcome and response, some of which are consistent across treatment. Fronto-limbic regions, in particular the hippocampus, prefrontal, anterior cingulate cortex, amygdala and insula, most frequently affect treatment outcome, although the directions of association may differ for distant treatment strategies. Notably, quantitative electroencephalographic (QEEG) is a technique used to investigate biomarkers for treatment outcome or response, but a meta-analysis revealed that QEEG is not reliable for predicting response to pharmacotherapy in MDD.

References

Alexander AL et al (2007) Diffusion tensor imaging of the brain 4(3):316–329

Alexopoulos GS et al (2012) Functional connectivity in the cognitive control network and the default mode network in late-life depression. 139(1):56

Almeida JR et al (2013) Pattern recognition analysis of anterior cingulate cortex blood flow to classify depression polarity. Br J Psychiatry 203(3):310–311

Amen DG et al (2017) Classification of depression, cognitive disorders, and co-morbid depression and cognitive disorders with perfusion SPECT neuroimaging **57**(1):253–266

Ando T, Dunn AJ (1999) Mouse tumor necrosis factor-alpha increases brain tryptophan concentrations and norepinephrine metabolism while activating the HPA axis in mice. NeuroImmunoModulation 6(5):319

Andrei M et al (2013) Insular dysfunction within the salience network is associated with severity of symptoms and aberrant inter-network connectivity in major depressive disorder 7(2):930

Arnone D (2019) Functional MRI findings, pharmacological treatment in major depression and clinical response. Prog Neuropsychopharmacol Biol Psychiatry 91:28–37

Ball TM, Stein MB, Paulus MP (2014) Toward the application of functional neuroimaging to individualized treatment for anxiety and depression. Depress Anxiety 31(11):920–933

Beijers L et al (2019) Data-driven biological subtypes of depression: systematic review of biological approaches to depression subtyping. Mol Psychiatry 24:888–900

Bick J et al (2019) Early parenting intervention and adverse family environments affect neural function in middle childhood. Biol Psychiatry 85(4):326–335

Breitenstein B, Scheuer S, Holsboer F (2014) Are there meaningful biomarkers of treatment response for depression? Drug Discov Today 19(5):539–561

Bunge SA, Kahn IJEoN (2009) Cognition: an overview of neuroimaging techniques 1063–1067

Carballedo A et al (2013) Brain-derived neurotrophic factor Val66Met polymorphism and early life adversity affect hippocampal volume. Am J Med Genet B Neuropsychiatr Genet 162B(2):183–190

Cardoner N et al (2013) Val66Met BDNF genotypes in melancholic depression: effects on brain structure and treatment outcome. Depress Anxiety 30(3):225–233

Catenadell'Osso M et al (2013) Inflammation, serotonin and major depression. Curr Drug Targets 14(5):571–577

Chen J et al (2012) Genotypic association of the DAOA gene with resting-state brain activity in major depression. Mol Neurobiol 46(2):361–373

Chen Y et al (2015) Aberrant connectivity within the default mode network in first-episode, treatment-naïve major depressive disorder 183:49–56

Chen G et al (2017) Intrinsic disruption of white matter microarchitecture in first-episode, drug-naive major depressive disorder: a voxel-based meta-analysis of diffusion tensor imaging. Prog Neuropsychopharmacol Biol Psychiatry 76:179–187

Chi KF, Korgaonkar M, Grieve SM (2015) Imaging predictors of remission to anti-depressant medications in major depressive disorder. J Affect Disord 186:134–144

Christmas DM, Potokar J, Davies SJ (2011) A biological pathway linking inflammation and depression: activation of indoleamine 2, 3-dioxygenase. Neuropsychiatric Dis Treat 7:431

Costafreda SG et al (2009a) Neural correlates of sad faces predict clinical remission to cognitive behavioural therapy in depression. Neuroreport 20(7):637–641

Costafreda SG et al (2009b) Prognostic and diagnostic potential of the structural neuroanatomy of depression. PLoS ONE 4(7):e6353

Dantzer R, Capuron L (2017) Inflammation-associated depression: evidence, mechanisms and implications. Springer International Publishing

Dantzer R et al (2008) From inflammation to sickness and depression: when the immune system subjugates the brain. Nat Rev Neurosci 9:46

Delany FM et al (2016) Depression, immune function, and early adrenarche in children. Psychoneuroendocrinology 63:228–234

Deng F et al (2017) Abnormal segments of right uncinate fasciculus and left anterior thalamic radiation in major and bipolar depression. Prog Neuro-Psychopharmacol & Biol Psychiatry 81

Dichter GS, Gibbs D, Smoski MJ (2015) A systematic review of relations between resting-state functional-MRI and treatment response in major depressive disorder. J Affect Disord 172:8–17

Dinarello CA (1996) Biologic basis for interleukin-1 in disease. Blood 87(6):2095–2147

Domschke K et al (2016) Magnetoencephalographic correlates of emotional processing in major depression before and after pharmacological treatment. Int J Neuropsychopharmacol 19(2)

Drysdale AT et al (2017) Resting-state connectivity biomarkers define neurophysiological subtypes of depression. Nat Med 23(1):28–38

Dunlop BW et al (2017) Functional connectivity of the subcallosal cingulate cortex and differential outcomes to treatment with cognitive-behavioral therapy or antidepressant medication for major depressive disorder. Am J Psychiatry 174(6):533–545

Dusi N et al (2015) Brain structural effects of antidepressant treatment in major depression. Curr Neuropharmacol 13(4):458–465

Eisenberger NI et al (2009) An fMRI study of cytokine-induced depressed mood and social pain: the role of sex differences. Neuroimage 47(3):881–890

Eisenberger NI et al (2010) Inflammation-induced anhedonia: endotoxin reduces ventral striatum responses to reward. Biol Psychiat 68(8):748–754

El-Hage W et al (2013) Mechanisms of antidepressant resistance. Front Pharmacol 4:146

Eschweiler GW et al (2000) Left prefrontal activation predicts therapeutic effects of repetitive transcranial magnetic stimulation (rTMS) in major depression. Psychiatry Research: Neuroimaging 99(3):161–172

Felger JC et al (2016) Inflammation is associated with decreased functional connectivity within corticostriatal reward circuitry in depression. Mol Psychiatry 21(10):1358–1365

Feder S et al (2017) Sample heterogeneity in unipolar depression as assessed by functional connectivity analyses is dominated by general disease effects. J Affect Disord 222:79–87

Flint J, Kendler KS (2014) The genetics of major depression. Neuron 81(3):484–503

Frodl T, Amico F (2014) Is there an association between peripheral immune markers and structural/functional neuroimaging findings? Prog Neuropsychopharmacol Biol Psychiatry 48(2):295–303

Frodl T et al (2004) Hippocampal and amygdala changes in patients with major depressive disorder and healthy controls during a 1-year follow-up. J Clin Psychiatry 65(4):492–499

Frodl TS et al (2008) Depression-related variation in brain morphology over 3 years: effects of stress? Arch Gen Psychiatry 65(10):1156–1165

Frodl T et al (2010) Childhood stress, serotonin transporter gene and brain structures in major depression. Neuropsychopharmacology 35(6):1383–1390

Frodl T et al (2012) Reduced expression of glucocorticoid-inducible genes GILZ and SGK-1: high IL-6 levels are associated with reduced hippocampal volumes in major depressive disorder. Translational Psychiatry 2(3):e88

Fu CH et al (2008) Neural responses to sad facial expressions in major depression following cognitive behavioral therapy. Biol Psychiat 64(6):505–512

Fujino J et al (2015) Anterior cingulate volume predicts response to cognitive behavioral therapy in major depressive disorder. J Affect Disord 174:397–399

Fung G et al (2015) Distinguishing bipolar and major depressive disorders by brain structural morphometry: a pilot study. Bmc Psychiatry 15(1):298

Gao S, Calhoun VD, Sui J (2018) Machine learning in major depression: from classification to treatment outcome prediction. CNS Neurosci Ther 24(11):1037–1052

Glover GH (2011) Overview of functional magnetic resonance imaging. Neurosurg Clin N Am 22(2):133–9, vii

Goldapple K et al (2004) Modulation of cortical-limbic pathways in major depression: treatment-specific effects of cognitive behavior therapy. Arch Gen Psychiatry 61(1):34–41

Gong Q, He Y (2015) Depression, neuroimaging and connectomics: a selective overview. Biol Psychiatry 77(3):223–235

Gong L et al (2018) Disrupted reward and cognitive control networks contribute to anhedonia in depression. J Psychiatr Res 103:61–68

Guo H et al (2014) Synergistic effect of 5-HT2A receptor gene and MAOA gene on the negative emotion of patients with depression. Clin Physiol Funct Imaging 34(4):277–281

Han KM et al (2014) Cortical thickness, cortical and subcortical volume, and white matter integrity in patients with their first episode of major depression. J Affect Disord 155:42–48

Han KM et al (2017) Local gyrification index in patients with major depressive disorder and its association with tryptophan hydroxylase-2 (TPH2) polymorphism. Hum Brain Mapp 38(3):1299–1310

Hannestad J et al (2012) Endotoxin-induced systemic inflammation activates microglia: [11C]PBR28 positron emission tomography in nonhuman primates. Neuroimage 63(1):232–239
Haroon E et al (2016) Conceptual convergence: increased inflammation is associated with increased basal ganglia glutamate in patients with major depression. Molecular Psychiatry 21(10):1351–1357
Harrison NA et al (2011) Inflammation causes mood changes through alterations in subgenual cingulate activity and mesolimbic connectivity. Biol Psychiat 66(5):407–414
Hayley S et al (2005) The pathogenesis of clinical depression: stressor- and cytokine-induced alterations of neuroplasticity. Neuroscience 135(3):659–678
Hestad KA et al (2017) The relationships among tryptophan, kynurenine, indoleamine 2,3-dioxygenase, depression, and neuropsychological performance. Front Psychol 8(1561)
Himmerich H et al (2006) Successful antidepressant therapy restores the disturbed interplay between TNF-α system and HPA axis. Biol Psychiat 60(8):882
Hirakawa N et al (2017) Right hemisphere pitch-mismatch negativity reduction in patients with major depression: An MEG study. J Affect Disord 215:225
Hosang GM et al (2014a) Interaction between stress and the BDNF Val66Met polymorphism in depression: a systematic review and meta-analysis 12(1):1–11
Hosang GM et al (2014b) Interaction between stress and the BDNF Val66Met polymorphism in depression: a systematic review and meta-analysis. BMC Med 12:7
Hsieh MH et al (2002) Hippocampal volume and antidepressant response in geriatric depression. Int J Geriatr Psychiatry 17(6):519–525
Hsu DT et al (2012) Variation in the corticotropin-releasing hormone receptor 1 (CRHR1) gene influences fMRI signal responses during emotional stimulus processing. J Neurosci 32(9):3253–3260
Huang TL, Lin CC (2015) Advances in biomarkers of major depressive disorder. Adv Clin Chem 68:177
Jaworska N et al (2016) The influence of 5-HTTLPR and Val66Met polymorphisms on cortical thickness and volume in limbic and paralimbic regions in depression: a preliminary study. BMC Psychiatry 16:61
Jefferson AL et al (2007) Inflammatory biomarkers are associated with total brain volume. The Framingham Heart Study 68(13):1032–1038
Jehn CF et al (2010) Association of IL-6, hypothalamus-pituitary-adrenal axis function, and depression in patients with cancer. Integrative Cancer Therapies 9(3):270–275
Jiang J et al (2017) Microstructural brain abnormalities in medication-free patients with major depressive disorder: a systematic review and meta-analysis of diffusion tensor imaging. J Psychiatry Neurosci 42(3):150–163
Jie NF et al (2015) Discriminating bipolar disorder from major depression based on SVM-FoBa: efficient feature selection with multimodal brain imaging data. IEEE Trans Auton Ment Dev 7(4):320–331
Kambeitz J et al (2017) Detecting neuroimaging biomarkers for depression: a meta-analysis of multivariate pattern recognition studies. Biol Psychiatry 82(5):330–338
Kaneko N et al (2006) Suppression of cell proliferation by interferon-alpha through interleukin-1 production in adult rat dentate gyrus. Neuropsychopharmacology 31(12):2619–2626
Kelly PA et al (2013) Cortical thickness, surface area, and gyrification abnormalities in children exposed to maltreatment: neural markers of vulnerability? Biol Psychiatry 74(11):845–852
Kennedy SH et al (2007) Differences in brain glucose metabolism between responders to CBT and venlafaxine in a 16-week randomized controlled trial. Am J Psychiatry 164(5):778–788
Konarski JZ et al (2009) Predictors of nonresponse to cognitive behavioural therapy or venlafaxine using glucose metabolism in major depressive disorder. J Psychiatry Neurosci 34(3):175–180
Korgaonkar MS et al (2014) Abnormal structural networks characterize major depressive disorder: a connectome analysis. Biol Psychiat 76(7):567–574
Kullmann JS et al (2013) Neural response to emotional stimuli during experimental human endotoxemia. Hum Brain Mapp 34(9):2217–2227

Lee BT et al (2009) Impact of the tryptophan hydroxylase 1 gene A218C polymorphism on amygdala activity in response to affective facial stimuli in patients with major depressive disorder. Genes Brain Behav 8(5):512–518

Lener MS, Iosifescu DV (2015) In pursuit of neuroimaging biomarkers to guide treatment selection in major depressive disorder: a review of the literature. Ann N Y Acad Sci 1344:50–65

Lener MS et al (2017) Glutamate and gamma-aminobutyric acid systems in the pathophysiology of major depression and antidepressant response to ketamine. Biol Psychiatry 81(10):886–897

Leuchter AF et al (2010) Biomarkers to predict antidepressant response. Curr Psychiatry Rep 12(6):553–562

Li M et al (2013a) SLC6A15 rs1545843 and depression: implications from brain imaging data. Am J Psychiatry 170(7):805

Li B et al (2013b) A treatment-resistant default mode subnetwork in major depression. Biol Psychiat 74(1):48–54

Librenza-Garcia D et al (2017) The impact of machine learning techniques in the study of bipolar disorder: a systematic review. Neurosci Biobehav Rev 80:538–554

Linden DE (2012) The challenges and promise of neuroimaging in psychiatry. Neuron 73(1):8–22

Liu CH et al (2015a) Alteration of spontaneous neuronal activity within the salience network in partially remitted depression. Brain Res 1599:93–102

Liu CH et al (2015b) Alteration of spontaneous neuronal activity within the salience network in partially remitted depression. 1599:93

Liu Y, Zhao J, Guo W (2018) Emotional Roles of Mono-Aminergic Neurotransmitters in Major Depressive Disorder and Anxiety Disorders. Front Psychol 9:2201

Lois G, Wessa M (2016) Differential association of default mode network connectivity and rumination in healthy individuals and remitted MDD patients. Soc Cogn Affect Neurosci 11(11):1792–1801

Lu Q et al (2013) Predicting depression based on dynamic regional connectivity: a windowed Granger causality analysis of MEG recordings. Brain Res 1535:52–60

Luborzewski A et al (2007) Metabolic alterations in the dorsolateral prefrontal cortex after treatment with high-frequency repetitive transcranial magnetic stimulation in patients with unipolar major depression. J Psychiatr Res 41(7):606–615

Macqueen G, Frodl T (2011) The hippocampus in major depression: evidence for the convergence of the bench and bedside in psychiatric research|[quest]. Molecular Psychiatry 16(3):252

MacQueen GM et al (2008) Posterior hippocampal volumes are associated with remission rates in patients with major depressive disorder. Biol Psychiatry 64(10):880–883

Maglanoc LA et al (2019) Data-driven clustering reveals a link between symptoms and functional brain connectivity in depression. Biol Psychiatry Cogn Neurosci Neuroimaging 4(1):16–26

Marchand WR (2010) Cortico-basal ganglia circuitry: a review of key research and implications for functional connectivity studies of mood and anxiety disorders. Brain Struct Funct 215(2):73–96

Marsland AL et al (2008) Interleukin-6 covaries inversely with hippocampal grey matter volume in middle-aged adults. Biol Psychiat 64(6):484

Masdeu JC (2011) Neuroimaging in psychiatric disorders. Neurotherapeutics 8(1):93–102

Mastorakos G, Chrousos GP, Weber JS (1993) Recombinant interleukin-6 activates the hypothalamic-pituitary-adrenal axis in humans. J Clin Endocrinol Metab 77(6):1690–1694

Mayer EA et al (2014) Gut microbes and the brain: paradigm shift in neuroscience. J Neurosci Off J Soc Neuroscic 34(46):15490–15496

Mayer EA, Tillisch K, Gupta A (2015) Gut/brain axis and the microbiota. J Clin Investig 125(3):926–938

McGrath CL et al (2013) Toward a neuroimaging treatment selection biomarker for major depressive disorder. JAMA Psychiatry 70(8):821–829

Milak MS et al (2009) Pretreatment regional brain glucose uptake in the midbrain on PET may predict remission from a major depressive episode after three months of treatment. Psychiatry Res 173(1):63–70

Miller JM et al (2013) Brain serotonin 1A receptor binding as a predictor of treatment outcome in major depressive disorder. Biol Psychiatry 74(10):760–767

Morón JA et al (2003) Mitogen-activated protein kinase regulates dopamine transporter surface expression and dopamine transport capacity. Journal of Neuroscience the Official Journal of the Society for Neuroscience 23(24):8480

Myint AM, Kim YK (2003) Cytokine-serotonin interaction through IDO: a neurodegeneration hypothesis of depression. Med Hypotheses 61(5):519–525

Myint A-M et al (2007) Kynurenine pathway in major depression: evidence of impaired neuroprotection. J Affect Disord 98(1–2):143–151

Nejad AB, Fossati P, Lemogne C (2013) Self-referential processing, rumination, and cortical midline structures in major depression. Front Hum Neurosci 7:666

Nugent AC et al (2016) Preliminary differences in resting state MEG functional connectivity pre- and post-ketamine in major depressive disorder. Psychiatry Res Neuroimaging 254:56–66

Olbrich S, Arns MJIRoP (2013) EEG biomarkers in major depressive disorder: Discriminative power and prediction of treatment response 25(5):604–618

Opel N et al (2019) Mediation of the influence of childhood maltreatment on depression relapse by cortical structure: a 2-year longitudinal observational study. The Lancet Psychiatry 6(4):318–326

Pemberton LA et al (1997) Quinolinic acid production by macrophages stimulated with IFN-gamma, TNF-alpha, and IFN-alpha. J Interf & Cytokine Res Off J Int Soc Interf & Cytokine Res 17(10):589–595

Peng D et al (2018) The metabolic factor kynurenic acid of kynurenine pathway predicts major depressive disorder. Front psychiatry 9:552

Pezawas L, Meyer-Lindenberg A (2010) Imaging genetics: Progressing by leaps and bounds. Neuroimage 53(3):801–803

Phillips ML et al (2003) Neurobiology of emotion perception I: the neural basis of normal emotion perception. Biol Psychiat 54(5):504–514

Phillips JL et al (2015a) Impact of monoamine-related gene polymorphisms on hippocampal volume in treatment-resistant depression. Acta Neuropsychiatr 27(6):353–361

Phillips ML et al (2015b) Identifying Predictors. Moderators, and Mediators of Antidepressant Response in Major Depressive Disorder: Neuroimaging Approaches. 172(2):124–138

Phillips ML et al (2015c) Identifying predictors, moderators, and mediators of antidepressant response in major depressive disorder: neuroimaging approaches. Am J Psychiatry 172(2):124–138

Pirnia T et al (2016) Electroconvulsive therapy and structural neuroplasticity in neocortical, limbic and paralimbic cortex. Transl Psychiatry 6(6):e832

Pizzagalli DA (2011) Frontocingulate dysfunction in depression: toward biomarkers of treatment response. Neuropsychopharmacology 36(1):183–206

Price RB et al (2017a) Parsing heterogeneity in the brain connectivity of depressed and healthy adults during positive mood. Biol Psychiatry 81(4):347–357

Price RB et al (2017b) Data-driven subgroups in depression derived from directed functional connectivity paths at rest. Neuropsychopharmacology 42(13):2623–2632

Qin J et al (2017) Reconfiguration of hub-level community structure in depressions: a follow-up study via diffusion tensor imaging. J Affect Disord 207:305–312

Qiu M et al (2018) Aberrant neural activity in patients with bipolar depressive disorder distinguishing to the unipolar depressive disorder: a resting-state functional magnetic resonance imaging study. Front psychiatry 9:238

Raison CL, Capuron L, Miller AH (2006) Cytokines sing the blues: inflammation and the pathogenesis of depression. Trends Immunol 27(1):24–31

Rajeev K et al (2015) Cardio-metabolic risk factors and cortical thickness in a neurologically healthy male population: results from the psychological, social and biological determinants of ill health (pSoBid) study. Neuroimage Clinical 9(3):e1

Redlich R et al (2014) Brain morphometric biomarkers distinguishing unipolar and bipolar depression. A voxel-based morphometry-pattern classification approach. JAMA Psychiatry 71(11): 1222–1230

Renteria ME et al (2017) Subcortical brain structure and suicidal behaviour in major depressive disorder: a meta-analysis from the ENIGMA-MDD working group. Transl Psychiatry 7(5):e1116

Richter-Levin G, Xu L (2018) How could stress lead to major depressive disorder? IBRO Rep 4:38–43

Ritchey M et al (2011) Neural correlates of emotional processing in depression: changes with cognitive behavioral therapy and predictors of treatment response. J Psychiatr Res 45(5):577–587

Rosa MJ et al (2015) Sparse network-based models for patient classification using fMRI. Neuroimage 105:493–506

Rosa CE et al (2017) Glutamatergic and neural dysfunction in postpartum depression using magnetic resonance spectroscopy. Psychiatry Res 265:18–25

Sacchet MD et al (2015) Support vector machine classification of major depressive disorder using diffusion-weighted neuroimaging and graph theory. Front Psychiatry 6:21

Saleh A et al (2017) Effects of early life stress on depression, cognitive performance and brain morphology. Psychol Med 47(1):171–181

Sambataro F et al (2014) Revisiting default mode network function in major depression: evidence for disrupted subsystem connectivity. 44(10):2041

Satomura Y et al (2019) Severity-dependent and -independent brain regions of major depressive disorder: A long-term longitudinal near-infrared spectroscopy study. J Affect Disord 243:249–254

Savitz J et al (2013) Inflammation and neurological disease-related genes are differentially expressed in depressed patients with mood disorders and correlate with morphometric and functional imaging abnormalities. Brain Behav Immun 31(1):161

Schmaal L et al (2016) Subcortical brain alterations in major depressive disorder: findings from the ENIGMA Major Depressive Disorder working group. Mol Psychiatry 21(6):806–812

Schmaal L et al (2017) Cortical abnormalities in adults and adolescents with major depression based on brain scans from 20 cohorts worldwide in the ENIGMA Major Depressive Disorder Working Group. Mol Psychiatry 22(6):900–909

Schoenfeld TJ et al (2017) Stress and loss of adult neurogenesis differentially reduce hippocampal volume. Biol Psychiatry 82(12):914–923

Setiawan E et al (2015) Role of translocator protein density, a marker of neuroinflammation, in the brain during major depressive episodes. JAMA Psychiatry 72(3):268–275

Sharpley CF, Agnew LL (2011) Cytokines and depression: findings, issues, and treatment implications. Rev Neurosci 22(3):295

Sheline YI et al (2010) Resting-state functional MRI in depression unmasks increased connectivity between networks via the dorsal nexus. Proc Natl Acad Sci U S A 107(24):11020–11025

Sheline YI et al (2012) Treatment course with antidepressant therapy in late-life depression. Am J Psychiatry 169(11):1185–1193

Shelton RC et al (2009) Elevated 5-HT 2A receptors in postmortem prefrontal cortex in major depression is associated with reduced activity of protein kinase A. Neuroscience 158(4):1406–1415

Shen T et al (2015) Increased cognition connectivity network in major depression disorder: a FMRI study. 12(2):227

Shirayama Y et al (2017) Myo-inositol, glutamate, and glutamine in the prefrontal cortex, hippocampus, and amygdala in major depression. Biol Psychiatry: Cogn Neurosci Neuroimaging 2(2):196–204

Siegle GJ, Carter CS, Thase ME (2006) Use of FMRI to predict recovery from unipolar depression with cognitive behavior therapy. Am J Psychiatry 163(4):735–738

Siegle GJ et al (2012) Toward clinically useful neuroimaging in depression treatment: prognostic utility of subgenual cingulate activity for determining depression outcome in cognitive therapy across studies, scanners, and patient characteristics. Arch Gen Psychiatry 69(9):913–924

Singleton A, Hardy J (2011) A generalizable hypothesis for the genetic architecture of disease: pleomorphic risk loci. Hum Mol Genet 20(R2):R158–R162

Smitha DFJEN (2009) Molecular tools for assessing human depression by positron emission tomography 19(9):611–628

Straub J et al (2017) Successful group psychotherapy of depression in adolescents alters fronto-limbic resting-state connectivity. J Affect Disord 209:135–139

Su L et al (2014) Cerebral metabolism in major depressive disorder: a voxel-based meta-analysis of positron emission tomography studies. 14(1):1–7

Suh JS et al (2019) Cortical thickness in major depressive disorder: a systematic review and meta-analysis. Prog Neuropsychopharmacol Biol Psychiatry 88:287–302

Symms M et al (2004) A review of structural magnetic resonance neuroimaging. J Neurol Neurosurg Psychiatry 75(9):1235–1244

Takizawa R et al (2014) Neuroimaging-aided differential diagnosis of the depressive state. Neuroimage 85(1):498–507

Tang S et al (2018) Abnormal amygdala resting-state functional connectivity in adults and adolescents with major depressive disorder: a comparative meta-analysis. EBioMedicine 36:436–445

Taylor WD et al (2011) One-year change in anterior cingulate cortex white matter microstructure: relationship with late-life depression outcomes. The American Journal of Geriatric Psychiatry 19(1):43–52

Taylor WD et al (2012) AGTR1 gene variation: association with depression and frontotemporal morphology. Psychiatry Res 202(2):104–109

Tekin S, Cummings JL (2002) Frontal–subcortical neuronal circuits and clinical neuropsychiatry: an update 53(2): 647–654

Tekin Erguzel T, Tas C, Cebi M (2015) A wrapper-based approach for feature selection and classification of major depressive disorder-bipolar disorders. Comput Biol Med 64:127–37

Tillisch K et al (2013) Consumption of fermented milk product with probiotic modulates brain activity. Gastroenterology 144(7):1394–1401

Tozzi L et al (2016) Single-nucleotide polymorphism of the FKBP5 gene and childhood maltreatment as predictors of structural changes in brain areas involved in emotional processing in depression. Neuropsychopharmacology 41(2):487–497

van der Schot AC et al (2009) Influence of genes and environment on brain volumes in twin pairs concordant and discordant for bipolar disorder. Arch Gen Psychiatry 66(2):142–151

Viviani B et al (2003) Interleukin-1β enhances NMDA receptor-mediated intracellular calcium increase through activation of the src family of kinases. J Neurosci 23(25):8692–8700

Wang Z et al (2012) Abnormal default-mode network in angiotensin converting enzyme D allele carriers with remitted geriatric depression. Behav Brain Res 230(2):325–332

Wang L et al (2014a) Overlapping and segregated resting-state functional connectivity in patients with major depressive disorder with and without childhood neglect. Hum Brain Mapp 35(4):1154–1166

Wang L et al (2014b) Short-term effects of escitalopram on regional brain function in first-episode drug-naive patients with major depressive disorder assessed by resting-state functional magnetic resonance imaging. Psychol Med 44(7):1417–1426

Wegener I et al (2015) Changes of explicitly and implicitly measured self-esteem in the treatment of major depression: evidence for implicit self-esteem compensation. Compr Psychiatry 58:57–67

Widge AS et al (2018) Electroencephalographic biomarkers for treatment response prediction in major depressive illness: a meta-analysis. Am J Psychiatry 176(1):44–56

Wilkin TJ et al (2009) Regimen simplification to Atazanavir-Ritonavir alone as maintenance antiretroviral therapy: final 48-week clinical and virologic outcomes. J Infect Dis 199(6):866–871

Wise T et al (2016) Voxel-Based Meta-Analytical Evidence of Structural Disconnectivity in Major Depression and Bipolar Disorder. Biol Psychiatry 79(4):293–302

Wu HQ, Rassoulpour A, Schwarcz R (2007) Kynurenic acid leads, dopamine follows: a new case of volume transmission in the brain? Journal of Neural Transmission 114(1):33–41

Wu MJ et al (2015) Prediction of pediatric unipolar depression using multiple neuromorphometric measurements: a pattern classification approach. J Psychiatr Res 62:84–91

Wohleb ES et al (2011) ß-Adrenergic receptor antagonism prevents anxiety-like behavior and microglial reactivity. J Neurosci 31(17):6277–6288
Yang W et al (2016) Abnormal brain activation during directed forgetting of negative memory in depressed patients. J Affect Disord 190:880–888
Yang XH et al (2017a) Anhedonia correlates with abnormal functional connectivity of the superior temporal gyrus and the caudate nucleus in patients with first-episode drug-naive major depressive disorder. J Affect Disord 218:284–290
Yang XH et al (2017b) White matter microstructural abnormalities and their association with anticipatory anhedonia in depression. Psychiatry Res 264:29–34
Yeh YW et al (2014) Incongruent reduction of serotonin transporter associated with suicide attempts in patients with major depressive disorder: a positron emission tomography study with 4-[18F]-ADAM. Int J Neuropsychopharmacol 18(3)
Zalcman S et al (1994) Cytokine-specific central monoamine alterations induced by interleukin-1, -2 and -6. Brain Res 643(1–2):40–49
Zeng LL et al (2012) Identifying major depression using whole-brain functional connectivity: a multivariate pattern analysis. 135(Pt 5):1498
Zhai ZW et al (2019) Childhood trauma moderates inhibitory control and anterior cingulate cortex activation during stress. Neuroimage 185:111–118
Zhang J et al (2011) Disrupted brain connectivity networks in drug-naive, first-episode major depressive disorder. Biol Psychiatry 70(4):334–342
Zhang X et al (2014) First-episode medication-naive major depressive disorder is associated with altered resting brain function in the affective network 9(1):e85241
Zhang K et al (2016a) Molecular, Functional, and Structural Imaging of Major Depressive Disorder. Neurosci Bull 32(3):273–285
Zhang X et al (2016b) Altered neuronal spontaneous activity correlates with glutamate concentration in medial prefrontal cortex of major depressed females: An fMRI-MRS study. J Affect Disord 201:153–161
Zhang FF et al (2018a) Brain structure alterations in depression: psychoradiological evidence. CNS Neurosci Ther 24(11):994–1003
Zhang H et al (2019) Intrinsic gray-matter connectivity of the brain in major depressive disorder. J Affect Disord 251:78–85
Zhang HF, Mellor D, Peng DH (2018b) Neuroimaging genomic studies in major depressive disorder: a systematic review. CNS Neurosci Ther 24(11):1020–1036
Zunszain PA, Hepgul N, Pariante CM (2012) Inflammation and depression. In: Behavioral neurobiology of depression and its treatment. Springer, pp 135–151

Chapter 4
Neuroimmune Advance in Depressive Disorder

Guoqing Zhao and Xiaohua Liu

Abstract Major depressive disorder (MDD) or depression is one of the most highly prevalent, chronic, and recurrent disorders, which is associated with a high burden of disease and substantial impairment in social functions. Both immune molecules and cells have been implicated in the pathophysiology and maintenance of MDD. Findings in animals and MDD patients have suggested that both pro- and anti-inflammatory cytokines are activated in the neuroinflammation which contribute to behavioral symptoms and changes in the course of depression. There is a growing body of evidence to support that neuroinflammation is a mediator for the communication among stress response, neuroendocrine, neurotransmission, neurogenesis, and gut microbiota. These communications have been known as risk factors in the pathogenesis of MDD. In the meantime, accumulating evidence has suggested that some interventions targeting the inflammatory processes may play an important role in the treatment of MDD.

Keywords Major depressive disorder · Neuroinflammation · Neurogenesis · Stress · Cytokines · Gut microbiota

4.1 Introduction

Major depressive disorder (MDD) is one of the most important psychiatric diseases with high prevalence and a leading contributor to global disability. Mounting pre-clinical and clinical evidence has demonstrated that peripheral and central immune

The original version of this chapter was revised: The reference Schedlowski et al. 2016 has been replaced with "Walsh et al. (2016)". The correction to this chapter is available at: https://doi.org/10.1007/978-981-32-9271-0_16

G. Zhao
Department of Psychology, Provincial Hospital Affiliated to Shandong University, Jinan, China

X. Liu (✉)
Division of Mood Disorders, Shanghai Mental Health Center, Shanghai Jiao Tong University School of Medicine, Shanghai, China
e-mail: drliuxiaohua@gmail.com

© Springer Nature Singapore Pte Ltd. 2019, corrected publication 2020
Y. Fang (ed.), *Depressive Disorders: Mechanisms, Measurement and Management*, Advances in Experimental Medicine and Biology 1180,
https://doi.org/10.1007/978-981-32-9271-0_4

dysfunction may contribute to the pathophysiology of MDD. By data mining, it was found that there was no overlap in the genes that comprise the immune/inflammation response-related molecules between depression and other mental diseases, such as schizophrenia and bipolar disorder (Kim et al. 2016a). It has been confirmed that bipolar disorders and depression showed different immune activations by proteomics analysis (Chen et al. 2015). These findings have suggested the unique immunological mechanisms of MDD.

4.2 The Innate Immune System and MDD

4.2.1 Immune Cells

Microglia

Microglia as major components of brain cells are involved in many pathological and physiological processes. Their function is much more diverse, which depends on the central nervous system (CNS) homeostasis. Being equipped by numerous receptors and cell surface molecules, microglia can perform bidirectional interactions with other cell types in the CNS (Szepesi et al. 2018). There is accumulating evidence showing that neurons inform microglia about their status and thus are capable of controlling microglial activation and motility, while conversely microglia also modulates neuronal activities.

In normal conditions, functions of microglia mainly contribute to the brain development and normal structural and functional processes of neurons. However, in the presence of pathogens and stress, microglia, as one essential effector and regulator in the CNS, can be activated and polarized to M1 or M2 phenotype. In the central nervous system, microglia as a principal immune mediator responding to and releasing immune signals in the brain is considered to be associated with pathological conditions, such as psychological stress, pathological aging, and chronic infections. In response to different stressors, microglial activity is shown to be changed and varied. Moreover, inflammation and microglial activation mediate individual differences in resilience and the risk of stress-induced MDD (Reus et al. 2018).

In the prefrontal cortex (PFC), microglia is activated by stress exposure and undergoes morphological and functional changes, which contribute to synaptic deficits and the development of anxiety- and depression-like behavior (Wohleb et al. 2018). Recently, the loss and dystrophy of hippocampal microglia induced by chronic unpredictable stress (CUS) have been reported to mediate the development of major depression in mice. Results showed that the cellular numbers, process lengths, soma areas, and activation markers of endogenous hippocampal microglia were markedly reduced in mice treated with different types of chronic stresses (Tong et al. 2017).

Also, the treatment of MDD was also illuminated to bring changes in a microglial marker in the brain of patients with MDD. In MDD patients receiving cognitive behavioral therapy (CBT) group, translocator protein total distribution volume (TSPO VT, a marker of microglial density and inflammation) was significantly reduced during the treatment period, and reductions in TSPO VT were correlated with the amelioration of depressive symptoms (Juarez-Orozco et al. 2018). This study may indicate reduced microglia activation contributes to the MDD recovery.

Microglia, along with astrocytes, prominently express various pro-inflammatory (such as tumor necrosis factor (TNF)-α and interleukin (IL)-1β) and anti-inflammatory cytokines (such as IL-1, IL-10, and transforming growth factor (TGF)-β1) in the brain and hence, any action of these cytokines can primarily be associated with microglial activity in the CNS. Studies show that there was a dynamic crosstalk between microglia and peripheral monocytes that underlies stress-induced neuroinflammation which may lead to perturbations in neuroimmune function and contribute to synaptic plasticity deficits that underlie behavioral symptoms of mental health disorders, such as depression (Wohleb and Delpech 2017).

During stress and pathologies, microglia play a significant role in the disruption of neuroplasticity and have detrimental effects on neuroprotection causing neuroinflammation and exacerbation of depression. Studies have suggested that molecular pathways to depression are linked through microglia-associated neuroinflammation and hippocampal degeneration. Recent evidence has confirmed that both over-expressed and under-expressed microglia can cause depression. While over-expressed microglia trigger the onset of depression through the neuroinflammatory pathway as mentioned before; under-expressed microglia could result in depression through hippocampal degeneration pathway. Research has focused on the interplay between cytokines and HPA axis-derived corticosteroids in regulating microglial activity. Chronic forms of stressors, for example, chronic unpredictable stress, chronic restraint stress, and chronic social defeat stress, have all led to depression through reduction in the number of hippocampal microglia (Tong et al. 2017).

Astrocytes

In the central nervous system, there are many multiple receptors expressed in astrocytes which also interact with neurons at the synapse. Astrocytes are also considered the principal immune mediators in the brain, responding to and releasing immune signals, while neurons also express genes and signaling systems that modulate CNS immune responses (Mayfield et al. 2013). Astroglia contributes to many neurological diseases, such as Alzheimer's disease, epilepsy, spinal cord injury, and ischemia (Valori et al. 2019). Similarly, astrocyte dysfunction is associated with the pathogenesis of MDD in both animal models and human brains. In depression, astrocytes show signs of degeneration and are decreased in numbers, which may lead to a misbalance in neurotransmission and aberrant synaptic connectivity. However, the mechanisms underlying these effects remain largely unknown.

Neuroimaging and postmortem studies of MDD subjects reveal that pathology in astrocytes may significantly contribute to this reduced volume and the involvement of the hippocampus in MDD. In major depression, astrocytes not only undergo morphological and functional atrophy but also have enzymatic abnormalities. Glutamine synthetase (GS) is mainly located in astroglia which catalyzes the ATP-dependent condensation of ammonia and glutamate to form glutamine, playing a central role in glutamate and glutamine homeostasis. Some research has found that GS expressing

ACs is prominently involved in glutamate-related disturbances in major depression, but not in bipolar disorder (Bernstein et al. 2015).

Antidepressant medications affect astroglial biochemistry and physiology, which may represent a part of the therapeutic action (Peng et al. 2015). A recent study has found that astrocyte displays a distinct role for menin in regulating neuroinflammation in depression (Leng et al. 2018). A finding from the recent study reveals astrocyte contributions to neuronal function in the hilus. The density of astrocytes in the hilus of hippocampus, but not CA1 or CA2/3, was significantly decreased only in depressed subjects not taking an antidepressant medication (Cobb et al. 2016).

Both transcranial direct current stimulation (tDCS) and repetitive transcranial magnetic stimulation (rTMS) have been reported to be effective in treating MDD. But, the detailed mechanism of action has not been fully understood. Recent results have demonstrated that the astrocytic calcium signaling is activated by tDCS which is a required component for treating chronic restraint stress-induced depressive mice (Monai and Hirase 2018). The notion that the inhibition of Sirt1/MAO-A expression in astrocytes in the PFC could contribute to the different anti-depressive effects of rTMS with different parameters also has been put forward (Peng et al. 2018).

T cells

T lymphocytes have been associated with the production of certain neurotrophic factors and cytokines in the brain in the condition of stress-induced depression. Chronic stress may influence the T cells via the activation of the NF-kB pathways (Silberman et al. 2005). Regulatory T cell (Treg) has been reported to be associated with depression, and some agents such as chronic high-dose captopril administration may induce depression-like behaviors via Treg reduction and microglial activation (Park et al. 2017). In MDD, T cell phenotype and T cell receptor (TCR) repertoire are distortive on several levels (Patas et al. 2018).

Understanding the activities of T cells may help MDD patients restore the neuroimmune homeostasis. T lymphocytes are also critical in the resilience to MDD. During depression, T lymphocytes have the essential roles in the neuroimmune network in balancing the adaptive and maladaptive immune responses (Toben and Baune 2015).

B cells

B cells have also been regarded to play a key role in inflammatory responses. The study which investigated homeostasis of specific B cell subsets in MDD has found that naive lgD(+)CD27(−) B cells and B cells with immune-regulatory capacities such as CD1d(+)CD5(+) B cells and CD24(+)CD38(hi) transitional B cells were reduced.

4.2.2 Immune Molecule

Although the pathogenesis of MDD has not yet been exactly elucidated in previous findings, the association between chronic inflammation and MDD has been confirmed. Accumulating evidence has confirmed that innate immune signals of chemokines and cytokines have been activated in the brain of MDD patients. Levels of some pro- and anti-inflammatory cytokines have been reported to be increased in MDD, such as the C-reactive protein (CRP), IL 1β, IL 6, and the TNF-α. The association of chemokine as well as chemokine receptors and MDD has been detected in MDD (Milenkovic 2019).

Chemokines

In many neuroimmune processes relevant to depression, recent findings have implicated that chemokines [small (8–12 kDa) chemotactic cytokines] play roles in direct chemotaxis induction, leukocyte and macrophage migration, and inflammatory response propagation. When infection happens, chemokines have the ability to call for the migration of immune cells to the infection site. In the brain, chemokines demonstrate a novel, brain-specific mechanisms including the modulation of neuroendocrine functions, chemotaxis, cell adhesion, and neuroinflammation. Recent studies support elevated levels of chemokines in MDD patients. Meta-analysis has found the association between the abnormalities of blood chemokines and depression. MDD patients had higher levels of blood CXCL4 and CXCL7 and lower levels of blood CCL4 (Leighton and Nerurkar 2018). However, other meta-analyses got the diverse results that depressed patients had higher concentrations of the chemokine MCP-1/CCL2 (Krishnadas et al. 2016).

In summary, chemokines and their receptors have been demonstrated an important role in the pathophysiology of depression. Based on the additional research in larger populations and longitudinal studies, chemokines and their receptors could become novel diagnostic markers or therapeutic targets for MDD.

IL 1β

Pro- and anti-inflammatory cytokines play different roles in the modulation of depression-like behaviors and the hormonal and neurotransmitter systems in rats. A study has confirmed that centrally administered IL-1β (a pro-inflammatory cytokine) is associated with activation of CNS inflammatory pathways and behavioral changes, and treatment with IL-4 could modulate IL-1β-induced depression-like behaviors (anhedonic responses) and increase social exploration and total activity (Park et al. 2015). Thus, inhibition of IL-1β may have a therapeutic benefit in the treatment of neuropsychiatric disorders. IL-1β is necessary for the oxidative and neuroinflammatory responses produced by lipopolysaccharide (LPS), when IL-1β knock down in the hippocampus significantly attenuated the memory deficits and anxiety- and depression-like behaviors induced by LPS in mice (Li et al. 2017a). Suppression of inflammation could then serve as a potential strategy in depression therapy. Evidence has demonstrated that curcumin protects against chronic stress-induced dysregulation of neuroplasticity, neuronal apoptosis, and depression-like behaviors via suppressing IL-1β pathway in rats (Fan et al. 2018a, b).

IL 6

IL-6 is a cytokine that can be induced by stress. Its proinflammatory functions may be mediated through the soluble IL-6 receptor/trans-signaling (Fonseka et al. 2015). During chronic stress exposures, IL-6 can signal via the gp130 and IL-6Rα receptors to trigger the JAK/STAT3 signaling pathways (Girotti et al. 2013). IL-6 may be involved in depression via its impacts on the HPA axis (Girotti et al. 2013).

Based on these mechanisms, potential antidepressant strategies have been proposed to selectively target the IL-6 trans-signaling pathways (Fonseka et al. 2015). During the first depressive episode in drug-naive patients with MDD, the serum IL-6 level was significantly higher and played a key role in the morphological changes in the PFC during the early stage of MDD (Kakeda et al. 2018). A meta-analysis of 34 relevant studies revealed that only IL-6 remained significantly higher in the elderly with depression (Ng et al. 2018). A study in healthy male volunteers demonstrated that endotoxin-induced increase of IL-6 in the CSF might be primarily linked to the appearance of depressive symptoms (Engler et al. 2017).

IL-6 might be a potential therapeutic target in mood disorders. A study in college women found that a brief mindfulness intervention predicted greater reductions in salivary IL-6 and TNF-α, and changes in IL-6 were sustained at a 3-month follow-up (Walsh et al. 2016).

IL-18

A study assessed serum cytokines levels of 64 depressive patients and found an increase in serum TNF-α and IL-6 concentrations but a decrease in IL-18 concentrations (Fan and Luo 2017). But, the opposite result, that is depression showed increased GR, IL-18, and OxLDL concentrations, was also reported in a recent study (Oglodek 2017).

TNF-α

TNF-α may be involved in the modulation of the hypothalamic–pituitary–adrenal (HPA) axis in systemic immune diseases (Straub et al. 2011). TNF-α also induces the alterations of the blood–brain barrier and inflammation in the brain. Studies conducted in humans have shown that TNF-α antagonist infliximab can significantly relieve depressive symptoms in patients with drug-resistant depressive disorders (Bekhbat et al. 2018). It may become the target for anti-inflammatory treatment for the prevention of depression.

In a new model of prolonged depression-like behavior, the research has found that TNF-α disrupts blood–brain barrier integrity (Cheng et al. 2018). Depression is accompanied by increased expression of the TNF, TNFRSF1A, and TNFRSF1B genes at both mRNA and protein levels, and the elevated expression of TNF-α, TNFRSF1A, and TNFRSF1B negatively correlates with cognitive efficiency: working memory, executive function, attention, auditory–verbal memory, effectiveness of learning processes, and verbal fluency (Bobinska et al. 2017). TNF-α antagonists contribute to alleviating depressive symptoms and improving cognitive deficits (Bortolato et al. 2015). TNF-α serves as a therapeutic target for management of MDD, and its inhibitor (Infliximab) shows the ability to improve the spatial and emotional memory impairments induced by chronic stress in rats by modulating inflammation in hippocampus (Fasick et al. 2015; Sahin et al. 2015; Ma et al. 2016). But, a recent meta-analysis of 10 case-control studies found no association of a single-nucleotide polymorphism in TNF-α G-308A gene with depression (Shin et al. 2017).

Interferon

Inflammation has an essential role in these processes at various systems' biology levels. Evidence referring to the close associations between interferon (IFN)-α and the pathophysiology of depression has also been found in a recent study (Su 2015). IFN-α could induce depression which has more somatic symptoms and fewer symptoms of mood, anxiety, and negative cognition (Su et al. 2019). During many inflammatory and neurodegenerative conditions, abundant studies have reported abnormality of IFN-I. Paeoniflorin might reverse abnormal inflammatory cytokine levels and the increased density of microglia in IFN-α-treated mice, thus have a preventive therapeutic potential in IFN-α-induced depression (Li et al. 2017b).

4.3 Stress, HPA, and Depression

Childhood adversity has been repeatedly and robustly linked to physical and mental illness across the lifespan. Both the animal and human literature might support that the elements of childhood adversity are salient to the developing organism: physical threat, disrupted caregiving, and unpredictable environmental conditions (Kuhlman et al. 2017). The biological pathways for child adversity might be the inflammatory arm of the immune system and the hypothalamic–pituitary–adrenal (HPA) axis. HPA axis is altered after stress. Stress-induced disturbance of the HPA axis is strongly associated with the incidence of depression. A heightened neuroinflammatory response plays a fundamental role in the dysfunction of the HPA axis. Therefore, when stress-induced HPA axis was prevented, depressive behavior would be improved by alleviating inflammation and oxidative stress in the hypothalamus (Zhao et al. 2017).

4.4 Neuroinflammation and Neurogenesis in Depression

Pro-inflammatory cytokines and neuroinflammation are important not only in inflammatory responses but also in neurogenesis and neuroprotection in major depression (Kim et al. 2016). Several neuroinflammation-associated conditions have been reported to impair neurogenesis; the neuroinflammation-induced depression would be associated with hippocampal neurogenesis dysfunction. In the hippocampus, the neuroinflammatory state showed an increased production of pro-inflammatory cytokines, including IL-1β, IL-6, and TNF-α, and inhibited cell proliferation and newborn cell survival (Tang et al. 2016). The decreased hippocampal neurogenesis in cell proliferation was significantly correlated with depression-like behavior.

FGF2

Fibroblast growth factor (FGF)2 has been demonstrated anti-depressive effect by regulation of microglia in a neuroinflammatory model of depression (Tang et al. 2018)

BDNF

Brain-derived neurotrophic factor (BDNF) has essential roles in many functions. Both basic scientific and clinical studies have indicated the importance of BDNF in the pathophysiology of MDD. The levels of BDNF in the human serum have been measured as a marker for the efficacy of antidepressants in certain psychiatric disorders. The results from animal studies have demonstrated that agents exert antidepressant-like effect which is associated with enhancement of BDNF expression and inhibition of neuroinflammation (Li et al. 2018; Ren et al. 2018).

TGF

In the last decade, it was found that cytokine gene polymorphisms of TGF-β in the promoter and/or coding regions of regulatory genes are likely to be used as disease-associated markers in depressive disorder (Mihailova et al. 2016). Reduced TGF-β1 plasma levels are found in MDD, which correlated with depression severity, and significantly contributed to treatment resistance in MDD. TGF-β1 signaling might serve as a new pharmacological target for the medication of MDD (Caraci et al. 2018).

4.5 Neuroinflammation, Gonadal Hormones, and Gut–Brain Axis

Gonadal Hormones

Activation of the innate immune system has profound effects on brain activation and behavior, inducing fatigue, worsened mood, and pain sensitivity. The vulnerability factors for women in the behavioral response to inflammation might contribute to the understanding of the etiology and maintenance of depression as well as other symptoms (Lasselin et al. 2018). And, this vulnerability of women to be depressed than men might be due to the sex differences in the neuroimmune consequences of stress (Bekhbat and Neigh 2018).

Gut Microbiota

The importance of the gut microbiota to mental health has been noted by psychiatrists in recent studies. The mechanisms of microbiota to brain communication are determined and investigated. Up to now, microbiota–brain communication is emerging as an important factor in brain development and function, and immune system might be the mediator between microbiome and behavior (Foster 2016). Gut microbiota may play a causal role in the development of features of depression. In a recent study, researchers transplanted fecal microbiota from a subgroup of depressed patients to a microbiota-deficient rat model by oral gavage. The results showed that depressed patients had decreased gut microbiota richness and diversity and the fecal microbiota transplantation could induce behavioral and physiological features characteristic of depression in the recipient animals (Kelly et al. 2016).

The inflammasome dysregulation may be implicated in depression and its activation under stress causes activation of cytokines, such as IL-1β and IL-18. Recently, it has been reported that caspase-1 matured by inflammasome altered the fecal microbiome composition and displayed anxiety- and depression-like behavior (Wong et al. 2016). Gut microbiome plays an important role in neurodevelopment and in the etiology of the depressive syndrome, including nutritional, immunological, and energy homeostasis approaches (Lima-Ojeda et al. 2017). Exactly, the relationship between the gut and the brain is bidirectional communication, which may be mediated by neuroimmune, neuroendocrine, and sensory neural pathways. This association between gut microbiome and depression is well supported by research that alterations in the gut microbiota populations of specific species might contribute to depression, and depressive states might induce modification of specific gut microbiota species (Winter et al. 2018).

The findings showed that microbial diversity and taxonomic compositions were significantly changed in patients with mood disorders. Short-chain fatty acids-producing bacterial genera were decreased, while pro-inflammatory genera and those involved in lipid metabolism were increased in patients with depressive episodes. Specific bacteria of the gut were associated with clinical characteristics, inflammatory profiles, metabolic markers, and pharmacological treatment. The brain–gut–microbiota axis might be a promising target for disease diagnosis and therapeutic interventions for mood disorders in the future (Huang et al. 2019).

4.6 Clinical Applications in Neuroinflammatory Depression

Up to now, the current practice of antidepressant medication prescription is a trial-and-failure process. Over 1/3 MDD patients fail to respond to two or more antidepressant treatments, most of which target monoamine neurotransmission. In such so-called treatment-resistant depression (TRD) patients, inflammatory biomarkers are expressed excessively in both the CNS and the peripheral blood. Although mounting evidence has supported a mutual relationship between inflammation and major depression, no anti-inflammation agent has been approved for depression medication.

A number of clinical trials have suggested that conventional antidepressants have the property of inhibiting inflammation response. Other agents, such as nonsteroidal anti-inflammatory drugs (NSAIDs), cytokine inhibitors, ketamine, polyunsaturated fatty acids, statins, and curcumin, have demonstrated immunomodulatory properties in the treatment of depression.

There are no valid and available clinical markers for current antidepressant medications, but inflammatory cytokines may serve as markers of depression and potential predictors of treatment response. A variety of pieces of evidence have indicated that inflammation is involved in the pathogenesis, course, and treatment of major depression. Findings from recent studies supported inflammation and immune regulation as potential drug targets in antidepressant treatment (Schmidt et al. 2016). The suppression of inflammation and microglial M1-polarization would display antidepressant effect for depressed patients (Kalkman and Feuerbach 2016). Some studies have focused on the novel personalized treatment targeted at inflammation (Jha and Trivedi 2018). For example, monoclonal antibodies against pro-inflammatory cytokines, have demonstrated antidepressant effects in patients with chronic inflammatory conditions who report significant depressive symptoms. The antidepressant effect of Ketamine has drawn great attention, especially in patients with TRD.

Ketamine, as a glutamate receptor antagonist, exerts a therapeutic effect on account of crosstalk between inflammation and glutamate system (Cui et al. 2019). Due to the heterogeneity of MDD, further studies are required to personalize antidepressant prescription for the subgroups of patients responsive to inflammatory agents.

4.7 Conclusion

In summary, understanding the role of the neuroimmune system and the inflammatory mechanisms in MDD may help in explaining symptoms of depression, identifying biomarkers for diagnosis, and exploring new therapeutic approaches. A variety of potential inflammatory biomarkers and agents have been implicated by research so far, but as yet their role in the pathophysiology of MDD requires further elaboration and clarification. All the studies both on animals and patients provide novel insights into the understanding of the mechanism of neuroinflammation-associated depression. Loads of evidence have revealed robust connections among stress, depression, and inflammatory pathways (Noto et al. 2014). These findings may provide the implications for the inflammatory networks among gut microbiota, neuroendocrine, and neurogenesis. The understandings of the existing immune-therapeutics research for depression have led to a novel and personalized antidepressant interventions.

Acknowledgements We thank Prof. Yiru Fang for many constructive opinions on the writing of this chapter.

References

Bekhbat M, Neigh GN (2018) Sex differences in the neuro-immune consequences of stress: focus on depression and anxiety. Brain Behav Immun 67:1–12

Bekhbat M, Chu K, Le NA, Woolwine BJ, Haroon E, Miller AH, Felger JC (2018) Glucose and lipid-related biomarkers and the antidepressant response to infliximab in patients with treatment-resistant depression. Psychoneuroendocrinology 98:222–229

Bernstein HG, Meyer-Lotz G, Dobrowolny H, Bannier J, Steiner J, Walter M, Bogerts B (2015) Reduced density of glutamine synthetase immunoreactive astrocytes in different cortical areas in major depression but not in bipolar I disorder. Front Cell Neurosci 9:273

Bobinska K, Galecka E, Szemraj J, Galecki P, Talarowska M (2017) Is there a link between TNF gene expression and cognitive deficits in depression? Acta Biochim Pol 64:65–73

Bortolato B, Carvalho AF, Soczynska JK, Perini GI, McIntyre RS (2015) The involvement of TNF-α in cognitive dysfunction associated with major depressive disorder: an opportunity for domain specific treatments. Curr Neuropharmacol 13:558–576

Caraci F, Spampinato SF, Morgese MG, Tascedda F, Salluzzo MG, Giambirtone MC, Caruso G, Munafo A, Torrisi SA, Leggio GM, Trabace L, Nicoletti F, Drago F, Sortino MA, Copani A (2018) Neurobiological links between depression and AD: the role of TGF-beta1 signaling as a new pharmacological target. Pharmacol Res 130:374–384

Chen J, Huang C, Song Y, Shi H, Wu D, Yang Y, Rao C, Liao L, Wu Y, Tang J, Cheng K, Zhou J, Xie P (2015) Comparative proteomic analysis of plasma from bipolar depression and depressive disorder: identification of proteins associated with immune regulatory. Protein Cell 6:908–911

Cheng Y, Desse S, Martinez A, Worthen RJ, Jope RS, Beurel E (2018) TNFα disrupts blood brain barrier integrity to maintain prolonged depressive-like behavior in mice. Brain Behav Immun 69:556–567

Cobb JA, O'Neill K, Milner J, Mahajan GJ, Lawrence TJ, May WL, Miguel-Hidalgo J, Rajkowska G, Stockmeier CA (2016) Density of GFAP-immunoreactive astrocytes is decreased in left hippocampi in major depressive disorder. Neuroscience 316:209–220

Cui W, Ning Y, Hong W, Wang J, Liu Z, Li MD (2019) Crosstalk between inflammation and glutamate system in depression: signaling pathway and molecular biomarkers for Ketamine's antidepressant effect. Mol Neurobiol 56:3484–3500

Engler H, Brendt P, Wischermann J, Wegner A, Rohling R, Schoemberg T, Meyer U, Gold R, Peters J, Benson S (2017) Selective increase of cerebrospinal fluid IL-6 during experimental systemic inflammation in humans: association with depressive symptoms. Mol Psychiatry 22:1448–1454

Fan N, Luo Y (2017) Altered serum levels of TNF-alpha, IL-6, and IL-18 in depressive disorder patients. Hum Psychopharmacol :e2588

Fan C, Song Q, Wang P, Li Y, Yang M, Liu B, Yu SY (2018a) Curcumin protects against chronic stress-induced dysregulation of neuroplasticity and depression-like behaviors via suppressing IL-1beta pathway in rats. Neuroscience 392:92–106

Fan C, Song Q, Wang P, Li Y, Yang M, Yu SY (2018b) Neuroprotective effects of curcumin on IL-1beta-induced neuronal apoptosis and depression-like behaviors caused by chronic stress in rats. Front Cell Neurosci 12:516

Fasick V, Spengler RN, Samankan S, Nader ND, Ignatowski TA (2015) The hippocampus and TNF: common links between chronic pain and depression. Neurosci Biobehav Rev 53:139–159

Fonseka TM, McIntyre RS, Soczynska JK, Kennedy SH (2015) Evidence to support peripheral and central IL-6 signaling targets to treat depression. Expert Opin Investig Drugs 24:991–992

Foster JA (2016) Gut microbiome and behavior: focus on neuroimmune interactions. Int Rev Neurobiol 131:49–65

Girotti M, Donegan JJ, Morilak DA (2013) Influence of hypothalamic IL-6/gp130 receptor signaling on the HPA axis response to chronic stress. Psychoneuroendocrinology 38:1158–1169

Huang TT, Lai JB, Du YL, Xu Y, Ruan LM, Hu SH (2019) Current understanding of gut microbiota in mood disorders: an update of human studies. Behav Sci (Basel, Switzerland) 10:98

Jha MK, Trivedi MH (2018) Personalized antidepressant selection and pathway to novel treatments: clinical utility of targeting inflammation. Mol Neurobiol 19

Juarez-Orozco LE, Kurtys E, Dierckx RA, Moriguchi-Jeckel CM, Doorduin J, Li H, Sagar AP, Keri S (2018) Translocator protein (18 kDa TSPO) binding, a marker of microglia, is reduced in major depression during cognitive-behavioral therapy. J Cereb Blood Flow Metablsm: Off J Int Soc Cereb Blood Flow Metabolism 83:1–7

Kakeda S, Watanabe K, Katsuki A, Sugimoto K, Igata N, Ueda I, Igata R, Abe O, Yoshimura R, Korogi Y (2018) Relationship between interleukin (IL)-6 and brain morphology in drug-naive, first-episode major depressive disorder using surface-based morphometry. Sci Rep 8:10054

Kalkman HO, Feuerbach D (2016) Antidepressant therapies inhibit inflammation and microglial M1-polarization. Pharmacol Ther 163:82–93

Kelly JR, Borre Y, O' Brien C, Patterson E, El Aidy S, Deane J, Kennedy PJ, Beers S, Scott K, Moloney G, Hoban AE, Scott L, Fitzgerald P, Ross P, Stanton C, Clarke G, Cryan JF, Dinan TG (2016) Transferring the blues: depression-associated gut microbiota induces neurobehavioral changes in the rat. J Psychiatr Res 82:109–118

Kim S, Hwang Y, Webster MJ, Lee D (2016a) Differential activation of immune/inflammatory response-related co-expression modules in the hippocampus across the major psychiatric disorders. Mol Psychiatry 21:376–385

Kim YK, Na KS, Myint AM, Leonard BE (2016b) The role of pro-inflammatory cytokines in neuroinflammation, neurogenesis and the neuroendocrine system in major depression. Prog Neuropsychopharmacol Biol Psychiatry 64:277–284

Krishnadas R, Johnman C, Graham GJ, Cavanagh J, Eyre HA, Air T, Pradhan A, Johnston J, Lavretsky H, Stuart MJ, Baune BT (2016) A meta-analysis of chemokines in major depression. Mol Psychiatry 68:1–8

Kuhlman KR, Chiang JJ, Horn S, Bower JE (2017) Developmental psychoneuroendocrine and psychoneuroimmune pathways from childhood adversity to disease. Neurosci Biobehav Rev 80:166–184

Lasselin J, Lekander M, Axelsson J, Karshikoff B (2018) Sex differences in how inflammation affects behavior: what we can learn from experimental inflammatory models in humans. Pharmaceuticals (Basel, Switzerland) 50:91–106

Leighton SP, Nerurkar L (2018) Chemokines in depression in health and in inflammatory illness: a systematic review and meta-analysis. Int J Mol Sci 23:48–58

Leng L, Zhuang K, Liu Z, Huang C, Gao Y, Chen G, Lin H, Hu Y, Wu D, Shi M, Xie W, Sun H, Shao Z, Li H, Zhang K, Mo W, Huang TY, Xue M, Yuan Z, Zhang X, Bu G, Xu H, Xu Q, Zhang J (2018) Menin deficiency leads to depressive-like behaviors in mice by modulating astrocyte-mediated neuroinflammation. Neuron 100(551–563):e557

Li M, Li C, Yu H, Cai X, Shen X, Sun X, Wang J, Zhang Y, Wang C (2017a) Lentivirus-mediated interleukin-1beta (IL-1beta) knock-down in the hippocampus alleviates lipopolysaccharide (LPS)-induced memory deficits and anxiety- and depression-like behaviors in mice. J Neuroinflammation 14:190

Li J, Huang S, Huang W, Wang W, Wen G, Gao L, Fu X, Wang M, Liang W, Kwan HY, Zhao X, Lv Z (2017b) Paeoniflorin ameliorates interferon-alpha-induced neuroinflammation and depressive-like behaviors in mice. Oncotarget 8:8264–8282

Li DD, Xie H, Du YF, Long Y, Reed MN, Hu M, Suppiramaniam V, Hong H, Tang SS (2018) Antidepressant-like effect of zileuton is accompanied by hippocampal neuroinflammation reduction and CREB/BDNF upregulation in lipopolysaccharide-challenged mice. J Affect Disord 227:672–680

Lima-Ojeda JM, Rupprecht R, Baghai TC (2017) "I Am I and My Bacterial Circumstances": linking Gut microbiome, neurodevelopment, and depression. Frontiers in psychiatry. 8:153

Ma K, Zhang H, Baloch Z (2016) Pathogenetic and therapeutic applications of tumor necrosis factor-α (TNF-α) in major depressive disorder: a systematic review. Int J Mol Sci 17

Mayfield J, Ferguson L, Harris RA (2013) Neuroimmune signaling: a key component of alcohol abuse. Curr Opin Neurobiol 23:513–520

Mihailova S, Ivanova-Genova E, Lukanov T, Stoyanova V, Milanova V, Naumova E (2016) A study of TNF-alpha, TGF-beta, IL-10, IL-6, and IFN-gamma gene polymorphisms in patients with depression. J Neuroimmunol 293:123–128

Milenkovic VM (2019) The role of chemokines in the pathophysiology of major depressive disorder. World J Biol Psychiatry: Off J World Fed Soc Biol Psychiatry 20

Monai H, Hirase H (2018) Astrocytes as a target of transcranial direct current stimulation (tDCS) to treat depression. Neurosci Res 126:15–21

Ng A, Tam WW, Zhang MW, Ho CS, Husain SF, McIntyre RS, Ho RC (2018) IL-1beta, IL-6, TNF-alpha and CRP in elderly patients with depression or Alzheimer's disease: systematic review and meta-analysis. Sci Rep 8:12050

Noto C, Rizzo LB, Mansur RB, McIntyre RS, Maes M, Brietzke E (2014) Targeting the inflammatory pathway as a therapeutic tool for major depression. NeuroImmunomodulation 21:131–139

Oglodek EA (2017) The role of PON-1, GR, IL-18, and OxLDL in depression with and without posttraumatic stress disorder. Pharmacol Rep: PR 69:837–845

Park HJ, Shim HS, An K, Starkweather A, Kim KS, Shim I (2015) IL-4 Inhibits IL-1beta-induced depressive-like behavior and central neurotransmitter alterations. Mediat Inflamm 2015:941413

Park HS, Han A, Yeo HL, Park MJ, You MJ, Choi HJ, Hong CW, Lee SH, Kim SH, Kim B, Kwon MS (2017) Chronic high dose of captopril induces depressive-like behaviors in mice: possible mechanism of regulatory T cell in depression. Oncotarget. 8:72528–72543

Patas K, Willing A, Demiralay C, Engler JB, Lupu A, Ramien C, Schafer T, Gach C, Stumm L, Chan K, Vignali M, Arck PC, Friese MA, Pless O, Wiedemann K, Agorastos A, Gold SM (2018) T cell phenotype and T cell receptor repertoire in patients with major depressive disorder. Front Immunol 9:291

Peng L, Verkhratsky A, Gu L, Li B (2015) Targeting astrocytes in major depression. Expert Rev Neurother 15:1299–1306

Peng ZW, Xue F, Zhou CH, Zhang RG, Wang Y, Liu L, Sang HF, Wang HN, Tan QR (2018) Repetitive transcranial magnetic stimulation inhibits Sirt1/MAO-A signaling in the prefrontal cortex in a rat model of depression and cortex-derived astrocytes. Mol Cell Biochem 442:59–72

Ren Z, Yan P, Zhu L, Yang H, Zhao Y, Kirby BP, Waddington JL, Zhen X (2018) Dihydromyricetin exerts a rapid antidepressant-like effect in association with enhancement of BDNF expression and inhibition of neuroinflammation. Psychopharmacology 235:233–244

Reus GZ, de Moura AB, Silva RH, Resende WR, Quevedo J (2018) Resilience dysregulation in major depressive disorder: focus on glutamatergic imbalance and microglial activation. Evid-Based Complement Altern Med: eCAM 16:297–307

Sahin TD, Karson A, Balci F, Yazir Y, Bayramgurler D, Utkan T (2015) TNF-alpha inhibition prevents cognitive decline and maintains hippocampal BDNF levels in the unpredictable chronic mild stress rat model of depression. Behav Brain Res 292:233–240

Schmidt FM, Kirkby KC, Lichtblau N (2016) Inflammation and immune regulation as potential drug targets in antidepressant treatment. Curr Neuropharmacol 14:674–687

Shin KH, Jeong HC, Choi DH, Kim SN, Kim TE (2017) Association of TNF-alpha G-308A gene polymorphism with depression: a meta-analysis. Neuropsychiatric disease and treatment. 13:2661–2668

Silberman DM, Zorrilla-Zubilete M, Cremaschi GA, Genaro AM (2005) Protein kinase C-dependent NF-kappaB activation is altered in T cells by chronic stress. Cell Mol Life Sci: CMLS 62:1744–1754

Straub RH, Buttgereit F, Cutolo M (2011) Alterations of the hypothalamic-pituitary-adrenal axis in systemic immune diseases—a role for misguided energy regulation. Clin Exp Rheumatol 29:S23–S31

Su KP (2015) Nutrition, psychoneuroimmunology and depression: the therapeutic implications of omega-3 fatty acids in interferon-alpha-induced depression. BioMedicine 5:21

Su KP, Lai HC, Peng CY, Su WP, Chang JP, Pariante CM (2019) Interferon-alpha-induced depression: comparisons between early- and late-onset subgroups and with patients with major depressive disorder. Brain, Behav, Immun 80:512–518

Szepesi Z, Manouchehrian O, Bachiller S, Deierborg T (2018) Bidirectional microglia-neuron communication in health and disease. Front Cell Neurosci 12:323

Tang MM, Lin WJ, Pan YQ, Guan XT, Li YC (2016) Hippocampal neurogenesis dysfunction linked to depressive-like behaviors in a neuroinflammation induced model of depression. Physiol Behav 161:166–173

Tang MM, Lin WJ, Pan YQ, Li YC (2018) Fibroblast growth factor 2 modulates hippocampal microglia activation in a neuroinflammation induced model of depression. Front Cell Neurosci 12:255

Toben C, Baune BT (2015) An act of balance between adaptive and maladaptive immunity in depression: a role for T lymphocytes. J Neuroimmune Pharmacol: Off J Soc Neuroimmune Pharmacology 10:595–609

Tong L, Gong Y, Wang P, Hu W, Wang J, Chen Z, Zhang W, Huang C (2017) Microglia loss contributes to the development of major depression induced by different types of chronic stresses. Neurochem Res 42:2698–2711

Valori CF, Guidotti G, Brambilla L, Rossi D (2019) Astrocytes: emerging therapeutic targets in neurological disorders. Trends Mol Med https://doi.org/10.1016/j.molmed.2019.04.010

Walsh E, Eisenlohr-Moul T, Baer R (2016) Brief mindfulness training reduces salivary IL-6 and TNF-α in young women with depressive symptomatology. J Consult Clin Psychol 84(10):887–897. https://doi.org/10.1037/ccp0000122

Winter G, Hart RA, Charlesworth RPG, Sharpley CF (2018) Gut microbiome and depression: what we know and what we need to know. Rev Neurosci 29:629–643

Wohleb ES, Delpech JC (2017) Dynamic cross-talk between microglia and peripheral monocytes underlies stress-induced neuroinflammation and behavioral consequences. Prog Neuropsychopharmacol Biol Psychiatry 79:40–48

Wohleb ES, Terwilliger R, Duman CH, Duman RS (2018) Stress-induced neuronal colony stimulating factor 1 provokes microglia-mediated neuronal remodeling and depressive-like behavior. Biol Psychiatry 83:38–49

Wong ML, Inserra A, Lewis MD, Mastronardi CA, Leong L, Choo J (2016) Inflammasome signaling affects anxiety- and depressive-like behavior and gut microbiome composition 21:797–805

Zhao XJ, Zhao Z, Yang DD, Cao LL, Zhang L, Ji J, Gu J, Huang JY, Sun XL (2017) Activation of ATP-sensitive potassium channel by iptakalim normalizes stress-induced HPA axis disorder and depressive behaviour by alleviating inflammation and oxidative stress in mouse hypothalamus. Brain Res Bull 130:146–155

Chapter 5
Neurophysiologic Advance in Depressive Disorder

Lin Xu and Rongrong Mao

Abstract Enormous efforts for near half-century have harvested a plenty of understanding on major depressive disorder (MDD), although the underlying mechanisms are still elusive. The available antidepressants are far from satisfaction due to long-delay action (LDA) of antidepressant efficacy and low response rates in MDD patients. Notably, discovery of a single low-dose ketamine-producing rapid-onset and sustained antidepressant efficacy has inspired new research direction. These new studies have revealed ketamine's NMDAR-dependent and NMDAR-independent mechanisms, most of which are well known to be the key bases of synaptic plasticity as well as learning and memory. In fact, animal models of MDD are all based on the principle of learning and memory, i.e., the change of a behavior, for which monoaminergic and glutamatergic systems are the major modulators and executors, respectively. Reconsidering MDD as an aberrant form of emotion-related learning and memory would endow us a clearer research direction for developing new techniques or ways to prevent, diagnose, and treat MDD.

Keywords Depression · Neurophysiology · Ketamine · Glutamatergic system · Neural plasticity · Psychological stress

5.1 Introduction

The monoamine hypotheses of MDD have been developed since the 50s of twentieth century. It is mainly because most of the available antidepressants enhance the central monoamine systems. However, these hypotheses met difficulty in explaining why the monoamine-based antidepressants produce antidepressant efficacy with long-delay

L. Xu
Key Laboratory of Animal Models and Human Disease Mechanisms and Laboratory of Learning and Memory, Kunming Institute of Zoology, The Chinese Academy of Sciences, Kunming, China

R. Mao (✉)
Department of Pathology and Pathophysiology, School of Basic Medical Science, Kunming Medical University, Kunming, China
e-mail: talktomaomao@163.com

© Springer Nature Singapore Pte Ltd. 2019

Y. Fang (ed.), *Depressive Disorders: Mechanisms, Measurement and Management*, Advances in Experimental Medicine and Biology 1180,
https://doi.org/10.1007/978-981-32-9271-0_5

action (LDA) for about 4–8 weeks. Furthermore, an assumption is even proposed that LDA is a prerequisite for animal models of MDD. For example, although forced swimming test (TST) is widely used to develop antidepressant, it is regarded as not a model of MDD. In contrast, chronic unpredictable mild stress (CUMS) was assumed to be a model of MDD because it demonstrated LDA of antidepressant. On one hand, translational study following the monoamine hypotheses is very successful, and tens of new antidepressants have been developed. On the other hand, this "success" is far from satisfaction because new antidepressants showed the efficacy, LDA, and side effects very similar to the original ones, due to similar mechanisms of action (SMOA).

It is thus a breaking-through finding that a single low dose of ketamine produces rapid-onset (about 2 h) and sustained (for a week) antidepressant efficacy. Ketamine is known as a dirty drug regulating many pharmacological targets including NMDAR, which are well known to be critical for synaptic plasticity as well as learning and memory. Ketamine is also known as the addictive drug "K powder" producing psychosis and drug addiction. Similarly, other addictive drugs such as opioids and cannabis were once prescribed to treat MDD and/or anxiety with rapid onset of antidepressant efficacy (within hours to days) in the early twentieth century. Consistent with this point of view, a self-medication hypothesis of drug addiction was proposed in the 90s for which it was assumed that peoples would like to take addictive drugs for enhancing mood.

Since long-term potentiation (LTP) was reported for the first time by Bliss TVP and his colleague in 1973, synaptic plasticity has attracted great attention for understanding the mechanisms of learning and memory. Ample evidence thereafter shows that NMDAR activation leads to Ca^{2+} influx and a cascade of intracellular signaling, and eventually causes LTP by trafficking more AMPAR onto the postsynaptic membrane surface. On the contrary, NMDAR can also lead to less membrane expression of AMPAR if Ca^{2+} influx is not enough, resulting in long-term depression (LTD) of the glutamate synaptic transmission.

It is therefore unbearable that NMDAR and AMPAR are widely discussed in the fields of MDD for understanding the mechanisms of action by ketamine, but the influence on learning and memory is almost ignored. Furthermore, other types of synaptic plasticity such as spike timing-dependent plasticity and input timing-dependent plasticity are governed by the similar mechanisms such as dependence of NMDAR and AMPAR. If simplifying an input of a neuron is composed of many NMDARs and AMPARs in a glutamatergic synapse, we would suggest that a part of the more than 1000 inputs of a neuron was possibly chosen to link the selected neurons by using synaptic plasticity, eventually leading to the formation of a neural circuit that governs certain function. Thus, ketamine regulates synaptic plasticity and neural circuit possibly to be the key mechanisms for understanding MDD. However, we would like to focus on why ketamine- but not monoamine-based drugs produce rapid-onset antidepressant efficacy.

5.2 NMDAR-Dependent Mechanisms of Action by Ketamine

NMDAR-dependent mechanisms have been the major research direction for studying learning and memory. However, it was demonstrated for the first time that certain antagonists to NMDAR can produce antidepressant-like efficacy in an inescapable stress model of MDD in 1990 (Trullas and Skolnick 1990). Ten years later, NMDAR has attracted much attention in the research fields of MDD because a clinical group led by Professor John Krystal in Yale medical school reported that a single sub-anesthetic dose of ketamine (0.5 mg/kg), known as a noncompetitive antagonist to NMDAR, produced rapid and sustained antidepressant efficacy in MDD patients (Berman et al. 2000). Ketamine showed a similar time course of antidepressant efficacy in treating the treatment-resistant MDD patients (Murrough et al. 2013; Zarate et al. 2006). These findings were further confirmed by recent clinical studies (Fava et al. 2018; Kishimoto et al. 2016). In addition, ketamine also exerts promising antidepressant efficacy in bipolar depression (Zarate et al. 2012b). In this line of clinical studies, it appears to be very important to find whether rapid-onset antidepressant efficacy by ketamine is dependent on NMDAR or not, because NMDAR-dependent mechanisms are easier to be relevant to learning and memory.

Over the past decades, diverse antagonists to NMDAR were developed, and some of them may have antidepressant efficacy. Those antagonists include the NMDAR competitive antagonists AP-7 and CGP 37849, the noncompetitive NMDAR antagonists ketamine, and lanicemine, the NR2B-containing NMDAR antagonists Ro25-6981, eliprodil, CP101-606, and MK-0657(Li et al. 2011b; Murrough et al., 2017), and the divalent cation blockers Mg^{2+} and Zn^{2+}. In contrast, the partial agonists at glycine site of NMDAR, known able to increase NMDAR activation, such as ACPC, GLYX-13, AV-101, D-cycloserine, and orcinoside (Chen et al. 2010) can also produce antidepressant efficacy (Burgdorf et al. 2013; Murrough et al. 2017; Wang et al. 2015). It is not understood yet that the pharmacological actions produced by antagonists and partial agonists to NMDAR are opposite but still lead to a similar antidepressant efficacy.

The above evidence suggests the relevance of NMDAR to antidepressant efficacy. A possible explanation is that direct action on NMDAR is responsible for rapid action of a low-dose ketamine (0.5 mg/kg), since ketamine is an antagonist to NMDAR (Aleksandrova et al. 2017; Autry et al. 2011) with a high affinity about 500 nM of the IC_{50}. This appears to be possible particularly related to the binding properties of ketamine on NMDAR, because the enantiomer R-ketamine shows faster and more sustained antidepressant efficacy than the enantiomer S-ketamine does, which has nearly fourfold greater affinity to NMDAR than R-ketamine (Hashimoto 2016; Zanos et al. 2016). Unfortunately, this still seems not consistent with the evidence that other antagonists to NMDAR such as MK-801, memantine, and the selective NR2B-containing NMDAR antagonist Ro25-6981 show less antidepressant efficiency or no effect at all (Newport et al. 2015; Zanos et al. 2016). For example, memantine is one of the approved drugs to treat Alzheimer's disease (AD) presumably by blocking overactivity of NMDAR. The difference between ketamine and memantine may attribute to NMDAR-mediated miniature EPSC, which can be blocked by the former but not the latter under the physiological Mg^{2+} levels (Gideons et al. 2014). Furthermore, this assumption is also not consistent with the antidepressant efficacy produced by the NMDAR partial agonists GLYX-13, D-cycloserine, and orcinoside (Burgdorf et al. 2013; Levin et al. 2015; Richter-Levin and Xu 2018; Wang et al. 2015). Strong evidence shows that the NR2B-containing NMDAR are mainly distributed in extra synapses to be critical for LTD (Liu et al. 2004). Ketamine may particularly target on this subtype of NMDAR to regulate LTD in the prefrontal cortex (Miller et al. 2014). Consistent with this finding, ketamine induces an endogenous LTD in the hippocampal CA1 region in freely moving rat (Duan et al. 2013). This evidence would be useful for understanding why ketamine must be low dose because higher doses of ketamine would block both the NMDAR-dependent LTP and LTD. Whether these partial agonists to NMDAR can similarly induce an endogenous LTD in the hippocampal CA1 region would be an interesting question. If it is the case, the antagonist ketamine and the partial agonists eventually lead to a similar effect on synaptic weight, LTD, which implicates that the mechanisms of action by ketamine may be still NMDAR-dependent even at low dose.

Alternative explanation is that ketamine or certain NMDAR antagonists can preferentially block the NMDAR on the presynaptic GABAergic neurons, where the inhibitory output to glutamatergic neurons is disinhibited, leading to potentiation of synaptic transmission of the glutamatergic synapses (Homayoun and Moghaddam 2007; Moghaddam et al. 1997; Stone et al. 2012) This point of view is highly consistent with the findings that ketamine or the NR2B-containing NMDAR antagonist Ro25-6981 increases extracellular glutamate levels (Moghaddam et al. 1997) and glutamate recycling (Chowdhury et al. 2017) in the prefrontal cortex. Nevertheless, the increased glutamatergic synaptic transmission would have eventually led to higher possibility of the postsynaptic NMDAR activation because depolarization of the membrane potentials needs to activate NMDAR. However, this interpretation appears to be not convincing yet because it is hardly understood why ketamine is so unique able to block the presynaptic NMDAR to cause disinhibition but memantine or MK-801 does not. It is noteworthy that recent achievements suggest that NMDAR in the lateral habenula (LHb) may play important roles in antidepressant-like efficacy in mice (Yang et al. 2018), leading to alterations of Kir4.1 function that governs neuronal firing pattern (Cui et al. 2018).

Up to date, we still need to take cautions that the NMDAR-dependent mechanisms of action by ketamine could not explain its antidepressant efficacy very well because rapid-onset and sustained action for over a week cannot be explained by the pharmacological properties on NMDAR since the half-life time of ketamine is only about 3 h. Perhaps, NMDAR-dependent mechanisms of action by ketamine are only an initial step that further influences synaptic plasticity as well as learning and memory to maintain the antidepressant efficacy for a longer period.

5.3 NMDAR-Independent Mechanisms of Action by Ketamine

NMDAR-independent hypotheses for rapid action of ketamine have been also developed. In particular, disinhibition through blocking the presynaptic NMDAR located on GABAergic neurons will bypass the postsynaptic NMDAR and lead to potentiation of the postsynaptic AMPAR-mediated EPSC. It is not known if this NMDAR-independent LTP-like process can be long-term maintained. This seems unlikely because ketamine is rapidly metabolized and the potentiation cannot be maintained for over 24 h. Nevertheless, stronger evidence is that the metabolism of racemic ketamine to (2R, 6R)-HNK is essential for antidepressant efficacy, but the HNK may directly potentiate the AMPAR-mediated current without affecting NMDAR (Carrier and Kabbaj 2013; Franceschelli et al. 2015). If HNK works in clinical trial, it may avoid the side effects of ketamine in blocking NMDAR. In fact, an earlier clinical research already examined the relationship between ketamine metabolites and psychotomimetic symptoms in MDD and bipolar depression patients, and discovered a positive link between plasma HNK metabolite levels and antidepressant response to ketamine (Zarate et al. 2012a). This is consistent with the findings from Todd D. Gould's group in which the antidepressant action of the metabolite HNK is NMDAR-independent involves early and sustained activation of AMPAR (Zanos et al. 2016). Furthermore, the ketamine-induced synaptic potentiation and antidepressant efficacy are absent in the knockout mice without the AMPAR subunit GluA2 (Nosyreva et al. 2013). The important role of AMPAR for the antidepressant efficacy of (2R, 6R)-HNK is associated with the upregulation of AMPAR in hippocampal synapses but can be blocked by the AMPAR antagonist NBQX (Yao et al. 2017; Zanos 2017; Zanos et al. 2016). Previous studies have demonstrated that pretreatment with NBQX can attenuate both ketamine-induced antidepressant efficacy and the levels of phosphorylated GluA1 of the AMPAR in the hippocampus (Maeng et al. 2008). Meanwhile, the AMPAR agonist CX546 pretreatment potentiates the antidepressant efficacy of ketamine (Zhou et al. 2014). It is believed that the enhanced gamma-band power induced by ketamine can reveal the cortical disinhibition-dependent on the activation of AMPAR (Zanos et al. 2016). Moreover, the group II metabotropic glutamate (mGlu2/3) receptor antagonists MGS0039 and LY341495 may also exert their rapid antidepressant efficacy through AMPAR activation (Karasawa et al. 2005; Koike and Chaki 2014).

These findings reveal that AMPAR activation is crucial for antidepressant efficacy of ketamine or its metabolites. However, it is worthy of keeping in mind, all these mechanisms mentioned in the above are well known in the research fields of synaptic plasticity as well as learning and memory. A slight difference is that the AMPAR-mediated potentiation here is not dependent on NMDAR, suggesting that widespread of the potentiation could happen in the brain. Furthermore, it is notable that fluoxetine and imipramine also induce upregulation of AMPAR and their antidepressant efficacy can also be blocked by NBQX, implicating that the mechanisms induced by monoamine-based antidepressants and ketamine or its metabolites would be the same (Freudenberg et al. 2015). This is in fact counterintuitive because the antidepressant efficacy of ketamine is rapid-onset but that of monoamine-based antidepressants is very slow.

Based on these AMPAR-dependent mechanisms, the AMPAR-positive allosteric modulators LY392098, LY451646, and Org 26576 (Aleksandrova et al. 2017), and the AMPAR agonists CX54620, and AMPA are aimed to be developed as a new generation of antidepressants. Furthermore, the AMPAR modulator Org 26576 in a small pilot clinical trial showed efficacy, tolerability, and cognitive enhancement in MDD patients (Nations et al. 2012). It is known that the phosphorylation of GluA1 and GluA2 regulates AMPAR trafficking and function, and thus the efficacy of ketamine or (2R, 6R)-HNK on the phosphorylation of GluA1 and GluA2 would have confirmed the role of AMPAR in the efficacy of the ketamine metabolite (Aleksandrova et al. 2017).

Nevertheless, these AMPAR-dependent mechanisms of action by ketamine are very similar to those of synaptic plasticity triggered by NMDAR-dependent activity. More importantly, these mechanisms seem very similar to those triggered by the monoamine-based antidepressants such as fluoxetine and imipramine, suggesting that the key mechanisms for rapid-onset action of ketamine or its metabolites remained still unknown.

5.4 The Common Mechanisms of Action by Antidepressant

The half-life time of ketamine is approximately 3 h (Wieber et al. 1975). However, its antidepressant efficacy is maintained for over a week. Particular mechanisms are required for the maintenance of the efficacy, implicating that synaptic plasticity and learning and memory are high possible, for which NMDAR and AMPAR are both critical. Evidence also suggests that ketamine can induce rapid synaptogenesis that may be able to maintain the antidepressant efficacy for over a week (Abdallah et al. 2016; Aleksandrova et al. 2017; Freudenberg et al. 2015). It is notable again that synaptic plasticity can lead to synaptogenesis to be the key mechanisms of learning and memory.

The intracellular signaling triggered by ketamine also sounds like that for learning and memory. Ketamine rapidly increases BDNF levels by deactivation of eukaryotic elongation factor 2 (eEF2) and activates the BDNF receptor TrkB in the hippocampus. BDNF is necessary for ketamine's sustained antidepressant efficacy although ketamine cannot increase BDNF levels 24 h after administration (Autry et al. 2011; Lepack et al. 2015). However, Todd D. Gould's group reported that both ketamine and (2R, 6R)-HNK induced a decrease in eEF2 phosphorylation in the hippocampus 1 h and 24 h after injection of the drugs, and increased hippocampal synaptoneurosomal BDNF at 24 h (Zanos et al. 2016). Furthermore, it is known that Met rs6265 allele induces deficits in BDNF processing and activity-dependent secretion. Both MDD patients carrying the Met rs6265 allele (Laje et al. 2012) and mice knocked in with human BDNF Val66met (rs6265) single-nucleotide polymorphism (Liu et al. 2012) do not respond to ketamine treatment, suggesting that increased BDNF synthesis is the downstream mechanisms of ketamine. The increased BDNF stimulates TrkB-Akt pathway that activates rapamycin (mTOR) signaling, resulting in increased proteins synthesis required for synapse maturation and formation including neurogenesis and dendritic spine growth (Zanos and Gould 2018). The activation of mTOR is not only found in both the prefrontal cortex and hippocampus of animals (Li et al. 2010; Paul et al. 2014; Zhou et al. 2014) but also in the peripheral blood of depressed patients (Denk et al. 2011; Yang et al. 2013) after ketamine administration. While the mTORC1 signaling is blocked, the antidepressant actions of ketamine are then blocked (Duman et al. 2016). These findings indicate that ketamine induced protein translation via mTOR activation. This mTOR activation is dependent on AMPAR but not on NMDAR (Advances in neurophysiological mechanisms underlying ketamine fast-acting antidepressant effects are summarized in Fig. 5.1).

The downstream mechanisms of action by ketamine appear to be very similar to those by classic antidepressants as well as those used by learning and memory. Because classic antidepressants are of LDA but ketamine is superior by rapid action, unique mechanisms of action by ketamine are unlikely to be those mentioned above. Nevertheless, these findings related to NMDAR-dependent and NMDAR-independent mechanisms of action by ketamine may overtake those of learning and memory to interrupt memory retrieval or to generate the drug-induced false memories so that the rapid-onset and sustained antidepressant efficacy could occur.

5.5 Aberrant Form of Learning and Memory for MDD

The possible mechanisms of action by ketamine seem to be NMDAR-dependent and NMDAR-independent, and the downstream mechanisms appear to be BDNF-TrkB-AKT-mTOR signaling, and eventually may lead to synaptogenesis, all of which are well known to be the fundamental basis of synaptic plasticity and learning and memory (Cohen-Cory 2002; Wang et al. 2003). It is due to this reason neural plasticity hypotheses of MDD have been developed (Duman et al. 2012; Liu et al. 2017; Wang et al. 2015). It is important to note that all these mechanisms are also found following the treatment of monoamine-based antidepressants. Thus, questions are still unanswered why ketamine produces rapid-onset and sustained antidepressant efficacy. One possible explanation is that MDD is caused by an aberrant form of learning

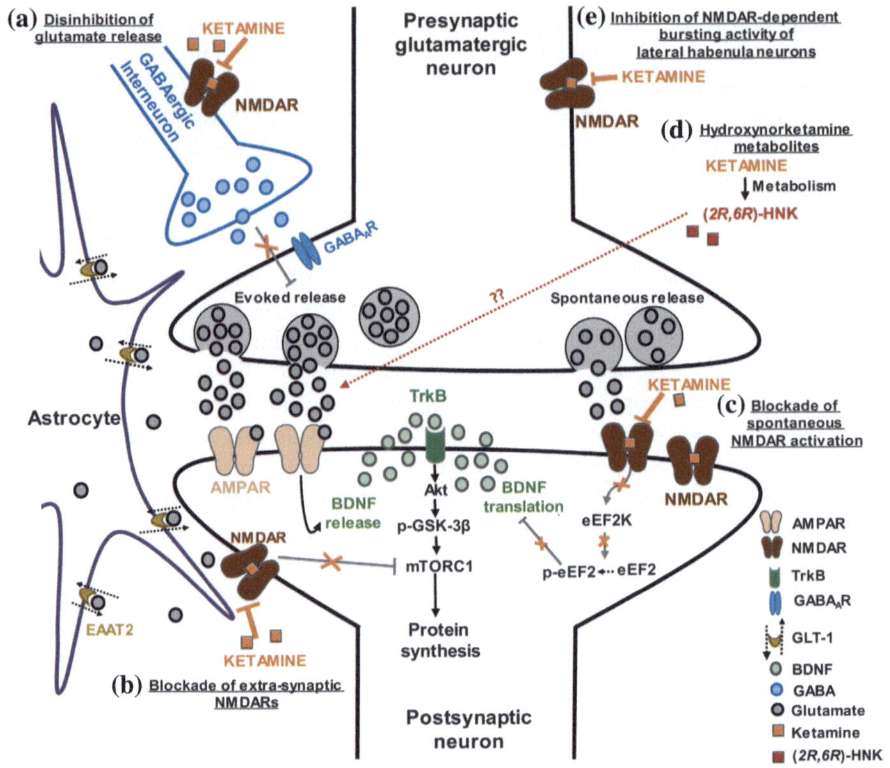

Fig. 5.1 Advances in neurophysiological mechanisms underlying ketamine fast-acting antidepressant effects. This figure is adapted from "Mechanisms of ketamine action as an antidepressant". Zanos, P., and Gould, T.D. (2018). Molecular Psychiatry 23, 801–811. (1) NMDAR inhibition-dependent mechanisms including presynaptic GABAergic neuron disinhibition (**a**), blockade of extra-synaptic NR2B containing NMDAR (**b**), blockade of spontaneous NMDAR activation at rest (**c**), and inhibition of NMDAR-dependent bursting firing in lateral habenula neurons (**e**). (2) Postsynaptic AMPAR activation (**d**), an NMDAR inhibition in dependent mechanism. (3) The convergent downstream mechanisms underlying various theories. The increased BDNF stimulates TrkB-Akt pathway that activates mTORC1 signaling, resulting in increased proteins synthesis required for synapse maturation and formation

and memory so that targeting on synaptic plasticity through NMDAR and AMPAR mechanisms would allow the negative memories to be blocked in the retrieval.

There are ample evidence suggesting that chronic stress or animal models of MDD can cause alterations of synaptic plasticity and synaptic structure such as atrophy of dendrite and spines in the prefrontal cortex and hippocampus (Kang et al. 2012; MacQueen et al. 2008), and also in the amygdala and nucleus accumbens (Qiao et al. 2016). In contrast, synaptogenesis can be triggered by Rac1 activation that is believed to be the key mechanism for forgetting. Thus, it is possible that atrophy of dendrite and spines in certain brain areas may be associated with the formation of negative memories but ketamine-induced synaptogenesis may cause forgetting so that depressive symptoms can be rapidly and temporally suppressed. The duration of antidepressant efficacy by ketamine would be dependent on how strong the ketamine-induced new synapses or how long the ketamine-induced synaptic plasticity can be maintained. Here, it is very important to address what type of synaptic plasticity was induced by ketamine.

It is well known that NMDAR-dependent LTP and LTD were changed by chronic stress or in animal models of MDD. Repeated treatment of monoamine-based antidepressants can rescue such changes, in particularly relevant to synaptic plasticity in the hippocampus (Jing et al. 2015; Zhang et al. 2014). These changes of synaptic plasticity in the hippocampus may allow the "negative memories" to be long-term stored (Citri and Malenka 2008). In fact, clinical investigations have revealed that stress is the major risk factor leading to the development of MDD. Ample evidence from Xu Lin's group from Kunming Institute of Zoology, the Chinese Academy of Sciences has suggested a concept that stress can impair LTP and facilitate LTD in the hippocampus (Cao et al. 2004; Rowan et al. 1998; Wang et al. 2006; Xiong et al. 2003, 2004; Xu et al. 1997, 1998; Yang et al. 2004b, 2007). It is assumed that the stress-limited LTP and stress-facilitated LTD may be a unique mechanism allowing the "negative memories" to be long-term stored more easily. Lost control of the retrieval of "negative memories" will lead to sustained depressed mood. Recent reports have demonstrated that optogenetic stimulation-induced LTD in the amygdala inactivates fear memory, while the stimulation-induced LTP activates it (Nabavi et al. 2014). This would be a reasonable explanation that retrieval of negative memory-associated depressed mood will be rapidly turned off if ketamine can induce LTD. It is in fact that sub-anesthetic dose of ketamine induced AMPAR trafficking-dependent LTD in the hippocampus (Duan et al. 2013). It is known that opioids and cannabis can also produce rapid action of antidepressant efficacy in MDD patients. It is remarkable that low dose of opioids or cannabis can also induce NMDAR-dependent LTD in the hippocampus (Han et al. 2012; Yang et al. 2004b). These findings strongly suggest that the drug-induced LTD in the hippocampus may have turned off the retrieval process of the negative memories and thus produces rapid action. Since the original negative memories are untouched, the depressive symptoms are still relapsed after 1 or 2 weeks from ketamine treatment. Consist with this view, Jing et al. (2015) from Xu L' group reported that despair memory may occur following forced swimming test with an endogenous LTP in the hippocampus, and ketamine can prevent such a process.

It is also the key questions in MDD what the mechanisms can cause people depressed. There is a very good hint that addictive drugs known to enhance the VTA functions and dopamine outputs all produce reinforcing and rewarding effects. It is possible that ketamine, opioids, and cannabis cause rewarding effects that counteract depressed mood in MDD patients. Back reasoning may predict that rewarding effects related to dopamine system are impaired in MDD patients. This point of view may be consistent with the finding that learned helplessness may relate to the excitatory synapses onto LHb neurons that relay excitation to the VTA dopamine neurons, and thus repeated electrical stimulation of LHb can effectively reduce depression-like behavior (Li et al. 2011a).

It is possible that the development of MDD is associated with stressful life events, which formed negative memories tagged with depressed mood. The underlying mechanisms are associated with NMDAR and AMPAR as well as monoaminergic systems. Blocking the retrieval of negative emotion and enhancing the rewarding effect would allow the depressive symptoms rapidly faded away temporally, while the NMDAR- and AMPAR-dependent mechanisms would allow the antidepressant efficacy to be maintained for a certain period.

5.6 New Diagnosis and Treatment for MDD

According to this aberrant form of learning and memory hypothesis of MDD, it would become clearer for understanding why electroconvulsion therapy (ETC) has been perhaps the best effective way to treat MDD because it causes amnesia. It is also the case for understanding why placebo effects are so strong in MDD patients because the mindset of the MDD patients would counteract the negative effects of life stress memories (Crum et al. 2013). It is unbearable today that we are still trying to understand MDD only on the level of receptors or transporters and the downstream signaling, but the only difference is from the monoamine systems moved to the glutamate ones (Duman et al. 2016; Hammen 2005; Richter-Levin and Xu 2018). The rapid-onset action of ketamine gives rise to us a chance for reconsidering the neuropathological basis of MDD and finding out how can we prevent the development of the disorder, and how can we use biological parameters to diagnose it, by which we can use them to monitor the antidepressant efficacy during treatment. There are two types of memory retrieval: specific recall and generalization. If the rapid-onset action of ketamine is due to enhancing mood and blocking retrieval of negative memories, the neuropathological basis of MDD is therefore possibly attributed to stressful life event-induced memories. This is consistent with clinical investigations for decades that psychological stress causes MDD (Cohen et al. 2007). We assumed that retrieval of negative memories is governed by generalization (Zhou et al. 2017) when peoples perceived novel situations but of certain similarity of inescapability to previous stressful life events, and their emotion was inevitably depressed unconsciously even without knowing what is the reason. This generalization process would become much faster if the stressful memories and current situations are somehow linked together (Cai et al. 2016; Silva 2017), leading to the negative emotion to be transferred rapidly from the past to the current (Zhou et al. 2017) by blocking generalization retrieval. If generalization retrieval can be blocked, the previous influence of stressful life events would hardly affect the current situations, implicating the prevention of the development of MDD.

A harder question is how we can use biological parameters to diagnose MDD and use them to monitor the antidepressant efficacy of treatment, and eventually to produce total remission on MDD once for all. It is important to note that not all types of stress but only inescapable stress can impair LTP, which facilitates LTD in the hippocampal CA1 region (Kim et al. 1996; Xu et al. 1997, 1998; Yang et al. 2004a). In short, inescapable stress causes despair emotion that was tagged with the stressful life experience. Consistent with this point of view, LTD is assumed to be induced by chronic life stress to be critical for the development of MDD (Duman and Aghajanian 2012; Duman et al. 2016; Manji et al. 2001). However, from our point of view, this is not the case because stressful life events can form long-lasting memories possibly only once. It must require fast learning mechanisms that are likely to be the so-called combinatory synaptic plasticity hypothesis, for which LTP and LTD are combined in the dendrite of the selected neurons that lead to little net changes in synaptic weight, with higher efficiency to encode and store the negative memories (Dudai 1996; Xu et al. 1997). Thus, a treatment if targeting both LTP and LTD under stressful situation may prevent the accumulation of negative memories. Consistent with this prediction, Xu L's laboratory found that TRPV1 activation in the hippocampus can prevent the effects of inescapable stress on LTP and LTD as well as spatial memory (Li et al. 2008). A growing body of evidence indicates that the role of central TRPV1 may be a good target to treat MDD (Madasu et al. 2015; Reyes-Mendez et al. 2018). The antidepressant duloxetine can increase neuronal survival by reducing oxidative stress through activation of TRPV1 receptor in the hippocampus (Demirdas et al. 2017; Reyes-Mendez et al. 2018). Furthermore, inactive of TRPV1 partially abolished the antidepressant effect of fluoxetine, while its effect was potentiated by activation of TRPV1 (Manna and Umathe 2012). These findings further support the stress-related neuropathological basis of MDD.

Testing generalization by the tasks of automatic negative thinking is possible to identify whether the patients had early life stress that leads to MDD in later life (Juruena et al. 2015; Liu et al. 2016; Radley et al. 2011), but this is still not a biological parameter. We believed that generalization of previous stress memories requires the activity of bilateral hippocampal CA1 region in the rostral–dorsal part of the hippocampus (Zhou et al. 2017). Recording the electric activity triggered by emotional generalization task may be a biological measurement for MDD.

Here, we assumed that strong negative memories are formed only under the situations of inescapable or uncontrollable stress, and re-encountering and perceiving new but similar situations with inescapability or uncontrollability will lead to generalization retrieval of the negative memories to cause depressed mood (Richter-Levin and Xu 2018, see Fig. 5.2). This hypothesis is consistent with the clinical observation that MDD patients exhibit overgeneralization of negative memories (Gotlib and Joormann 2010). Alternatively, MDD patients may favor negative but not positive memories for the cautions in response to novel situations due to the cognition pattern shaped by previous bad experience (Dillon and Pizzagalli 2018).

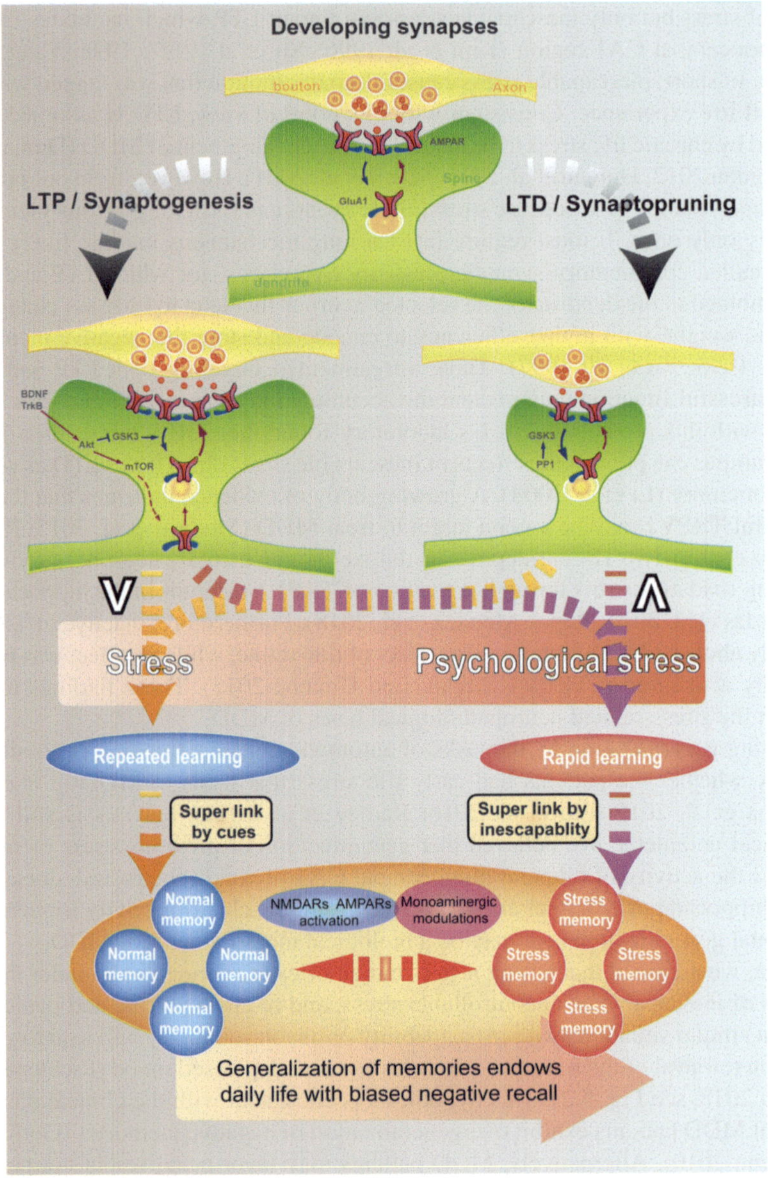

Fig. 5.2 The neurophysiological mechanisms mediating psychological stress may lead to major depressive disorder. This figure is adapted from "How could stress lead to major depressive disorder?" Richter-Levin, G., and Xu, L. (2018). IBRO reports 4, 38–43. Synaptic LTP/synaptogenesis or LTD/synaptopruning builds up normal neural circuits. Both LTP and LTD participate the repeated learning under controllable stress. In contrast, the distinct combination of LTP and LTD is engaged in the rapid learning related to psychological stress, which may be the mechanisms of generalization on the biased negative cognition in depression. NMDAR and AMPAR as well as monoaminergic systems modulate these processes. ∨: LTP or LTD; ∧: Combined LTP and LTD

5.7 Conclusions

The discovery of iproniazid led to the revolution of neuropharmacology that has resulted in more and better monoamine-based antidepressants today. We would expect that the discovery of ketamine-producing rapid-onset and sustained antidepressant efficacy for a week will lead to another revolution. However, this goal looks still a matter of big challenge because repeated use of ketamine cannot avoid the tolerance and addictive effects. The metabolites of ketamine if without these two mechanisms are unlikely to produce rapid-onset action, because other addictive drugs can also produce rapid-onset action on MDD. A better way is to allow MDD patients to learn that new situations have nothing to do with the past bad experience, by enhancing NMDAR activation through the partial agonist at the glycine site of NMDAR. It is very much hopeful that the better ability in learning and memory would be the better way to cure MDD once for all.

References

Abdallah CG, Adams TG, Kelmendi B, Esterlis I, Sanacora G, Krystal JH (2016) Ketamine's mechanism of action: a path to rapid-acting antidepressants. Depress Anxiety 33:689–697

Aleksandrova LR, Phillips AG, Wang YT (2017) Antidepressant effects of ketamine and the roles of AMPA glutamate receptors and other mechanisms beyond NMDA receptor antagonism. J Psychiatr Neurosci 42:222–229

Autry AE, Adachi M, Nosyreva E, Na ES, Los MF, Cheng PF, Kavalali ET, Monteggia LM (2011) NMDA receptor blockade at rest triggers rapid behavioural antidepressant responses. Nature 475:91

Berman RM, Cappiello A, Anand A, Oren DA, Heninger GR, Charney DS, Krystal JH (2000) Antidepressant effects of ketamine in depressed patients. Biol Psychiatry 47:351–354

Burgdorf J, Zhang X-L, Nicholson KL, Balster RL, Leander JD, Stanton PK, Gross AL, Kroes RA, Moskal JR (2013) GLYX-13, a NMDA receptor glycine-site functional partial agonist, induces antidepressant-like effects without ketamine-like side effects. Neuropsychopharmacology 38:729–742

Cai DJ, Aharoni D, Shuman T, Shobe J, Biane J, Song W, Wei B, Veshkini M, La-Vu M, Lou J et al (2016) A shared neural ensemble links distinct contextual memories encoded close in time. Nature 534:115–118

Cao J, Chen NH, Xu TL, Xu L (2004) Stress-facilitated LTD induces output plasticity through a synchronized-spikes and spontaneous unitary discharges the CA1 region of the hippocampus. Neurosci Res 49:229–239

Carrier N, Kabbaj M (2013) Sex differences in the antidepressant-like effects of ketamine. Neuropharmacology 70:27–34

Chen J, Xu L, Zhou J, Lv J, Mao RR, Tian M, Zhou QX, Zhang XM, Shen Y, Jiang ZY et al (2010) The application of 5-methyl-1,3-benzenediol or derivatives thereof in the preparation of medicines and functional foods for treatment or prevention of depression. China Patent Number CN101742992A

Chowdhury GMI, Zhang J, Thomas M, Banasr M, Ma X, Pittman B, Bristow L, Schaeffer E, Duman RS, Rothman DL et al (2017) Transiently increased glutamate cycling in rat PFC is associated with rapid onset of antidepressant-like effects. Mol Psychiatry 22:120–126

Citri A, Malenka RC (2008) Synaptic plasticity: multiple forms, functions, and mechanisms. Neuropsychopharmacology 33:18–41
Cohen-Cory S (2002) The developing synapse: construction and modulation of synaptic structures and circuits. Science 298:770–776
Cohen S, Janicki-Deverts D, Miller GE (2007) Psychological stress and disease. JAMA 298:1685–1687
Crum AJ, Salovey P, Achor S (2013) Rethinking stress: the role of mindsets in determining the stress response. J Personal Soc Psychol 104:716–733
Cui Y, Yang Y, Ni Z, Dong Y, Cai G, Foncelle A, Ma S, Sang K, Tang S, Li Y et al (2018) Astroglial Kir4.1 in the lateral habenula drives neuronal bursts in depression. Nature 554:323–327
Demirdas A, Naziroglu M, Ovey IS (2017) Duloxetine reduces oxidative stress, apoptosis, and Ca^{2+} entry through modulation of TRPM2 and TRPV1 channels in the hippocampus and dorsal root ganglion of rats. Mol Neurobiol 54:4683–4695
Denk MC, Rewerts C, Holsboer F, Erhardt-Lehmann A, Turck CW (2011) Monitoring ketamine treatment response in a depressed patient via peripheral Mammalian target of rapamycin activation. Am J Psychiatry 168:751–752
Dillon DG, Pizzagalli DA (2018) Mechanisms of memory disruption in depression. Trends Neurosci 41:137–149
Duan TT, Tan JW, Yuan Q, Cao J, Zhou QX, Xu L (2013) Acute ketamine induces hippocampal synaptic depression and spatial memory impairment through dopamine D1/D5 receptors. Psychopharmacology 228:451–461
Dudai Y (1996) Consolidation: fragility on the road to the engram. Neuron 17:367–370
Duman RS, Aghajanian GK (2012) Synaptic dysfunction in depression: potential therapeutic targets. Science 338:68–72
Duman RS, Li N, Liu R-J, Duric V, Aghajanian G (2012) Signaling pathways underlying the rapid antidepressant actions of ketamine. Neuropharmacology 62:35–41
Duman RS, Aghajanian GK, Sanacora G, Krysta JH (2016) Synaptic plasticity and depression: new insights from stress and rapid-acting antidepressants. Nat Med 22:238–249
Fava M, Freeman MP, Flynn M, Judge H, Hoeppner BB, Cusin C, Ionescu DF, Mathew SJ, Chang LC, Iosifescu DV et al (2018) Double-blind, placebo-controlled, dose-ranging trial of intravenous ketamine as adjunctive therapy in treatment-resistant depression (TRD). Mol Psychiatry
Franceschelli A, Sens J, Herchick S, Thelen C, Pitychoutis PM (2015) Sex differences in the rapid and the sustained antidepressant-like effects of ketamine in stress-naive and "depressed" mice exposed to chronic mild stress. Neuroscience 290:49–60
Freudenberg F, Celikel T, Reif A (2015) The role of alpha-amino-3-hydroxy-5-methyl-4-isoxazolepropionic acid (AMPA) receptors in depression: central mediators of pathophysiology and antidepressant activity? Neurosci Biobehav R 52:193–206
Gideons ES, Kavalali ET, Monteggia LM (2014) Mechanisms underlying differential effectiveness of memantine and ketamine in rapid antidepressant responses. Proc Natl Acad Sci USA 111:8649–8654
Gotlib IH, Joormann J (2010) Cognition and depression: current status and future directions. Annu Rev Clin Psychol 6:285–312
Hammen C (2005). Stress and depression. Annu Rev Clin Psychol 293–319
Han J, Kesner P, Metna-Laurent M, Duan T, Xu L, Georges F, Koehl M, Abrous DN, Mendizabal-Zubiaga J, Grandes P et al (2012) Acute cannabinoids impair working memory through astroglial CB1 receptor modulation of hippocampal LTD. Cell 148:1039–1050
Hashimoto K (2016) R-ketamine: a rapid-onset and sustained antidepressant without risk of brain toxicity. Psychol Med 46:2449–2451
Homayoun H, Moghaddam B (2007) NMDA receptor hypofunction produces opposite effects on prefrontal cortex Interneurons and pyramidal neurons. J Neurosci 27:11496–11500
Jing L, Duan TT, Tian M, Yuan Q, Tan JW, Zhu YY, Ding ZY, Cao J, Yang YX, Zhang X et al (2015) Despair-associated memory requires a slow-onset CA1 long-term potentiation with unique underlying mechanisms. Sci Rep 5

Juruena MF, Baes CV, Menezes IC, Graeff FG (2015) Early life stress in depressive patients: role of glucocorticoid and mineralocorticoid receptors and of hypothalamic-pituitary-adrenal axis activity. Curr Pharm Des 21:1369–1378

Kang HJ, Voleti B, Hajszan T, Rajkowska G, Stockmeier CA, Licznerski P, Lepack A, Majik MS, Jeong LS, Banasr M et al (2012) Decreased expression of synapse-related genes and loss of synapses in major depressive disorder. Nat Med 18:1413–1417

Karasawa J, Shimazaki T, Kawashima N, Chaki S (2005) AMPA receptor stimulation mediates the antidepressant-like effect of a group II metabotropic glutamate receptor antagonist. Brain Res 1042:92–98

Kim JJ, Foy MR, Thompson RF (1996) Behavioral stress modifies hippocampal plasticity through N-methyl-D-aspartate receptor activation. Proc Natl Acad Sci USA 93:4750–4753

Kishimoto T, Chawla JM, Hagi K, Zarate CA, Kane JM, Bauer M, Correll CU (2016) Single-dose infusion ketamine and non-ketamine N-methyl-D-aspartate receptor antagonists for unipolar and bipolar depression: a meta-analysis of efficacy, safety and time trajectories. Psychol Med 46:1459–1472

Koike H, Chaki S (2014) Requirement of AMPA receptor stimulation for the sustained antidepressant activity of ketamine and LY341495 during the forced swim test in rats. Behav Brain Res 271:111–115

Laje G, Lally N, Mathews D, Brutsche N, Chemerinski A, Akula N, Kelmendi B, Simen A, McMahon FJ, Sanacora G et al (2012) Brain-derived neurotrophic factor Val66Met polymorphism and antidepressant efficacy of ketamine in depressed patients. Biol Psychiatry 72:E27–E28

Lepack AE, Fuchikami M, Dwyer JM, Banasr M, Duman RS (2015) BDNF release is required for the behavioral actions of ketamine. Int J Neuropsychoph 18

Levin R, Dor-Abarbanel AE, Edelman S, Durrant AR, Hashimoto K, Javitt DC, Heresco-Levy U (2015) Behavioral and cognitive effects of the N-methyl-D-aspartate receptor co-agonist D-serine in healthy humans: initial findings. J Psychiatr Res 61:188–195

Li NX, Lee B, Liu RJ, Banasr M, Dwyer JM, Iwata M, Li XY, Aghajanian G, Duman RS (2010) mTOR-dependent synapse formation underlies the rapid antidepressant effects of NMDA antagonists. Science 329:959–964

Li B, Piriz J, Mirrione M, Chung CH, Proulx CD, Schulz D, Henn F, Malinow R (2011a). Synaptic potentiation onto habenula neurons in the learned helplessness model of depression. Nature 470:535–539

Li N, Liu R-J, Dwyer JM, Banasr M, Lee B, Son H, Li X-Y, Aghajanian G, Duman RS (2011b) Glutamate N-methyl-D-aspartate receptor antagonists rapidly reverse behavioral and synaptic deficits caused by chronic stress exposure. Biol Psychiatry 69:754–761

Li HB, Mao RR, Zhang JC, Cao YJ, Xu L (2008) Antistress effect of TRPV1 channel on synaptic plasticity and spatial memory. Biol Psychiatry 64:286–292

Liu SY, Zeng YC, Lei XL (2016) Effects of electron-impurity scattering on density of states in silicene: impurity bands and band-gap narrowing. Phys Rev B 94

Liu RJ, Lee FS, Li XY, Bambico F, Duman RS, Aghajanian GK (2012) Brain-derived neurotrophic factor Val66Met allele impairs basal and ketamine-stimulated synaptogenesis in prefrontal cortex. Biol Psychiatry 71:996–1005

Liu W, Ge TT, Leng YS, Pan ZX, Fan JE, Yang W, Cui RJ (2017) The role of neural plasticity in depression: from hippocampus to prefrontal cortex. Neural Plast 11

Liu LD, Wong TP, Pozza MF, Lingenhoehl K, Wang YS, Sheng M, Auberson YP, Wang YT (2004) Role of NMDA receptor subtypes in governing the direction of hippocampal synaptic plasticity. Science 304:1021–1024

MacQueen GM, Yucel K, Taylor VH, Macdonald K, Joffe R (2008) Posterior hippocampal volumes are associated with remission rates in patients with major depressive disorder. Biol Psychiatry 64:880–883

Madasu MK, Roche M, Finn DP (2015) Supraspinal transient receptor potential subfamily V member 1 (TRPV1) in pain and psychiatric disorders. Mod Trends Pharmacopsychiatry 30:80–93

Maeng S, Zarate CA Jr, Du J, Schloesser RJ, McCammon J, Chen G, Manji HK (2008) Cellular mechanisms underlying the antidepressant effects of ketamine: role of alpha-amino-3-hydroxy-5-methylisoxazole-4-propionic acid receptors. Biol Psychiatry 63:349–352

Manji HK, Drevets WC, Charney DS (2001) The cellular neurobiology of depression. Nat Med 7:541–547

Manna SSS, Umathe SN (2012) A possible participation of transient receptor potential vanilloid type 1 channels in the antidepressant effect of fluoxetine. Eur J Pharmacol 685:81–90

Miller OH, Yang L, Wang CC, Hargroder EA, Zhang Y, Delpire E, Hall BJ (2014) GluN2B-containing NMDA receptors regulate depression-like behavior and are critical for the rapid antidepressant actions of ketamine. eLife 3:e03581

Moghaddam B, Adams B, Verma A, Daly D (1997) Activation of glutamatergic neurotransmission by ketamine: a novel step in the pathway from NMDA receptor blockade to dopaminergic and cognitive disruptions associated with the prefrontal cortex. J Neurosci 17:2921–2927

Murrough JW, Abdallah CG, Mathew SJ (2017) Targeting glutamate signalling in depression: progress and prospects. Nat Rev Drug Discovery 16:472–486

Murrough JW, Iosifescu DV, Chang LC, Al Jurdi RK, Green CE, Perez AM, Iqbal S, Pillemer S, Foulkes A, Shah A et al (2013) Antidepressant efficacy of ketamine in treatment-resistant major depression: a two-site randomized controlled trial. Am J Psychiatry 170:1134–1142

Nabavi S, Fox R, Proulx CD, Lin JY, Tsien RY, Malinow R (2014) Engineering a memory with LTD and LTP. Nature 511:348-+

Nations KR, Dogterom P, Bursi R, Schipper J, Greenwald S, Zraket D, Gertsik L, Johnstone J, Lee A, Pande Y et al (2012) Examination of Org 26576, an AMPA receptor positive allosteric modulator, in patients diagnosed with major depressive disorder: an exploratory, randomized, double-blind, placebo-controlled trial. J Psychopharmacol 26:1525–1539

Newport DJ, Carpenter LL, McDonald WM, Potash JB, Tohen M, Nemeroff CB, Novel APACRTF (2015) Ketamine and other NMDA antagonists: early clinical trials and possible mechanisms in depression. Am J Psychiatry 172:950–966

Nosyreva E, Szabla K, Autry AE, Ryazanov AG, Monteggia LM, Kavalali ET (2013) Acute suppression of spontaneous neurotransmission drives synaptic potentiation. J Neurosci 33:6990–7002

Paul RK, Singh NS, Khadeer M, Moaddel R, Sanghvi M, Green CE, O'Loughlin K, Torjman MC, Bernier M, Wainer IW (2014) (R, S)-Ketamine metabolites (R, S)-norketamine and (2S, 6S)-hydroxynorketamine increase the mammalian target of rapamycin function. Anesthesiology 121:149–159

Qiao H, Li MX, Xu C, Chen HB, An SC, Ma XM (2016) Dendritic spines in depression: what we learned from animal models. Neural Plast 26

Radley JJ, Kabbaj M, Jacobson L, Heydendael W, Yehuda R, Herman JP (2011) Stress risk factors and stress-related pathology: neuroplasticity, epigenetics and endophenotypes. Stress 14:481–497

Reyes-Mendez ME, Castro-Sanchez LA, Dagnino-Acosta A, Aguilar-Martinez I, Perez-Burgos A, Vazquez-Jimenez C, Moreno-Galindo EG, Alvarez-Cervera FJ, Gongora-Alfaro JL, Navarro-Polanco RA et al (2018) Capsaicin produces antidepressant-like effects in the forced swimming test and enhances the response of a sub-effective dose of amitriptyline in rats. Physiol Behav 195:158–166

Richter-Levin G, Xu L (2018) How could stress lead to major depressive disorder? IBRO Rep 4:38–43

Rowan MJ, Anwyl R, Xu L (1998) Stress and long-term synaptic depression. Mol Psychiatry 3:472–474

Silva AJ (2017) Memory's intricate web. Sci Am 317:30–37

Stone JM, Dietrich C, Edden R, Mehta MA, De Simoni S, Reed LJ, Krystal JH, Nutt D, Barker GJ (2012) Ketamine effects on brain GABA and glutamate levels with 1H-MRS: relationship to ketamine-induced psychopathology. Mol Psychiatry 17:664–665

Trullas R, Skolnick P (1990) Functional antagonists at the NMDA receptor complex exhibit antidepressant actions. Eur J Pharmacol 185:1–10

Wang J, Jing L, Toledo-Salas JC, Xu L (2015) Rapid-onset antidepressant efficacy of glutamatergic system modulators: the neural plasticity hypothesis of depression. Neurosci Bull 31:75–86

Wang MN, Yang Y, Dong ZF, Cao J, Xu L (2006) NR2B-containing N-methyl-D-aspartate subtype glutamate receptors regulate the acute stress effect on hippocampal long-term potentiation/long-term depression in vivo. Neuro Rep 17:1343–1346

Wang QH, Liu LD, Pei L, Ju W, Ahmadian G, Lu J, Wang YS, Liu F, Wang YT (2003) Control of synaptic strength, a novel function of Akt. Neuron 38:915–928

Wieber J, Gugler R, Hengstmann JH, Dengler HJ (1975) Pharmacokinetics of ketamine in man. Anaesthesist 24:260–263

Xiong WY, Yang YX, Cao J, Wei HM, Liang CL, Yang SC, Xu L (2003) The stress experience dependent long-term depression disassociated with stress effect on spatial memory task. Neurosci Res 46:415–421

Xiong WY, Wei HM, Xiang XJ, Cao J, Dong ZF, Wang YF, Xu TL, Xu L (2004) The effect of acute stress on LTP and LTD induction in the hippocampal CA1 region of anesthetized rats at three different ages. Brain Res 1005:187–192

Xu L, Anwyl R, Rowan MJ (1997) Behavioural stress facilitates the induction of long-term depression in the hippocampus. Nature 387:497–500

Xu L, Holscher C, Anwyl R, Rowan MJ (1998) Glucocorticoid receptor and protein/RNA synthesis-dependent mechanisms underlie the control of synaptic plasticity by stress. Proc Natl Acad Sci USA 95:3204–3208

Yang CH, Huang CC, Hsu KS (2004a) Behavioral stress modifies hippocampal synaptic plasticity through corticosterone-induced sustained extracellular signal-regulated kinase/mitogen-activated protein kinase activation. J Neurosci 24:11029–11034

Yang Y, Zheng XG, Wang YF, Cao J, Dong ZF, Cai JX, Sui N, Xu L (2004b) Stress enables synaptic depression in CA1 synapses by acute and chronic morphine: possible mechanisms for corticosterone on opiate addiction. J Neurosci 24:2412–2420

Yang C, Zhou ZQ, Gao ZQ, Shi JY, Yang JJ (2013) Acute increases in plasma Mammalian target of rapamycin, glycogen synthase kinase-3 beta, and eukaryotic elongation factor 2 phosphorylation after ketamine treatment in three depressed patients. Biol Psychiatry 73:E35–E36

Yang JL, Hou CL, Ma N, Liu J, Zhang Y, Zhou JS, Xu L, Li LJ (2007) Enriched environment treatment restores impaired hippocampal synaptic plasticity and cognitive deficits induced by prenatal chronic stress. Neurobiol Learn Mem 87:257–263

Yang Y, Cui YH, Sang KN, Dong YY, Ni ZY, Ma SS, Hu HL (2018) Ketamine blocks bursting in the lateral habenula to rapidly relieve depression. Nature 554:317–322

Yao N, Skiteva O, Zhang X, Svenningsson P, Chergui K (2017) Ketamine and its metabolite (2R, 6R)-hydroxynorketamine induce lasting alterations in glutamatergic synaptic plasticity in the mesolimbic circuit. Mol Psychiatry

Zanos P (2017) NMDAR inhibition-independent antidepressant actions of ketamine metabolites. Eur Neuropsychopharm 27:S528–S528

Zanos P, Gould TD (2018) Mechanisms of ketamine action as an antidepressant. Mol Psychiatry 23:801–811

Zanos P, Moaddel R, Morris PJ, Georgiou P, Fischell J, Elmer GI, Alkondon M, Yuan PX, Pribut HJ, Singh NS et al (2016) NMDAR inhibition-independent antidepressant actions of ketamine metabolites. Nature 533:481-+

Zarate CA, Brutsche N, Laje G, Luckenbaugh DA, Venkata SLV, Ramamoorthy A, Moaddel R, Wainer IW (2012a) Relationship of ketamine's plasma metabolites with response, diagnosis, and side effects in major depression. Biol Psychiatry 72:331–338

Zarate CA, Brutsche NE, Ibrahim L, Franco-Chaves J, Diazgranados N, Cravchik A, Selter J, Marquardt CA, Liberty V, Luckenbaugh DA (2012b) Replication of ketamine's antidepressant efficacy in bipolar depression: a randomized controlled add-on trial. Biol Psychiatry 71:939–946

Zarate CA, Singh JB, Carlson PJ, Brutsche NE, Ameli R, Luckenbaugh DA, Charney DS, Manji HK (2006) A randomized trial of an N-methyl-D-aspartate antagonist in treatment-resistant major depression. Arch Gen Psychiat 63:856–864

Zhang Y, Mao RR, Chen ZF, Tian M, Tong DL, Gao ZR, Huang M, Li X, Xu X, Zhou WH et al (2014). Deep-brain magnetic stimulation promotes adult hippocampal neurogenesis and alleviates stress-related behaviors in mouse models for neuropsychiatric disorders. Mol Brain 7

Zhou W, Wang N, Yang C, Li XM, Zhou ZQ, Yang JJ (2014) Ketamine-induced antidepressant effects are associated with AMPA receptors-mediated upregulation of mTOR and BDNF in rat hippocampus and prefrontal cortex. Eur Psychiatry 29:419–423

Zhou H, Xiong GJ, Jing L, Song NN, Pu DL, Tang X, He XB, Xu FQ, Huang JF, Li LJ et al (2017) The interhemispheric CA1 circuit governs rapid generalisation but not fear memory. Nat Commun 8:2190

Chapter 6
Biological Rhythms Advance in Depressive Disorder

Wu Hong and Qinting Zhang

Abstract Most processes of human body, such as brain function, are regulated by biological rhythms. Disturbance of biological rhythms impairs mood, behavior, cognition, sleep, and social activity and may lead to mental disorders. Disturbed rhythms are widely observable in patients with major depressive disorders (MDD) and make risk of onset, comorbidity, response of antidepressants, recurrence, cognition, social function, and complications of physical health. Therefore, it is crucial to assess and manage focus on biological rhythms for patients with MDD. There are several validated ways of assessing the biological rhythms, including 24 h fluctuations in cortisol or melatonin, sleep monitoring, actigraphy, and self-report scales. Chronotherapy, such as cognitive-behavioral therapy, interpersonal and social rhythm therapy, sleep deprivation, and bright light therapy was widely reported for treatment in patients with MDD. Monoamine antidepressants and lithium are attributed to regulation of biological rhythm. And some rhythm-regulated agents have been shown efficacy of antidepressant. Considering the crucial clinical significance of disturbed biological rhythms in MDD, we describe the mechanisms, clinical features, measurements, and treatments of the biological rhythms in patients with MDD.

Keywords Depressive disorder · Biological rhythms · Circadian · Chronotherapy

---

W. Hong
Division of Mood Disorders, Shanghai Mental Health Center, Shanghai Jiao Tong University School of Medicine, Shanghai, China
e-mail: drhongwu@126.com

Q. Zhang (✉)
Shanghai Key Laboratory of Forensic Medicine, Shanghai Forensic Service Platform, Academy of Forensic Science, Shanghai, China
e-mail: zhangqinting@126.com

© Springer Nature Singapore Pte Ltd. 2019
Y. Fang (ed.), *Depressive Disorders: Mechanisms, Measurement and Management*, Advances in Experimental Medicine and Biology 1180,
https://doi.org/10.1007/978-981-32-9271-0_6

6.1 Introduction

It is well known that biological rhythms influence various aspects in human's biochemical and physiological behaviors, which critically influence human's well-being and health. Endogenous biological rhythms are mainly regulated by suprachiasmatic nuclei and pineal gland (Takahashi et al. 2008). Accumulative clinical and biological studies show that disturbance of biological rhythms impairs mood, behavior, cognition, sleep, and social activity and may lead to mental disorders (Hickie and Rogers 2011; Jones and Benca 2015).

Disruption of biological rhythms is very common in mood disorders (Nechita et al. 2015; Lunsford-Avery 2017), including child and adolescent, women during pregnant and old patients. Disrupted rhythms in mood disorders have been widely reported (Bellivier et al. 2015; Dallaspezia and Benedetti 2015). Patients with major depressive disorders (MDD) manifest various biological functions such as hormone secretion (including melatonin and cortisol), blood pressure, and body temperature (Jones and Benca 2015; Schnell et al. 2014). It has been even suggested that the disruption of biological rhythms is the core characteristics of MDD.

The role of disturbance of biological rhythms plays in the risk of onset, early course, comorbidity, response of antidepressants, recurrence, cognition, social function, and physical health complications of patients with MDD (Hickie and Rogers 2011; Jones and Benca 2015; Albrecht 2013). Recent researches (Morris et al. 2007; Courtet and Olié 2012; Bahk et al. 2014; Berdynaj et al. 2016), including the famous sequenced treatment alternatives to relieve depression(STAR-D)discovered that dysfunction of rhythms, such as disruption of sleep-cycle, evening-type rhythm, daytime oversleep, daily and seasonal mood change, predict severe symptoms, less response to antidepressants, more residual symptoms, lower remission, high risk of suicide, more likely pay attention to negative life events, less concerning to positive event, lower cognition, and lower social function (Robillard et al. 2016; Rastad et al. 2017; Soterio-Pires et al. 2017). In addition, disturbed rhythms can indicate mood and psychotic episodes in patients who already have psychiatric disorders (McA et al. 2017).

Therefore, it is crucial to pay attention to assessment and management focus on biological rhythms for patients with MDD. However, the attention about biological rhythms of MDD is not sufficient. There are several validated ways of assessing the biological rhythms, including 24 h fluctuations in cortisol or melatonin, sleep monitoring, actigraphy, and self-report scales. Chronotherapy, such as cognitive-behavioral therapy, interpersonal and social rhythm therapy, sleep deprivation, and bright light therapy, was widely discovered having efficacy of antidepressant. For example, sleep deprivation, which keeps patients awake all or part of the night, has been reported having rapid antidepressant action in half of patients with MDD by numerous studies (Bunney and Bunney 2013). However, when supplemented with both medication and sleep phase advance by bright light therapy (SPA), the effect may be continued for up to several months. Monoamine antidepressants and lithium are attributed to the regulation of biological rhythm. And some rhythm-regulated agents have been shown efficacy of antidepressant.

Considering the crucial clinical significance of disturbed biological rhythms in MDD, we describe the mechanisms, clinical features, measurements, and treatments of the biological rhythms in patients with MDD.

6.2 Potential Mechanisms of the Relation Between Biological Rhythms and MDD

The mechanisms of the relation between the disrupted biological rhythms and MDD remain undefined (Brown et al. 2018). A great deal of hypotheses have been suggested to explain the potential mechanisms of the relation between circadian rhythms and MDD (McClung 2013). Sleep is vitally important for proper maturation of circuit and long-term memory formation in animals and humans (Kragh et al. 2018; Marshall et al. 2006; Havekes et al. 2015), and many synaptic pruning and refinement processes occur during sleep. Therefore, disrupted circadian rhythms affect the synaptic pruning and neural circuits' maturation (Goldstone et al. 2018), which may be linked to the development of psychiatric disorders (Selemon and Zecevic 2015). Constant darkness can induce a depressive phenotype and apoptosis of monoaminergic neurons, particularly noradrenergic neurons in the locus coeruleus, and these effects are reversed by the norepinephrine reuptake inhibitor, desipramine. In addition, short light-dark cycle (22 h) also can induce a depressive state and showed desynchronized oscillations in the SCN, increased norepinephrine and dopamine levels, and serotonin turnover in the prefrontal cortex (Ben-Hamo et al. 2016).

Behavioral variations in circadian rhythms have been conceptualized in the form of chronotypes, which is determined by genetic variations in clock genes and environmental factors. Recently, several clock genes including Bmall, Clock, Perl, Per2, Per 3, Cryl, Cry 2, and Timeless, and their protein products express feedback loops to regulate the circadian rhythms. Genetic studies have shown that genes controlling circadian rhythms are involved in susceptibility to MDD. The circadian genes interact with various systems and processes, including monoaminergic and glutamatergic transmission, HPA axis function, metabolism, and immune function (Ketchesin et al. 2018). Increased risk of depression and significant changes in dopamine-related gene expression in the ventral tegmental area were found in CLOCK knockout mice (Mukherjee et al. 2010). Depressive-like behavior and a reduction in cortical norepinephrine level were found in Bmal1 knockout rats (Bering et al. 2018). Polymorphisms of *Per3*, *Cry1*, *Clock, and Npas2* were demonstrated to be associated with MDD (Shi et al. 2016; Hua et al. 2014). Expression levels of the *Bmal1*, *NR1D1*, *Per1*, *Per2*, *Per3*, *Dbp*, *sirtuin 1* (*Sirt1*), MDD *Dec1,* and *Dec2* were reported to be changed in the brains of MDD patients (Li et al. 2013). The epigenetic effects of circadian system were also found in patients with MDD. For instance, abnormal processing of pre-miR-182 which controls circadian system was described in MDD (Saus et al. 2010).

Disruption of circadian genes may be implicated in the clinical and course variables in patients with depression (Soria et al. 2010). For instance, the disruption of CLOCK, *PER3, CRY1*, and *NR1D1* have been associated with age at onset, sleep problems, diurnal activity patterns, information processing, illness recurrence, response to antidepressants treatment (Ma et al. 2018), and response to specific circadian-based therapies, such as light therapy or sleep deprivation (Benedetti et al. 2010).

Clock genes are also widely reported involving in the regulation of the antidepressants, which are well-known implication in the pathophysiology and treatment of MDD (Soria et al. 2010). Furthermore, the expression of clock genes in hippocampus and striatum of mouse brain regions which have been implicated in mood regulation are altered after being prescribed fluoxetine. Moreover, lithium, which is also an augmentation agent of antidepressant treatment in MDD, is known to regulate circadian rhythms. What is more, ketamine, being famous because of its rapid antidepressants effect, was found regulating the clock genes (Orozco-Solis et al. 2017). And non-pharmacological therapy based on the modulating the circadian rhythms, such as light therapy, sleep deprivation, and interpersonal and social rhythm therapy, has been demonstrated the effect of antidepressant.

Melatonin, which is secreted from pineal gland controlled by the SCN, is regulated by light, and is used as an index of the SCN rhythm rising in the evening under dim light, and then falling to virtually undetectable levels in the morning. Disturbances in the amplitude and rhythm of melatonin secretion have been described in patients with MDD. Decreased levels of nocturnal melatonin were widely reported in MDD (Brown et al. 2018) although there are also contrary reports and a report of increased serum melatonin (Bumb et al. 2016). Heterogeneity of MDD subforms may be related to the differences.

Overactivation hypothalamic pituitary adrenal (HPA) axis, which was regulated by SCN, of patients with MDD has long been known. As early as 1973, disrupted 24 h rhythms of cortisol were reported in patients with MDD (Sachar et al. 1973). Rhythms serum cortisol levels have been proposed as a diagnostic procedure for MDD (Bilello et al. 2015). Recently, cortisol awakening response, a rise in cortisol that occurs 30 min after waking, has been raised concern (Elder et al. 2014). Exaggerated response was reported in patients with MDD and other various conditions, which is interpreted as a response to stress and leads to alterations of immune system. High level of cortisol may be sustained throughout the day in depressed patients with psychotic and suicidal symptoms. Peak level of cortisol may be phase-advanced by 2–3 h in the morning. Selective serotonin reuptake inhibitors (SSRIs), such as escitalopram, might exert its antidepressants therapeutic effect in women in part through regulation of diurnal cortisol rhythm (Ronaldson et al. 2018). In remission, cortisol levels in evening may decrease to normal levels although not consistently.

6.3 Clinical Features of Biological Rhythms' Disturbances in MDD

Disturbed of biological rhythms is very common in MDD, and indicators of disturbed rhythm are observable in patients with major depression including disturbances of sleep pattern, changes of diurnal/seasonal mood variation, low energy, or loss of normal daily variations.

6.3.1 Disturbances of Sleep Pattern

Disturbances of sleep pattern represent obvious feature of depression (Fava 2004). A substantial body of research demonstrates up to 50–90% of patients with MDD complain impairment of their sleep disruption (Levitan 2007; Hickie et al. 2013) including difficulty in falling asleep, frequent nocturnal awakening, early morning awakening, staying asleep, disturbing dreams, and prolong sleep. The complaints of sleep disturbance are also reflected objectively in altered sleep architecture, including short sleep duration, prolonged sleep latency, sleep shallowness, shortened REM latency, increased rapid eye movements (especially in the first sleep cycle), delay of REM sleep (Benca et al. 1992; Buysse et al. 1997; Lam 2006), increased sleep fragmentation, reduced slow-wave activity and number of slow waves, and high comorbidity with sleep-related breathing disorders.

6.3.2 Diurnal/Seasonal Mood Variation

The main character of diurnal mood variation is "heavily morning lightly night." On the contrary, there are also some patients who present reversed diurnal variation with heavy mood at night (Joyce et al. 2005). Data of STAR-D indicates that there are about 21.6% of patients of MDD present diurnal mood variation (Morris et al. 2007). The diurnal mood variation may be related to changes in biological processes and metabolisms in the brain. Activation of back neural network which is related to mood control was indicated to be related to improvement of mood in the evening (Germain et al. 2007).

The main character seasonal mood variation is the correlation between depression and one season, especially in winter. The criteria of seasonal affective disorder (SAD) have been clear in diagnostic and statistical manual of mental disorders (DSM-5). For example, depressive episodes occur regularly during winter and remit during the subsequent spring and summer period. SAD is a disorder with apparent chronobiological abnormalities and seasonal disturbance of biological rhythms (Levitan 2007). Besides the seasonal mood variation, the abnormalities of biological rhythms in SAD patients also include quantitative changes, sleep disturbances, and phase delays in melatonin/cortisol secretion patterns, and increase in the minima of the nocturnal body temperature, as well as a phase delay of its 24 h rhythm.

6.3.3 Variations of Daily Activity

The variations of daily activity of patients with MDD include variation of appetite, decreasing activity, low energy, decreasing interest, and lack of motivation and fatigue. Patients with MDD present variation of appetite, mainly manifested decreased appetite and minority shows increased appetite.

The pattern of daily activity can be divided into morning type, evening type, and intermediate type. Population with morning type like early wake-up, more active in the morning, and falling sleep early. On the contrary, population with evening type like late wake-up, more activity in the afternoon and evening. The majority of the population is intermediate type. Circumstantial evidence in the last decade indicated evening type is a potential factor associated with MDD. Patients with evening-type preference correlate with more severe depressive symptoms and are associated with increased odds of a mood disorder.

6.4 Measurement of Biological Rhythms in MDD

Considering the crucial clinical significance of disrupted biological rhythms in MDD, it is important for measuring and accurately detecting for potential disrupt biological rhythms.

There are several validated ways of assessing the biological rhythms, including 24 h fluctuations in cortisol or melatonin, core body temperature monitoring, sleep monitoring, actigraphy (for example, typically at least two weeks), and self-report scales.

6.4.1 Physiological Markers of Biological Rhythm

There are various endogenous physiological markers of biological rhythm that can be used for patients with MDD. The most widely used method of measuring markers of circadian rhythmicity involves profiling the daily fluctuations in endogenous melatonin and cortisol levels.

Melatonin, which is secreted from pineal gland, is proposed as a diagnostic marker for biological rhythm. Levels of melatonin from blood, saliva, and urine, have been used as an index of the SCN rhythm, which typically rises prior to habitual bedtime, reach the peak in early morning hours roughly around 3:00 AM, and gradually decrease throughout the day during wakefulness to nearly undetectable levels. Immense body of evidence has demonstrated disrupted nocturnal melatonin including phase advance and phase delay and decreased levels in patients with MDD (Brown et al. 2018; Bumb et al. 2016; Hickie et al. 2013; Khaleghipour et al. 2012; Naismith et al. 2012; Bouwmans et al. 2015), although there are also contrary findings

(Bouwmans et al. 2015) of increasing levels of melatonin in MDD (Bumb et al. 2016). Heterogeneity of MDD may lead to different findings. Patients who spend less time in darkness because of sleep difficulty may present decreased levels of melatonin.

Cortisol, the fluctuate of which in saliva, urine, and blood, has been used as a marker of biological rhythm (Bilello et al. 2015). Cortisol levels typically reach to peak in the morning roughly around 6:00 AM and decrease throughout the day. The peak of cortisol release in morning is just well matched to the feelings of energy and capacity to cope with physical stressors. Altered patterns of cortisol release are in line with potential phase advance or loss of synchrony with another circadian rhythm. Therefore, the disrupted patterns of cortisol release appear to correlate with feelings of lack of energy and fatigue. Disrupted 24 h pattern of cortisol was first reported in patients with MDD in 1973. Recently, cortisol awakening response was attended (Elder et al. 2014), which could represent an inherent change in HPA regulation as the 24 h rhythm of cortisol (Nicolaides et al. 2014). Exaggerated responses were found in those with MDD and those at risk although exaggerated responses are also seen in some other conditions.

6.4.2 Electrophysiologic Index

With the development of electronic technique, more and more electric devices were used to monitor the sleep quality and daily activity patterns, including polysomnography (PSG), actigraphy, and smartphone. Importantly, such measures are only indirect indicators of biological rhythm, which cannot be used to confirm the presence of internal desynchrony.

Polysomnography (PSG) is the most classical and gold standard electric measures, but it is costly and time-consuming, often requires an adaptation night and time-intensive scoring by sleep technicians. Therefore, PSG is not suitable for daily monitoring. Actigraphy measures are somewhat well-validated, noninvasive and more convenient (at least in the short term) method of measuring multiday ambulatory activity rhythms assessment, which is a device like a watch, on the whist. Actigraphy has been widely used in studies for monitoring sleep and circadian, and shown high levels of agreement with PSG. Other new technologies, such as smartphone applications, permit home monitoring of extended periods of sleep–wake cycle behavior and patterns of physical activity.

6.4.3 Scales and Questionnaires

Various scales and questionnaires were developed to identify biological rhythm and individual differences in circadian preferences, which is suitable for prolonged reporting of sleep timing, latency, quality and duration, daily mood, and daytime physical activity. Compared to biological assessments, scales and questionnaires are useful and accessible as an indirect measure of intrinsic biological function rhythms. Most of the scales and questionnaires showed excellent internal consistency, test–retest reliability, and convergent validity, with adequate levels of construct C reported.

The Pittsburgh sleep quality index (PSQI), which is a self-reported scale, has been widely reported closely correlated with the latency and duration of sleep and circadian variables (phase markers), and weak correlations with objective PSG and actigraphy. The morningness–eveningness questionnaire (MEQ) and the composite scale of morningness (CSM) are the most frequently used self-report questionnaires for assessing individual preference rhythm typologies. Individuals are categorized into specific circadian typologies according to a composite score, with lower scores favoring an eveningness orientation and higher scores indicative of a morningness orientation.

Chronotype questionnaires, such as MEQ and CSM, are useful and feasible assessment tools for indirect measuring the intrinsic circadian function rhythms (Rastad et al. 2017). The MEQ is the most utilized and more frequently used measurement for evaluating rhythm type in patients with MDD. The MEQ is also positively correlated with circadian markers such as melatonin and core body temperature.

6.4.4 Physiological Index

Core body temperature monitoring is used for assessing biological rhythm. Core body temperature is 24 h measured from ingested wireless probe or rectal probe in controlled laboratory setting. Thermoregulation has a critical role in biological function, as little as 1 °C change can significantly change circadian gene expression (Saini et al. 2012). In healthy controls, core body temperature rises in early morning hours and decreases at night. But the diurnal rhythm was reported flattened in depressed patients, as nocturnal temperatures remain high (Avery et al. 1999). Decrease in nocturnal temperature and restoration of normal rhythms is usually associated with remission.

Cardiovascular monitoring (including heart rate and blood pressure) is also used for assessing biological rhythm. Reduced heart rate circadian amplitude (Kario et al. 2001) and increased nocturnal blood pressure are associated with depressive symptoms. Cardiovascular monitoring is continuous or repeated measure of blood pressure and heart rate parameters across 24 h in controlled laboratory settings.

6.5 Chronotherapy for MDD

Chronotherapy was widely reported for treatment in patients with MDD. In comparison with large number of pharmaceutical trials focusing on regulation of monoamine systems, chronotherapy does not figure prominently in major therapeutic guidelines for MDD (except for agomelatine). Evidence base for nonpharmacologic circadian interventions focusing on cognitive, interpersonal, and related behavioral factors is only modest. The effects of chronotherapies may be personalized individual. For example, MDD patients with more disturbed biological rhythms are more responsive to chronotherapy (Kragh et al. 2018).

6.5.1 Nonpharmacologic Treatment

6.5.1.1 Cognitive-Behavioral Therapy (CBT)

The antidepressant effects of CBT have been widely reported by substantial clinical trials and have been recommended to be used for patients with MDD by a lot of guidelines (Robillard et al. 2016; Bumb et al. 2016; Elder et al. 2014). In addition, circumstantial evidence demonstrated effectiveness of cognitive-behavioral therapy (CBT) on insomnia, and the effectiveness of CBT is more durable than that of pharmacotherapy. CBT specified for modifying the disrupted biological rhythms of MDD patients involves psychoeducation with circadian systems, cognitive restructuring (focusing on negative automatic thoughts about sleeping), and behavioral techniques (Hickie et al. 2013; Boland et al. 2019) (Table 6.1).

6.5.1.2 Light Treatment in the Morning

Light treatment is mostly used as a method of chronotherapy, to readjust the biological rhythm system. Light treatment is the most common and convenient therapy in this category. There are a lot of forms of light treatment; natural and artificial light are both used. Artificial light is the most common (Nussbaumer et al. 2015). The mechanism of light treatment is maybe related to adjust the circadian rhythm. For example, circadian rhythms will be advanced after bright light therapy in morning, and will be delayed after bright light therapy in the evening. Accumulative studies (Hamers et al. 2017) have demonstrated that bright light therapy can be combined used with antidepressant for patients with MDD.

6.5.1.3 Interpersonal and Social Rhythm Therapy

Interpersonal and social rhythm therapy (IPSRT) was first proposed by Ellen Frank in 2000 (Frank et al. 2000). The aims of IPSRT are regularizing daily activity rhythms (having regular bed, wake, and mealtimes, switching to a more regular work schedule, and incorporating a regular daily exercise session) and managing biological or psychosocial factors susceptible of dysregulating biological rhythms. Circumstantial evidence (Haynes et al. 2005, 2016a, b; Corruble et al. 2016) have shown that IPSRT was helpful for reducing the severity of depression, suicidal ideation, and hopelessness, accelerating remission, reducing recurrence, regularizing social rhythms, and improving social function in patients with MDD.

Table 6.1 CBT specified for modifying the disrupted biological rhythms of MDD patients

1. Psychoeducation with circadian systems	Explanation of circadian systems: Understanding the mechanism of circadian regulation and the processes through which circadian disturbances can be initiated and maintained; Understanding of the normal circadian and synchronization with hormonal, immune, body temperature, and other key physiological elements; Linking circadian disturbances to onset and relapse of mood disorders; Promoting awareness of how daily behaviors and environmental factors influence circadian rhythms (that is, sleep hygiene); Emphasis on setting the rhythm through morning rising, appropriately timed light exposure, regularity of activity cycles, daytime physical activity, bedtime schedules, and nighttime practices
2. Cognitive restructuring	Identifying and adjusting negative beliefs contributing to sleep; Understanding the influence of sleep and daytime functioning; Establishing realistic expectations about sleep
3. Set specific behavioral elements, Structured behavioral changes	Techniques to prevent evening/nighttime ruminations; Set daily activity schedules; Emphasize light exposure in morning; Stimulus control; reestablish positive associations between the bedroom and sleeping; Bed restriction therapy; Techniques used for insomnia include progressive muscle relaxation, diaphragmatic breathing, autogenic training, and imagery training

6.5.1.4 Sleep Deprivation

Sleep deprivation (SD), which keeps patients awake part or all of the night, was first being described in the 1970s (Yamaguchi et al. 1978). Numerous studies have reported the acute antidepressant effect of SD (Bunney and Bunney 2013; Dallaspezia and Benedetti 2015). But the clinical usefulness of SD has been questioned because of the shortage of short duration of its antidepressant effects. Indeed, the therapeutic response can vanish as fast as it develops; up to 80% of SD responders relapse (Giedke and Schwärzler 2002). New clinical strategies have been developed to prolong the effects of SD. For example, combining different chronobiological therapies (e.g., SD plus light therapy) or combining antidepressants and mood stabilizers was recommended.

In addition, it is important to note that sleep loss can trigger mania/hypomania episodes, epilepsy, and unsuspected medical conditions, e.g., undetected severe cardiovascular diseases.

Therefore, SD is not the first-line treatment, but SD has been widely used because of its rapid antidepressant efficacy (Bunney and Bunney 2013).

6.5.1.5 Blue Blocking (BB) Glasses in the Evening

The circadian timing system is regulated by light exposed through the mechanism of suppression of melatonin. Retinal ganglion cells containing the photopigment melanopsin are the most important for regulating the timing system, which is most sensitive to blue wavelengths (around 480 nm). Consequently, wearing blue-blocking glasses can preserve normal nighttime melatonin production in subjects exposed to light and improve sleep (Brown et al. 2018).

Wearing blue-blocking (BB) glasses in the evening has been found of improving sleep quality and depressive symptoms, although there are also some negative results (Esaki et al. 2017).

Therefore, BB glasses may be useful for patients with MDD who want to reduce hypnotic drugs or who are resistant to drug treatment.

6.5.2 Pharmacologic Treatment

Regulation of biological rhythm is one of the mechanisms of antidepressant therapy. Studies have shown that the antidepressant effects of monoamine antidepressants and lithium are attributed to regulation of biological rhythm (Abe et al. 2000; Uz et al. 2005; Sprouse et al. 2006). Some medications of rhythm have been shown having the effect of antidepressant.

6.5.2.1 Antidepressants

The effect in the management of circadian of monoamine-based antidepressant remains controversial. The tricyclic agents (TCAs) are still widely prescribed for their sedative properties, while SSRI and selective norepinephrine reuptake inhibitor (SNRI) agents disturb sleep during the early phases of therapy. Antidepressants, those with classical serotoninergic properties, are expected to increase wakefulness. Especially, those with more obvious noradrenergic properties have been used (with daytime or morning administration) to promote daytime activity and arousal and reduce fatigue.

Compared to monoamine-based antidepressants, the melatonin-based agents offer any specific benefits in sleep architecture, sleep–wake cycle, circadian, or mood disorders' perspective. Agomelatine, a melatonergic medication, is one of the new categories of antidepressants. Agomelatine's antidepressant actions are attributed to its chronobiotic actions mediated by activating MT1 and MT2 receptors present in the suprachiasmatic nucleus, as well as to its effects on the blockade of 5-HT2c receptors. Agomelatine's efficacy of antidepressant treatment has been established in many studies and meta-analyses (Stahl et al. 2010; Zajecka et al. 2010; Martinotti et al. 2012; Huang et al. 2014; Kennedy et al. 2014; Laux et al. 2017). Importantly, compared with SSRIs, agomelatine has the advantage of sleep promotion and improvement of sleep architecture (Quera-Salva et al. 2010).

6.5.2.2 Melatonin

Melatonin has been widely studied in the treatment of sleep disorders, and accumulated evidence has demonstrated the possible role for melatonin influencing mood (Fava et al. 2012). Although some studies were positive, meta-analyses did not show significant effect of melatonin on effects of antidepressant (De Crescenzo et al. 2017). Therefore, melatonin is only recommended for combined strategy.

6.5.2.3 Lithium

Lithium is the inhibitor of glycogen synthase kinase 3β(GSK-3β), which involved in the circadian regulation of the SCN. Therefore, lithium modulates biological rhythms (possibly by lengthening the circadian period or delaying endogenous rhythms) (Abe et al. 2000; Campbell et al. 1989; Kaladchibachi et al. 2007). For example, lithium can enhance the therapeutic effects of combined sleep deprivation and phase advance in people with bipolar disorders (Benedetti et al. 2001), decrease retinal sensitivity to light, and could possibly influence melatonin's sensitivity to light.

6.5.2.4 Other Medicine

Hypnotics/sedatives, those with delayed circadian rhythms, can be used in the short term to help realign the sleep phase to a regular schedule.

Stimulant-wakefulness agents, like modafinil, can be used for patients with daytime fatigue, low energy, reduced locomotor activity, and daytime sleeping to increase the wake period.

6.6 Conclusion

On the viewpoint of relationship between biological rhythms and MDD, more attention to recognition and measure on circadian should be paid. Chronotherapy should be selected to be prescribed for patients with significantly disrupted rhythms. The relationships and the potential mechanism of biological rhythms and MDD are still unclear. In the future, research toward the measurement and treatment focusing on biological rhythms of MDD which should be developed.

References

Abe M, Herzog ED, Block GD (2000) Lithium lengthens the circadian period of individual suprachiasmatic nucleus neurons. NeuroReport 11(14):3261–3264

Albrecht U (2013) Circadian clocks and mood-related behaviors. Handb Exp Pharmacol 217:227–239. https://doi.org/10.1007/978-3-642-25950-0_9

Avery DH, Shah SH, Eder DN et al (1999) Nocturnal sweating and temperature in depression. Acta Psychiatr Scand 100(4):295–301

Bahk YC, Han E, Lee SH (2014) Biological rhythm differences and suicidal ideation in patients with major depressive disorder. J Affect Disord 168:294–297. https://doi.org/10.1016/j.jad.2014.07.001

Bellivier F, Geoffroy PA, Etain B et al (2015) Sleep- and circadian rhythm-associated pathways as therapeutic targets in bipolar disorder. Expert Opin Ther Targets 19(6):747–763. https://doi.org/10.1517/14728222.2015.1018822

Benca RM, Obermeyer WH, Thisted RA et al (1992) Sleep and psychiatric disorders: a meta-analysis. Arch Gen Psychiatry 49(8):651–668; discussion 669–670

Benedetti F, Barbini B, Campori E et al (2001) Sleep phase advance and lithium to sustain the antidepressant effect of total sleep deprivation in bipolar depression: new findings supporting the internal coincidence model? J Psychiatr Res 35(6):323–329

Benedetti F, Barbini B, Bernasconi A et al (2010) Acute antidepressant response to sleep deprivation combined with light therapy is influenced by the catechol-O-methyltransferase Val(108/158)Met polymorphism. J Affect Disord 121(1–2):68–72. https://doi.org/10.1016/j.jad.2009.05.017

Ben-Hamo M, Larson TA, Duge LS et al (2016) Circadian forced desynchrony of the master clock leads to phenotypic manifestation of depression in rats. eNeuro 3(6). https://doi.org/10.1523/eneuro.0237-16.2016

Berdynaj D, Boudissa SN, Grieg MS et al (2016) Effect of chronotype on emotional processing and risk taking. Chronobiol Int 33(4):406–418. https://doi.org/10.3109/07420528.2016.1146739

Bering T, Carstensen MB, Wörtwein G et al (2018) The circadian oscillator of the cerebral cortex: molecular, biochemical and behavioral effects of deleting the arntl clock gene in cortical neurons. Cereb Cortex 28(2):644–657. https://doi.org/10.1093/cercor/bhw406

Bilello JA, Thurmond LM, Smith KM et al (2015) MDDScore: confirmation of a blood test to aid in the diagnosis of major depressive disorder. J Clin Psychiatry 76(2):e199–e206. https://doi.org/10.4088/JCP.14m09029

Boland EM, Bertulis K, Leong SH et al (2019) Preliminary support for the role of reward relevant effort and chronotype in the depression/insomnia comorbidity. J Affect Disord 242:220–223. https://doi.org/10.1016/j.jad.2018.08.057

Bouwmans ME, Bos EH, Booij SH et al (2015) Intra- and inter-individual variability of longitudinal daytime melatonin secretion patterns in depressed and non-depressed individuals. Chronobiol Int 32(3):441–446. https://doi.org/10.3109/07420528.2014.973114

Brown GM, McIntyre RS, Rosenblat J et al (2018) Depressive disorders: processes leading to neurogeneration and potential novel treatments[J]. Prog Neuropsychopharmacol Biol Psychiatry 80(Pt C):189–204. https://doi.org/10.1016/j.pnpbp.2017.04.023

Bumb JM, Enning F, Mueller JK et al (2016) Differential melatonin alterations in cerebrospinal fluid and serum of patients with major depressive disorder and bipolar disorder. Compr Psychiatry 68:34–39. https://doi.org/10.1016/j.comppsych.2016.03.005

Bunney BG, Bunney WE (2013) Mechanisms of rapid antidepressant effects of sleep deprivation therapy: clock genes and circadian rhythms. Biol Psychiatry 73(12):1164–1171. https://doi.org/10.1016/j.biopsych.2012.07.020

Buysse DJ, Frank E, Lowe KK et al (1997) Electroencephalographic sleep correlates of episode and vulnerability to recurrence in depression. Biol Psychiatry 41(4):406–418. https://doi.org/10.1016/S0006-3223(96)00041-8

Campbell SS, Gillin JC, Kripke DF et al (1989) Lithium delays circadian phase of temperature and REM sleep in a bipolar depressive: a case report. Psychiatry Res 27(1):23–29

Corruble E, Swartz HA, Bottai T et al (2016) Telephone-administered psychotherapy in combination with antidepressant medication for the acute treatment of major depressive disorder. J Affect Disord 190:6–11. https://doi.org/10.1016/j.jad.2015.07.052

Courtet P, Olié E (2012) Circadian dimension and severity of depression. Eur Neuropsychopharmacol 22(Suppl 3):S476–S481. https://doi.org/10.1016/j.euroneuro.2012.07.009

Dallaspezia S, Benedetti F (2015a) Chronobiology of bipolar disorder: therapeutic implication. Curr Psychiatry Rep 17(8):606. https://doi.org/10.1007/s11920-015-0606-9

Dallaspezia S, Benedetti F (2015b) Sleep deprivation therapy for depression. Curr Top Behav Neurosci 25:483–502. https://doi.org/10.1007/7854_2014_363

De Crescenzo F, Lennox A, Gibson JC et al (2017) Melatonin as a treatment for mood disorders: a systematic review. Acta Psychiatr Scand 136(6):549–558. https://doi.org/10.1111/acps.12755

Elder GJ, Wetherell MA, Barclay NL et al (2014) The cortisol awakening response–applications and implications for sleep medicine. Sleep Med Rev 18(3):215–224. https://doi.org/10.1016/j.smrv.2013.05.001

Esaki Y, Kitajima T, Takeuchi I et al (2017) Effect of blue-blocking glasses in major depressive disorder with sleep onset insomnia: a randomized, double-blind, placebo-controlled study. Chronobiol Int 34(6):753–761. https://doi.org/10.1080/07420528.2017.1318893

Fava M (2004) Daytime sleepiness and insomnia as correlates of depression. J Clin Psychiatry 65(Suppl 16):27–32

Fava M, Targum SD, Nierenberg AA et al (2012) An exploratory study of combination buspirone and melatonin SR in major depressive disorder (MDD): a possible role for neurogenesis in drug discovery. J Psychiatr Res 46(12):1553–1563. https://doi.org/10.1016/j.jpsychires.2012.08.013

Frank E, Swartz HA, Kupfer DJ (2000) Interpersonal and social rhythm therapy: managing the chaos of bipolar disorder. Biol Psychiatry 48(6):593–604

Germain A, Nofzinger EA, Meltzer CC et al (2007) Diurnal variation in regional brain glucose metabolism in depression. Biol Psychiatry 62(5):438–445. https://doi.org/10.1016/j.biopsych.2006.09.043

Giedke H, Schwärzler F (2002) Therapeutic use of sleep deprivation in depression. Sleep Med Rev 6(5):361–377

Goldstone A, Willoughby AR, de Zambotti M et al (2018) The mediating role of cortical thickness and gray matter volume on sleep slow-wave activity during adolescence. Brain Struct Funct 223(2):669–685. https://doi.org/10.1007/s00429-017-1509-9

Hamers PC, Evenhuis HM, Hermans H (2017) A multicenter randomized controlled trial for bright light therapy in adults with intellectual disabilities and depression: study protocol and obstacle management. Res Dev Disabil 60:96–106. https://doi.org/10.1016/j.ridd.2016.10.012

Havekes R, Meerlo P, Abel T (2015) Animal studies on the role of sleep in memory: from behavioral performance to molecular mechanisms. Curr Top Behav Neurosci 25:183–206. https://doi.org/10.1007/7854_2015_369

Haynes PL, Ancoli-Israel S, McQuaid J (2005) Illuminating the impact of habitual behaviors in depression. Chronobiol Int 22(2):279–297

Haynes PL, Gengler D, Kelly M (2016a) Social rhythm therapies for mood disorders: an update. Curr Psychiatry Rep 18(8):75. https://doi.org/10.1007/s11920-016-0712-3

Haynes PL, Kelly M, Warner L et al (2016b) Cognitive behavioral social rhythm group therapy for veterans with posttraumatic stress disorder, depression, and sleep disturbance: results from an open trial. J Affect Disord 192:234–243. https://doi.org/10.1016/j.jad.2015.12.012

Hickie IB, Rogers NL (2011) Novel melatonin-based therapies: potential advances in the treatment of major depression. Lancet 378(9791):621–631. https://doi.org/10.1016/S0140-6736(11)60095-0

Hickie IB, Naismith SL, Robillard R et al (2013) Manipulating the sleep-wake cycle and circadian rhythms to improve clinical management of major depression. BMC Med 11:79. https://doi.org/10.1186/1741-7015-11-79

Hua P, Liu W, Chen D et al (2014) Cry1 and Tef gene polymorphisms are associated with major depressive disorder in the Chinese population. J Affect Disord 157:100–103. https://doi.org/10.1016/j.jad.2013.11.019

Huang KL, Lu WC, Wang YY et al (2014) Comparison of agomelatine and selective serotonin reuptake inhibitors/serotonin-norepinephrine reuptake inhibitors in major depressive disorder: a meta-analysis of head-to-head randomized clinical trials. Aust NZ J Psychiatry 48(7):663–671. https://doi.org/10.1177/0004867414525837

Jones SG, Benca RM (2015) Circadian disruption in psychiatric disorders. Sleep Med Clin 10(4):481–493. https://doi.org/10.1016/j.jsmc.2015.07.004

Joyce PR, Porter RJ, Mulder RT et al (2005) Reversed diurnal variation in depression: associations with a differential antidepressant response, tryptophan: large neutral amino acid ratio and serotonin transporter polymorphisms. Psychol Med 35(4):511–517

Kaladchibachi SA, Doble B, Anthopoulos N et al (2007) Glycogen synthase kinase 3, circadian rhythms, and bipolar disorder: a molecular link in the therapeutic action of lithium. J Circadian Rhythms 5:3. https://doi.org/10.1186/1740-3391-5-3

Kario K, Schwartz JE, Davidson KW et al (2001) Gender differences in associations of diurnal blood pressure variation, awake physical activity, and sleep quality with negative affect: the work site blood pressure study. Hypertension 38(5):997–1002

Kennedy SH, Avedisova A, Giménez-Montesinos N et al (2014) A placebo-controlled study of three agomelatine dose regimens (10 mg, 25 mg, 25–50 mg) in patients with major depressive disorder. Eur Neuropsychopharmacol 24(4):553–563. https://doi.org/10.1016/j.euroneuro.2014.01.006

Ketchesin KD, Becker-Krail D, McClung CA (2018) Mood-related central and peripheral clocks. Eur J Neurosci. https://doi.org/10.1111/ejn.14253

Khaleghipour S, Masjedi M, Ahade H et al (2012) Morning and nocturnal serum melatonin rhythm levels in patients with major depressive disorder: an analytical cross-sectional study. Sao Paulo Med J 130(3):167–172

Kragh M, Larsen ER, Martiny K et al (2018) Predictors of response to combined wake and light therapy in treatment-resistant inpatients with depression. Chronobiol Int 35(9):1209–1220. https://doi.org/10.1080/07420528.2018.1468341

Lam RW (2006) Sleep disturbances and depression: a challenge for antidepressants. Int Clin Psychopharmacol 21(Suppl 1):S25–S29. https://doi.org/10.1097/01.yic.0000195658.91524.61

Laux G, Barthel B, Hajak G et al (2017) Pooled analysis of four non-interventional studies: effectiveness and tolerability of the antidepressant agomelatine in daily practice. Adv Ther 34(4):895–914. https://doi.org/10.1007/s12325-017-0485-z

Levitan RD (2007) The chronobiology and neurobiology of winter seasonal affective disorder. Dialogues Clin Neurosci 9(3):315–324

Li JZ, Bunney BG, Meng F et al (2013) Circadian patterns of gene expression in the human brain and disruption in major depressive disorder. Proc Natl Acad Sci USA 110(24):9950–9955. https://doi.org/10.1073/pnas.1305814110

Lunsford-Avery JR, BDSB G, Brietzke E et al (2017) Adolescents at clinical-high risk for psychosis: circadian rhythm disturbances predict worsened prognosis at 1-year follow-up. Schizophr Res189:37–42. https://doi.org/10.1016/j.schres.2017.01.051

Ma HY, Liu ZF, Xu YF et al (2018) The association study of CLOCK gene polymorphisms with antidepressant effect in Chinese with major depressive disorder. Per Med. https://doi.org/10.2217/pme-2018-0123

Marshall L, Helgadóttir H, Mölle M et al (2006) Boosting slow oscillations during sleep potentiates memory. Nature 444(7119):610–613. https://doi.org/10.1038/nature05278

Martinotti G, Sepede G, Gambi F et al (2012) Agomelatine versus venlafaxine XR in the treatment of anhedonia in major depressive disorder: a pilot study. J Clin Psychopharmacol 32(4):487–491. https://doi.org/10.1097/JCP.0b013e31825d6c25

McClung CA (2013) How might circadian rhythms control mood? Let me count the ways. Biol Psychiatry 74(4):242–249. https://doi.org/10.1016/j.biopsych.2013.02.019

Melo MCA, Abreu RLC, Linhares Neto VB et al (2017) Chronotype and circadian rhythm in bipolar disorder: a systematic review. Sleep Med Rev 34:46–58. https://doi.org/10.1016/j.smrv.2016.06.007

Morris DW, Rush AJ, Jain S et al (2007) Diurnal mood variation in outpatients with major depressive disorder: implications for DSM-V from an analysis of the sequenced treatment alternatives to relieve depression study data. J Clin Psychiatry 68(9):1339–1347

Mukherjee S, Coque L, Cao JL et al (2010) Knockdown of clock in the ventral tegmental area through RNA interference results in a mixed state of mania and depression-like behavior. Biol Psychiatry 68(6):503–511. https://doi.org/10.1016/j.biopsych.2010.04.031

Naismith SL, Hermens DF, Ip TK et al (2012) Circadian profiles in young people during the early stages of affective disorder. Transl Psychiatry 2:e123. https://doi.org/10.1038/tp.2012.47

Nechita F, Pîrlog MC, ChiriŢă AL (2015) Circadian malfunctions in depression—neurobiological and psychosocial approaches. Rom J Morphol Embryol 56(3):949–955

Nicolaides NC, Charmandari E, Chrousos GP et al (2014) Circadian endocrine rhythms: the hypothalamic-pituitary-adrenal axis and its actions. Ann NY Acad Sci 1318:71–80. https://doi.org/10.1111/nyas.12464

Nussbaumer B, Kaminski-Hartenthaler A, Forneris CA et al (2015) Light therapy for preventing seasonal affective disorder. Cochrane Database Syst Rev (11):CD011269. https://doi.org/10.1002/14651858.cd011269.pub2

Orozco-Solis R, Montellier E, Aguilar-Arnal L et al (2017) A circadian genomic signature common to ketamine and sleep deprivation in the anterior cingulate cortex. Biol Psychiatry 82(5):351–360. https://doi.org/10.1016/j.biopsych.2017.02.1176

Quera-Salva MA, Lemoine P, Guilleminault C (2010) Impact of the novel antidepressant agomelatine on disturbed sleep-wake cycles in depressed patients. Hum Psychopharmacol 25(3):222–229. https://doi.org/10.1002/hup.1112

Rastad C, Wetterberg L, Martin C (2017) Patients' experience of winter depression and light room treatment. Psychiatry J 2017:6867957. https://doi.org/10.1155/2017/6867957

Robillard R, Hermens DF, Lee RS et al (2016) Sleep-wake profiles predict longitudinal changes in manic symptoms and memory in young people with mood disorders. J Sleep Res 25(5):549–555. https://doi.org/10.1111/jsr.12413

Ronaldson A, Carvalho LA, Kostich K et al (2018) The effects of six-day SSRI administration on diurnal cortisol secretion in healthy volunteers. Psychopharmacology 235(12):3415–3422. https://doi.org/10.1007/s00213-018-5050-1

Sachar EJ, Hellman L, Roffwarg HP et al (1973) Disrupted 24-hour patterns of cortisol secretion in psychotic depression. Arch Gen Psychiatry 28(1):19–24

Saini C, Morf J, Stratmann M et al (2012) Simulated body temperature rhythms reveal the phase-shifting behavior and plasticity of mammalian circadian oscillators. Genes Dev 26(6):567–580. https://doi.org/10.1101/gad.183251.111

Saus E, Soria V, Escaramís G et al (2010) Genetic variants and abnormal processing of pre-miR-182, a circadian clock modulator, in major depression patients with late insomnia. Hum Mol Genet 19(20):4017–4025. https://doi.org/10.1093/hmg/ddq316

Schnell A, Albrecht U, Sandrelli F (2014) Rhythm and mood: relationships between the circadian clock and mood-related behavior. Behav Neurosci 128(3):326–343. https://doi.org/10.1037/a0035883

Selemon LD, Zecevic N (2015) Schizophrenia: a tale of two critical periods for prefrontal cortical development. Transl Psychiatry 5:e623. https://doi.org/10.1038/tp.2015.115

Shi SQ, White MJ, Borsetti HM et al (2016) Molecular analyses of circadian gene variants reveal sex-dependent links between depression and clocks. Transl Psychiatry 6:e748. https://doi.org/10.1038/tp.2016.9

Soria V, Martínez-Amorós E, Escaramís G et al (2010) Differential association of circadian genes with mood disorders: CRY1 and NPAS2 are associated with unipolar major depression and CLOCK and VIP with bipolar disorder. Neuropsychopharmacology 35(6):1279–1289. https://doi.org/10.1038/npp.2009.230

Soterio-Pires JH, Hirotsu C, Kim LJ et al (2017) The interaction between erectile dysfunction complaints and depression in men: a cross-sectional study about sleep, hormones and quality of life. Int J Impot Res 29(2):70–75. https://doi.org/10.1038/ijir.2016.49

Sprouse J, Braselton J, Reynolds L (2006) Fluoxetine modulates the circadian biological clock via phase advances of suprachiasmatic nucleus neuronal firing. Biol Psychiatry 60(8):896–899. https://doi.org/10.1016/j.biopsych.2006.03.003

Stahl SM, Fava M, Trivedi MH et al (2010) Agomelatine in the treatment of major depressive disorder: an 8-week, multicenter, randomized, placebo-controlled trial. J Clin Psychiatry 71(5):616–626. https://doi.org/10.4088/JCP.09m05471blu

Takahashi JS, Hong HK, Ko CH et al (2008) The genetics of mammalian circadian order and disorder: implications for physiology and disease. Nat Rev Genet 9(10):764–775. https://doi.org/10.1038/nrg2430

Uz T, Ahmed R, Akhisaroglu M et al (2005) Effect of fluoxetine and cocaine on the expression of clock genes in the mouse hippocampus and striatum. Neuroscience 134(4):1309–1316. https://doi.org/10.1016/j.neuroscience.2005.05.003

Yamaguchi N, Maeda K, Kuromaru S (1978) Sleep deprivation therapy for depression and diurnal rhythm of serum cortisol. Horumon to Rinsho 26(5):457–463

Zajecka J, Schatzberg A, Stahl S et al (2010) Efficacy and safety of agomelatine in the treatment of major depressive disorder: a multicenter, randomized, double-blind, placebo-controlled trial. J Clin Psychopharmacol 30(2):135–144. https://doi.org/10.1097/JCP.0b013e3181d420a7

Chapter 7
Advances in Molecular and Circuitry Mechanisms of Depressive Disorder—A Focus on Lateral Habenula

Hailan Hu

Abstract Depression is a devastating disorder with a combination of diverse symptoms such as low self-esteem, lack of motivation, anhedonia, loss of appetite, low energy, and discomfort without a clear cause. Depression has been suggested to be the result of maladaptive changes in specific brain circuits. Recently, the lateral habenula (LHb) has emerged as a key brain region in the pathophysiology of depression. Increasing evidence from rodent, nonhuman primate, and human studies indicates that the aberrant activity of the LHb is associated with depressive symptoms such as helplessness, anhedonia, and excessive negative focus. Revealing the molecular, cellular, and circuit properties of the LHb will help explain how abnormalities in LHb activity are linked to depressive disorders and shed light on developing novel strategies for depression treatment.

Keywords Depression · Lateral habenula · LHb circuits · Novel antidepressant treatments

7.1 Introduction

The lateral habenula (LHb) is a phylogenetically well-preserved brain structure (Aizawa et al. 2011) and has long been recognized as a key region mediating communication between the forebrain and monoaminergic systems in the midbrain and hindbrain (Herkenham 1979; Herkenham and Nauta 1979). Many functions are subject to LHb regulation, including pain, behavioral flexibility, maternal behavior, circadian rhythm, sleep, addiction, anxiety, and depression (Zahm and Root 2017; Li

H. Hu (✉)
Center for Neuroscience and Department of Psychiatry of First Affiliated Hospital, Zhejiang University School of Medicine, Hangzhou, China
e-mail: huhailan@zju.edu.cn

Interdisciplinary Institute of Neuroscience and Technology, Qiushi Academy for Advanced Studies, Zhejiang University, Hangzhou, China

NHC and CAMS Key Laboratory of Medical Neurobiology, Mental Health Center, Zhejiang University, Hangzhou, China

© Springer Nature Singapore Pte Ltd. 2019
Y. Fang (ed.), *Depressive Disorders: Mechanisms, Measurement and Management*, Advances in Experimental Medicine and Biology 1180,
https://doi.org/10.1007/978-981-32-9271-0_7

et al. 2016; Baker and Mizumori 2017; Matthews-Felton et al. 1995; Mendoza 2017; Aizawa et al. 2013; Meye et al. 2017; Proulx et al. 2014). Recently, several important findings have proposed that the LHb participates in the processing of negative valence information (Proulx et al. 2014; Matsumoto and Hikosaka 2007; Hong et al. 2011; Matsumoto and Hikosaka 2009). Mounting evidence from animal and human studies also suggests that aberrant LHb activity is associated with depressive symptoms such as helplessness (Li et al. 2011; Yang et al. 2008) and lack of pleasure (anhedonia) (Li et al. 2013). Circuitry-wise, the LHb acts as a relay station that interconnects the limbic forebrain with depression-related monoaminergic centers including the ventral tegmental area (VTA) and raphe (Zahm and Root 2017; Proulx et al. 2014; Hikosaka et al. 2008). In this article, we review the output regions downstream of the LHb, as well as the source of inputs into the LHb, which may mediate different negative symptoms of depression. We also summarize major findings addressing the molecular and cellular mechanisms by which the LHb neurons become aberrantly overactive and how that may lead to depression.

7.1.1 Activity of LHb Neurons in Normal and Depressive Conditions

The ability to predict reward and punishment and to adjust behavior accordingly is essential for animal survival (Proulx et al. 2014). The LHb is believed to play a central role in encoding negative valence signals and promoting behavioral aversion. Recordings in macaque monkeys show that LHb neurons encode negative reward prediction error (RPE), whereby they increase firing when animals fail to receive an expected reward ("disappointed") or receive a cue predicting aversive stimuli (Matsumoto and Hikosaka 2007, 2009). Moreover, the LHb is activated by different stressors and negative emotional stimuli (Wirtshafter et al. 1994), suggesting that it may mediate behavioral responses underlying stress and aversion under normal physiological conditions. However, aberrant LHb activity may lead to detrimental consequences, such as depression, as evidenced below.

The LHb shows higher metabolic activity in various animal models of depression (Caldecott-Hazard et al. 1988; Shumake et al. 2003; Shumake and Gonzalez-Lima 2003, 2013). In both acute learned helplessness (aLH) and congenital learned helplessness (cLH) rats, VTA-projecting habenula neurons demonstrate enhanced synaptic activity (Li et al. 2011). Interestingly, such elevated LHb firing and depressive behaviors can be reversed following antidepressant treatment (Li et al. 2011).

Due to the small size of the LHb, human functional imaging is challenging and still inadequate; however, several studies have provided evidence for habenula hyperactivity in depressed human subjects (Lawson et al. 2017; Morris et al. 1999). Tryptophan depletion treatment, which usually deteriorates symptoms in depressive patients, increases the cerebral blood flow in the habenular region (Morris et al. 1999).

Moreover, depressive patients show abnormal phasic habenular responses elicited by cues that predict upcoming punishment (Lawson et al. 2014).

Taken together, increasing evidence from both animal and human studies suggests that LHb hyperactivity is associated with symptoms of depression.

7.1.2 Output Regions of the LHb

The primary output regions from the lateral habenula are midbrain aminergic centers, which include the dopaminergic (DA) VTA and substantia nigra pars compacta (SNc), serotonergic (5-HT) dorsal and median raphe (DRN, MRN), and GABAergic rostromedial tegmental nucleus (RMTg, also known as tail-VTA).

The "amine hypothesis of depression" postulates that depression results from a deficiency of the monoamine neurotransmitters (Disorders et al. 2014). Consistent with this hypothesis, the LHb is known to exert a powerful inhibitory influence over the midbrain DA and 5-HT neurons (Matsumoto and Hikosaka 2007; Hong et al. 2011; Stamatakis and Stuber 2012; Park 1987; Gao et al. 1990; Varga et al. 2003; Tian and Uchida 2015). As the main output neurons of the LHb are glutamatergic (Li et al. 2011; Stamatakis and Stuber 2012; Aizawa et al. 2012; Lammel et al. 2012), this inhibition over midbrain monoaminergic neurons may be achieved either through relay at the GABAergic RMTg (Hong et al. 2011; Stamatakis and Stuber 2012; Sego et al. 2014) or local interneurons (Zhou et al. 2017) (Fig. 7.1). Ultrastructural analysis shows that within the VTA, LHb axons target GABAergic and dopaminergic neurons to a comparable extent (52% and 45%, respectively) (Omelchenko et al. 2009), suggesting that the robust inhibition of DA cells by the LHb is unlikely to be fully accounted for by selective innervation of VTA GABA neurons. The RMTg is a region that receives dense projections from the LHb and sends strong GABAergic projections to midbrain monoaminergic nuclei (Jhou et al. 2009; Kaufling et al. 2009). Axon-sparing lesions of the RMTg and recordings of DA neuron firing have shown that the robust inhibition of DA cells evoked by the LHb is mainly through the RMTg (Brown et al. 2017). At the behavioral level, optogenetic activation of the LHb-VTA or LHb-RMTg pathway produces conditioned place avoidance (Stamatakis and Stuber 2012; Lammel et al. 2012).

The DRN receives a moderate direct projection from the medial part of the LHb (LHbM) and a robust projection from the RMTg (Sego et al. 2014). The glutamatergic inputs from the LHb target both 5-HT and GABAergic neurons in the DR, which provide feedforward inhibition to 5-HT neurons (Zhou et al. 2017).

Collectively, the circuit configuration downstream of the LHb suggests that during depression, hyperactivation of the LHb may suppress DA and 5-HT neurons through the GABAergic RMTg neurons and local interneurons.

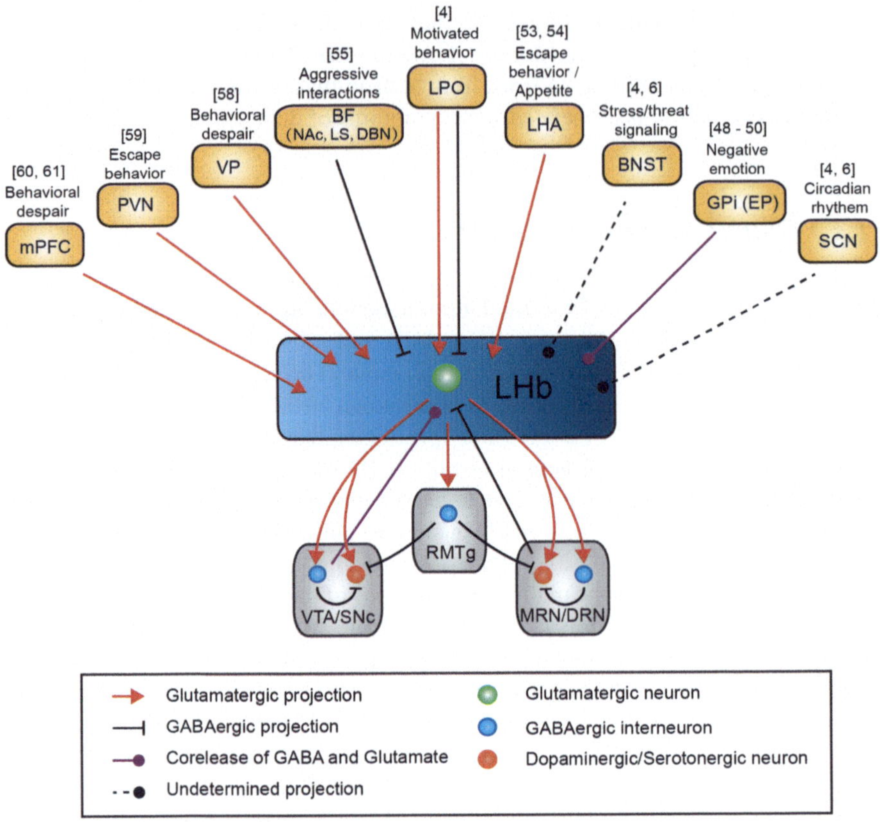

Fig. 7.1 Summary of the input and output circuitry of the LHb. Projections to the lateral habenula (LHb), shown in brown, include the paraventricular nucleus (PVN), basal forebrain [BF, including the nucleus accumbens (NAc), lateral septum, and diagonal band nuclei (DBN)], lateral hypothalamic area (LHA), lateral preoptic area (LPO), ventral pallidum (VP), globus pallidus (GPi), medial prefrontal cortex (mPFC), suprachiasmatic nucleus (SCN), and bed nucleus of the stria terminalis (BNST). Potential functional role of each input pathway is noted. The references are indicated on the top of the figure. The main output of the LHb, shown in red, is glutamatergic. The LHb neurons preferentially form synapses on GABAergic or dopaminergic neurons in the VTA and substantia nigra pars compacta (SNc) (Gao et al. 1990; Omelchenko et al. 2009), GABAergic or serotoninergic neurons in the dorsal raphe nucleus (DRN), and median raphe nucleus (MSN) (Sego et al. 2014; Zhou et al. 2017), as well as GABAergic neurons in the rostromedial tegmental nucleus (RMTg) (Jhou et al. 2009; Kaufling et al. 2009; Brown et al. 2017). The GABAergic RMTg inhibits DA and 5HT neurons (Jhou et al. 2009; Kaufling et al. 2009; Brown et al. 2017). The VTA and DRN/MSN also send reciprocal feedback inputs into the LHb (Root et al. 2014; Stamatakis et al. 2013; Taylor et al. 2014; Yoo et al. 2016; Yetnikoff et al. 2015)

7.1.3 Presynaptic Inputs Controlling LHb Activity

The primary regions sending input into the LHb are the various limbic and basal ganglion structures, which include the internal segment of the globus pallidus (GPi), lateral hypothalamic area (LHA), basal forebrain, ventral pallidum (VP), paraventricular nucleus (PVN), and lateral preoptic area (LPO) (Zahm and Root 2017; Kowski et al. 2008) (Fig. 7.1). These pathways contain both glutamatergic and GABAergic axons and sometimes cholinergic axons (Hikosaka 2010). In addition, another major input comes as a reciprocal feedback from the monoaminergic centers, including the VTA (Lammel et al. 2012; Root et al. 2014; Stamatakis et al. 2013; Taylor et al. 2014; Yoo et al. 2016) and raphe (Yetnikoff et al. 2015). These various presynaptic partners of the LHb convey different components of positive or negative emotional state to regulate LHb activity and modulate normal and depressive behaviors.

The GPi, as well as its non-primate homologue, the entopeduncular nucleus (EPN), is a major output station of basal ganglia and sends a direct projection to the LHb (Shabel et al. 2014; Shabel et al. 2012). Habenula-projecting globus pallidus (GPh) neurons are important for the evaluation of action outcome, encoding whether an outcome is better or worse than expected (Stephenson-Jones et al. 2016). Optical stimulation of the terminal of the EPN (GPh) $\rightarrow$ LHb pathway is aversive (Shabel et al. 2012). Although outputs from the EPN are mainly GABAergic, the efferents to the LHb co-release both GABA and glutamate (Shabel et al. 2014). Interestingly, at the EPN-LHb synapses, the ratio of GABA/glutamate neurotransmission is reduced in animal models of depression and increased by antidepressant treatment (Shabel et al. 2014). Thus, tipping the balance of this co-transmission at the EPN-LHb synapses may be important for regulating LHb activity and mood.

The co-transmission of GABA and glutamate also exists in the VTA-LHb projections (Root et al. 2014; Taylor et al. 2014; Yoo et al. 2016). Single axons from the mesohabenular neurons co-express vesicular GABA transporter (VGaT) and vesicular glutamate transporter 2 (VGluT2), and form both symmetric and asymmetric synapses on the same LHb neurons (Root et al. 2014). In vivo light stimulation of the mesohabenular terminals can either inhibit or excite LHb neurons, possibly dependent on the activity pattern or membrane potential of individual postsynaptic LHb neurons (Root et al. 2014). The LHb-targeting VTA neurons include a tyrosine hydroxylase (TH)-positive population. Although it is not clear whether this population releases dopamine or not (Lammel et al. 2015; Stuber et al. 2015), it is clear that they release GABA and suppress LHb output (Stamatakis et al. 2013). In vivo activation of this pathway prompts reward-related phenotypes dependent on intra-LHb GABA-A receptor signaling (Stamatakis et al. 2013). These findings reveal that neurotransmissions from the VTA to the LHb are highly heterogeneous. It will be of interest to examine whether these co-release mechanisms can be a target for the treatment of depression, as in the case of EPN-LHb synapses.

Loss of appetite is a common symptom of depression. A direct projection from the LHA to the LHb is suggested to negatively regulate food consumption (Stamatakis et al. 2016). Optogenetic activation of LHA-LHb glutamatergic fibers results in an increase in the firing of most LHb neurons as well as avoidance behavior, whereas optogenetic inhibition of this pathway produces a real-time place preference and an acute increase in the consumption of liquid caloric reward (Stamatakis et al. 2016). The same LHA-LHb circuit has also been shown to mediate footshock-driven escape behavior (Lecca et al. 2017).

Abnormality in social behavior is another hallmark of depression. A GABAergic input from the basal forebrain [BF, including the nucleus accumbens (NAc), lateral septum, and diagonal band nuclei (DBN)] to the LHb can bidirectionally control the valence of aggressive interactions (Golden et al. 2016). Optical silencing the GABAergic BF-LHb terminals of aggressors leads to an increase in LHb neuronal firing and abolishes conditioned place preference (CPP) to the intruder-paired context. Conversely, activation of the terminals of non-aggressors decreases LHb neuronal firing and promotes CPP to the intruder-paired context. These results reveal a critical role of the BF-LHb circuit in modulating aggression reward during social behaviors (Golden et al. 2016).

The VP is an important convergent region within the motivational and reward circuitry and has been implicated in depression (Haber et al. 1985; Smith et al. 2009). In social-defeat-stress-induced depressive mice, activity of parvalbumin (PV) neurons in the VP is significantly increased and can be reversed by antidepressant treatment (Knowland et al. 2017). Both optogenetic and chemogenetic silencing of $PV^{VP\text{-}LHb}$ neurons attenuates behavioral despair, whereas the same manipulation of $PV^{VP\text{-}VTA}$ neurons rescues social withdrawal. Therefore, distinct VP-PV neuronal projections to the LHb and VTA may contribute to different aspects of depressive-like behaviors (Knowland et al. 2017).

A group of vasopressin-expressing neurons in the PVN of the hypothalamus also sends glutamatergic input into the LHb, targeting the medial LHb (LHbM). Water deprivation suppresses LHb output through this pathway, leading to decreased freezing during innate fear and immobility in behavioral despair assessment (Zhang et al. 2016).

Circuit tracing studies also show inputs from the medial prefrontal cortex (mPFC) (Warden et al. 2012; Ye et al. 2016), suprachiasmatic nucleus (SCN), and bed nucleus of the stria terminalis (BNST) into the LHb (Zahm and Root 2017). Optogenetic activation of LHb-projecting mPFC axons induces a rapid decrease in mobility in the forced swim test (Warden et al. 2012). However, the functions of these pathways need further research.

7.1.4 Molecular and Cellular Mechanism of LHb Hyperactivation During Depression

In order to search for the molecular mechanism underlying habenular hyperactivity, Li et al. (2013) initiated an unbiased high-throughput proteomic screen and identified several proteins showing altered expression in the LHb of cLH rats

(Li et al. 2013). Among these, the β form of calcium/calmodulin-dependent protein kinase II (β-CaMKII) is significantly upregulated in animal models of depression and downregulated by an antidepressant (Li et al. 2013). Increasing β-, but not α-, CaMKII in the LHb enhances the synaptic efficacy of LHb neurons, possibly by facilitating the synaptic delivery of glutamate receptor GluR1, and produces profound depressive-like behaviors. Notably, loss-of-function manipulations of β-CaMKII are antidepressant (Fig. 7.2) (Li et al. 2013).

Apart from excitatory glutamatergic synapses, the inhibitory GABAergic transmission at the LHb also undergoes changes after aversive experience. Salvatore et al. found that footshock exposure upregulates the activity of protein phosphatase 2A (PP2A), which then triggers GABAB receptor (GABABR) and G protein-gated

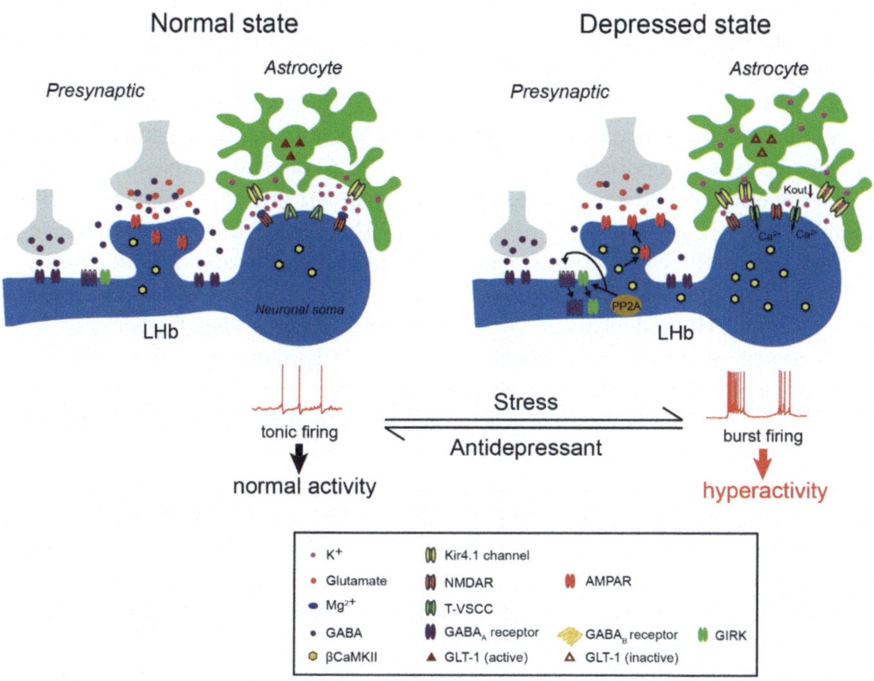

Fig. 7.2 Summary of the molecular mechanism underlying aberrant activity of LHb neurons during depression. Major molecular changes at several subcellular locations have been suggested to account for LHb hyperactivity in depression. At presynaptic sites, GABA/glutamate at the EPN-LHb synapse is shifted toward reduced GABA (Shabel et al. 2014). At the postsynaptic sites of excitatory synapses, upregulation of βCaMKII by stress can lead to increased membrane trafficking of GluR1 and increased synaptic efficacy (Li et al. 2013). At the postsynaptic sites of inhibitory synapses, activation of PP2A by footshock triggers internalization of GABAB1 and GIRK2, causing an increase in neuronal excitability (Lecca et al. 2016). At the interface between neuronal soma and astrocytic end feet, upregulation of astrocytic Kir4.1 causes enhanced K+ buffering, decreased extracellular potassium concentration (Kout), and increased NMDAR-dependent burst firing of LHb neurons (Cui et al. unpublished data). Blockade of NMDARs by ketamine stops burst firing and relieves depression (Yang et al. unpublished data)

inwardly rectifying potassium (GIRK) channel internalization, leading to rapid and persistent weakening of GABAB-activated GIRK-mediated (GABAB-GIRK) currents (Lecca et al. 2016). Accordingly, pharmacological inhibition of PP2A restores both GABAB-GIRK signaling and neuronal excitability, and ameliorates depressive-like symptoms (Lecca et al. 2016).

In addition to these postsynaptic mechanisms, presynaptic changes have also been reported, which include increased release probability of glutamatergic synapses onto LHb neurons (Li et al. 2011) and an altered GABA/glutamate ratio at EPN-LHb synapses (Shabel et al. 2014).

At the cellular level, several synaptic plasticity mechanisms have been described in the LHb. mGluR-dependent long-term depression (LTD) at the excitatory (eLTD) and inhibitory (iLTD) synapses both exist and can bidirectionally tune LHb output (Valentinova and Mameli 2016). mGluR-eLTD occurs presynaptically through a cannabinoid receptor 1(CB1R)-dependent decrease in glutamate release, whereas mGluR-iLTD occurs postsynaptically through a PKC-dependent reduction of b2-containing GABAA-R function (Valentinova and Mameli 2016). Effects of stress on bidirectional synaptic plasticity of long-term potentiation (LTP) and LTD at the LHb have also been reported (Park et al. 2017a, b). Acute stress exposure facilitates LTP by lowering the induction threshold (Park et al. 2017a), while impairing CB1R-dependent LTD (Park et al. 2017b) in the LHb. Thus, stress disturbs the balance of LTP and LTD in the LHb neurons, leading to further potentiation of LHb activity.

The above studies have focused on aberrant synaptic plasticity and synaptic transmission of the LHb in depression etiology. Several recent studies further revealed a critical contribution of LHb neuronal firing pattern, regulated by neural–glial interactions, to depression and the rapid antidepressant effects of ketamine. When characterizing additional candidates from the proteomic screen (Matsumoto and Hikosaka 2009), Cui et al. discovered an astrocyte-specific potassium channel, Kir4.1, which is upregulated in the LHb and causes increased burst firing of LHb neurons in depressive-like animals (Cui et al. unpublished data). This NMDAR-dependent bursting of LHb neurons turns out to be a potential target of ketamine (Yang, Y et al. & Hu, H, unpublished data). Another glial protein, glial glutamate transporter GLT-1, when inhibited in the LHb, was found to increase neuronal excitability and cause depressive-like phenotypes (Cui et al. 2014).

Collectively, it is likely that neurons and glia, synaptic inputs, and intrinsic membrane properties of LHb neurons interact intimately to give rise to enhanced LHb output in depression.

7.2 Conclusions and Perspective

The last 10 years have witnessed explosive growth in our understanding of the function of the LHb and its role in depression. According to current evidence, the hyperactivity of the LHb plays a critical role in depression etiology, and its suppression may relieve depressive symptoms. Based on such evidence, deep brain stimulation has already been applied to suppress neural activity of the LHb in treatment-resistant patients and has induced remission of depressive symptoms (Sartorius and Henn 2007; Sartorius et al. 2010). In the future, continuous efforts are needed to understand how the LHb becomes hyperactive in depression. Specifically, what kind of plasticity mechanisms lead to prolonged alteration of LHb neural activity in depression? How do stress and neuromodulators regulate the burst firing patterns of LHb? What molecular substrates can be targeted for selective suppression of this hyperactivity? Understanding these questions will help design effective pharmacological and interventional treatment for depression.

Acknowledgements This chapter was modified from the paper reported by our group in "current opinion in neurobiology" journal (Yang, Y. et al. (2018) Lateral habenula in the pathophysiology of depression. Curr. Opin. Neurobiol. 48, 90–96). The related contents are reused with permission. This work was supported by the Natural Science Foundation of China (#91432108, #31225010 and #81527901) and 111 projects to H.H.

Conflict of Interest Statement Nothing declared.

References

Aizawa H, Amo R, Okamoto H (2011) Phylogeny and ontogeny of the habenular structure. Front Neurosci 5:138

Aizawa H, Cui WP, Tanaka K, Okamoto H (2013) Hyperactivation of the habenula as a link between depression and sleep disturbance. Front Hum Neurosci 7

Aizawa H, Kobayashi M, Tanaka S, Fukai T, Okamoto H (2012) Molecular characterization of the subnuclei in rat habenula. J Comp Neurol 520:4051–4066

Baker PM, Mizumori SJY (2017) Control of behavioral flexibility by the lateral habenula. Pharmacol Biochem Behav 162:62–68

Brown PL, Palacorolla H, Brady D, Riegger K, Elmer GI, Shepard PD (2017) Habenula-induced inhibition of midbrain dopamine neurons is diminished by lesions of the rostromedial tegmental nucleus. J Neurosci 37:217–225

Caldecott-Hazard S, Mazziotta J, Phelps M (1988) Cerebral correlates of depressed behavior in rats, visualized using 14C-2-deoxyglucose autoradiography. J Neurosci 8:1951–1961

Cui W, Mizukami H, Yanagisawa M, Aida T, Nomura M, Isomura Y, Takayanagi R, Ozawa K, Tanaka K, Aizawa H (2014) glial dysfunction in the mouse habenula causes depressive-like behaviors and sleep disturbance. J Neurosci 34:16273–16285

Gao DM, Jeaugey L, Pollak P, Benabid AL (1990) Intensity-dependent nociceptive responses from presumed dopaminergic-neurons of the substantia-nigra, pars compacta in the rat and their modification by lateral habenula inputs. Brain Res 529:315–319

Golden SA, Heshmati M, Flanigan M, Christoffel DJ, Guise K, Pfau ML, Aleyasin H, Menard C, Zhang H, Hodes GE et al (2016) Basal forebrain projections to the lateral habenula modulate aggression reward. Nature 534:688–692

Haber SN, Groenewegen HJ, Grove EA, Nauta WJH (1985) Efferent connections of the ventral pallidum—evidence of a dual striato pallidofugal pathway. J Comp Neurol 235:322–335

Herkenham M (1979) The afferent and efferent connections of the ventromedial thalamic nucleus in the rat. J Comp Neurol 183:487–517

Herkenham M, Nauta WJ (1979) Efferent connections of the habenular nuclei in the rat. J Comp Neurol 187:19–47

Hikosaka O (2010) The habenula: from stress evasion to value-based decision-making. Nat Rev Neurosci 11:503–513

Hikosaka O, Sesack SR, Lecourtier L, Shepard PD (2008) Habenula: crossroad between the basal ganglia and the limbic system. J Neurosci 28:11825–11829

Hong S, Jhou TC, Smith M, Saleem KS, Hikosaka O (2011) Negative reward signals from the lateral habenula to dopamine neurons are mediated by rostromedial tegmental nucleus in primates. J Neurosci 31:11457–11471

Jhou TC, Geisler S, Marinelli M, Degarmo BA, Zahm DS (2009) The mesopontine rostromedial tegmental nucleus: a structure targeted by the lateral habenula that projects to the ventral tegmental area of Tsai and substantia nigra compacta. J Comp Neurol 513:566–596

Kaufling J, Veinante P, Pawlowski SA, Freund-Mercier MJ, Barrot M (2009) Afferents to the GABAergic tail of the ventral tegmental area in the rat. J Comp Neurol 513:597–621

Knowland D, Lilascharoen V, Pacia CP, Shin S, Wang EH, Lim BK (2017) Distinct ventral pallidal neural populations mediate separate symptoms of depression. Cell 170(284–297):e18

Kowski AB, Geisler S, Krauss M, Veh RW (2008) Differential projections from subfields in the lateral preoptic area to the lateral habenular complex of the rat. J Comp Neurol 507:1465–1478

Lammel S, Lim BK, Ran C, Huang KW, Betley MJ, Tye KM, Deisseroth K, Malenka RC (2012) Input-specific control of reward and aversion in the ventral tegmental area. Nature 491:212–217

Lammel S, Steinberg EE, Foldy C, Wall NR, Beier K, Luo LQ, Malenka RC (2015) Diversity of transgenic mouse models for selective targeting of midbrain dopamine neurons. Neuron 85:429–438

Lawson RP, Nord CL, Seymour B, Thomas DL, Dayan P, Pilling S, Roiser JP (2017) Disrupted habenula function in major depression. Mol Psychiatry 22:202–208

Lawson RP, Seymour B, Loh E, Lutti A, Dolan RJ, Dayan P, Weiskopf N, Roiser JP (2014) The habenula encodes negative motivational value associated with primary punishment in humans. Proc Natl Acad Sci USA 111:11858–11863

Lecca S, Pelosi A, Tchenio A, Moutkine I, Lujan R, Herve D, Mameli M (2016) Rescue of GABAB and GIRK function in the lateral habenula by protein phosphatase 2A inhibition ameliorates depression-like phenotypes in mice. Nat Med 22:254–261

Lecca S, Meye FJ, Trusel M, Tchenio A, Harris J, Schwarz MK, Burdakov D, Georges F, Mameli M (2017) Aversive stimuli drive hypothalamus-to-habenula excitation to promote escape behavior. Elife 6

Li JC, Li Y, Zhang BL, Shen XF, Zhao H (2016) Why depression and pain often coexist and mutually reinforce: role of the lateral habenula. Exp Neurol 284:106–113

Li B, Piriz J, Mirrione M, Chung C, Proulx CD, Schulz D, Henn F, Malinow R (2011) Synaptic potentiation onto habenula neurons in the learned helplessness model of depression. Nature 470:535–539

Li K, Zhou T, Liao L, Yang Z, Wong C, Henn F, Malinow R, Yates JR 3rd, Hu H (2013) betaCaMKII in lateral habenula mediates core symptoms of depression. Science 341:1016–1020

Matsumoto M, Hikosaka O (2007) Lateral habenula as a source of negative reward signals in dopamine neurons. Nature 447:1111

Matsumoto M, Hikosaka O (2009) Representation of negative motivational value in the primate lateral habenula. Nat Neurosci 12:77–84

Matthews-Felton T, Corodimas KP, Rosenblatt JS, Morrell JI (1995) Lateral habenula neurons are necessary for the hormonal onset of maternal behavior and for the display of postpartum estrus in naturally parturient female rats. Behav Neurosci 109:1172–1188

Mendoza J (2017) Circadian neurons in the lateral habenula: clocking motivated behaviors. Pharmacol Biochem Behav 162:55–61

Meye FJ, Trusel M, Soiza-Reilly M, Mameli M (2017) Neural circuit adaptations during drug withdrawal—spotlight on the lateral habenula. Pharmacol Biochem Behav 162:87–93

Morris JS, Smith KA, Cowen PJ, Friston KJ, Dolan RJ (1999) Covariation of activity in habenula and dorsal raphe nuclei following tryptophan depletion. Neuroimage 10:163–172

Omelchenko N, Bell R, Sesack SR (2009) Lateral habenula projections to dopamine and GABA neurons in the rat ventral tegmental area. Eur J Neurosci 30:1239–1250

Ostergaard SD (2014) Mood disorders and antidepressants. Stahl's essential psychopharmacology. Acta Psychiatr Scand 130:74–75

Park MR (1987) Monosynaptic inhibitory postsynaptic potentials from lateral habenula recorded in dorsal raphe neurons. Brain Res Bull 19:581–586

Park H, Rhee J, Park K, Han JS, Malinow R, Chung C (2017a) Exposure to stressors facilitates long-term synaptic potentiation in the lateral habenula. J Neurosci 37:6021–6030

Park H, Rhee J, Lee S, Chung C (2017b) Selectively impaired endocannabinoid-dependent long-term depression in the lateral habenula in an animal model of depression. Cell Rep 20:289–296

Proulx CD, Hikosaka O, Malinow R (2014) Reward processing by the lateral habenula in normal and depressive behaviors. Nat Neurosci 17:1146–1152

Root DH, Mejias-Aponte CA, Zhang S, Wang HL, Hoffman AF, Lupica CR, Morales M (2014) Single rodent mesohabenular axons release glutamate and GABA. Nat Neurosci 17:1543–1551

Sartorius A, Henn FA (2007) Deep brain stimulation of the lateral habenula in treatment resistant major depression. Med Hypotheses 69:1305–1308

Sartorius A, Kiening KL, Kirsch P, von Gall CC, Haberkorn U, Unterberg AW, Henn FA, Meyer-Lindenberg A (2010) Remission of major depression under deep brain stimulation of the lateral habenula in a therapy-refractory patient. Biol Psychiatry 67:e9–e11

Sego C, Goncalves L, Lima L, Furigo IC, Donato J Jr, Metzger M (2014) Lateral habenula and the rostromedial tegmental nucleus innervate neurochemically distinct subdivisions of the dorsal raphe nucleus in the rat. J Comp Neurol 522:1454–1484

Shabel SJ, Proulx CD, Trias A, Murphy RT, Malinow R (2012) Input to the lateral habenula from the basal ganglia is excitatory, aversive, and suppressed by serotonin. Neuron 74:475–481

Shabel SJ, Proulx CD, Piriz J, Malinow R (2014) Mood regulation. GABA/glutamate co-release controls habenula output and is modified by antidepressant treatment. Science 345:1494–1498

Shumake J, Gonzalez-Lima F (2003) Brain systems underlying susceptibility to helplessness and depression. Behav Cogn Neurosci Rev 2:198–221

Shumake J, Gonzalez-Lima F (2013) Functional opposition between habenula metabolism and the brain reward system. Front Hum Neurosci 7:662

Shumake J, Edwards E, Gonzalez-Lima F (2003) Opposite metabolic changes in the habenula and ventral tegmental area of a genetic model of helpless behavior. Brain Res 963:274–281

Smith KS, Tindell AJ, Aldridge JW, Berridge KC (2009) Ventral pallidum roles in reward and motivation. Behav Brain Res 196:155–167

Stamatakis AM, Stuber GD (2012) Activation of lateral habenula inputs to the ventral midbrain promotes behavioral avoidance. Nat Neurosci 15:1105–1107

Stamatakis AM, Jennings JH, Ung RL, Blair GA, Weinberg RJ, Neve RL, Boyce F, Mattis J, Ramakrishnan C, Deisseroth K et al (2013) A unique population of ventral tegmental area neurons inhibits the lateral habenula to promote reward. Neuron 80:1039–1053

Stamatakis AM, Van Swieten M, Basiri ML, Blair GA, Kantak P, Stuber GD (2016) Lateral hypothalamic area glutamatergic neurons and their projections to the lateral habenula regulate feeding and reward. J Neurosci 36:302–311

Stephenson-Jones M, Yu K, Ahrens S, Tucciarone JM, van Huijstee AN, Mejia LA, Penzo MA, Tai LH, Wilbrecht L, Li B (2016) A basal ganglia circuit for evaluating action outcomes. Nature 539:289–293

Stuber GD, Stamatakis AM, Kantak PA (2015) Considerations when using cre-driver rodent lines for studying ventral tegmental area circuitry. Neuron 85:439–445

Taylor SR, Badurek S, Dileone RJ, Nashmi R, Minichiello L, Picciotto MR (2014) GABAergic and glutamatergic efferents of the mouse ventral tegmental area. J Comp Neurol 522:3308–3334

Tian J, Uchida N (2015) Habenula lesions reveal that multiple mechanisms underlie dopamine prediction errors. Neuron 87:1304–1316

Valentinova K, Mameli M (2016) mGluR-LTD at excitatory and inhibitory synapses in the lateral habenula tunes neuronal output. Cell Rep 16:2298–2307

Varga V, Kocsis B, Sharp T (2003) Electrophysiological evidence for convergence of inputs from the medial prefrontal cortex and lateral habenula on single neurons in the dorsal raphe nucleus. Eur J Neurosci 17:280–286

Warden MR, Selimbeyoglu A, Mirzabekov JJ, Lo M, Thompson KR, Kim SY, Adhikari A, Tye KM, Frank LM, Deisseroth K (2012) A prefrontal cortex-brainstem neuronal projection that controls response to behavioural challenge. Nature 492:428–432

Wirtshafter D, Asin KE, Pitzer MR (1994) Dopamine agonists and stress produce different patterns of Fos-like immunoreactivity in the lateral habenula. Brain Res 633:21–26

Yang LM, Hu B, Xia YH, Zhang BL, Zhao H (2008) Lateral habenula lesions improve the behavioral response in depressed rats via increasing the serotonin level in dorsal raphe nucleus. Behav Brain Res 188:84–90

Ye L, Allen WE, Thompson KR, Tian Q, Hsueh B, Ramakrishnan C, Wang AC, Jennings JH, Adhikari A, Halpern CH et al (2016) Wiring and molecular features of prefrontal ensembles representing distinct experiences. Cell 165:1776–1788

Yetnikoff L, Cheng AY, Lavezzi HN, Parsley KP, Zahm DS (2015) Sources of input to the rostromedial tegmental nucleus, ventral tegmental area, and lateral habenula compared: a study in rat. J Comp Neurol 523:2426–2456

Yoo JH, Zell V, Gutierrez-Reed N, Wu J, Ressler R, Shenasa MA, Johnson AB, Fife KH, Faget L, Hnasko TS (2016) Ventral tegmental area glutamate neurons co-release GABA and promote positive reinforcement. Nat Commun 7:13697

Zahm DS, Root DH (2017) Review of the cytology and connections of the lateral habenula, an avatar of adaptive behaving. Pharmacol Biochem Behav 162:3–21

Zhang LM, Hernandez VS, Vazquez-Juarez E, Chay FK, Barrio RA (2016) Thirst is associated with suppression of habenula output and active stress coping: is there a role for a non-canonical vasopressin-glutamate pathway? Front Neural Circuits 10

Zhou L, Liu MZ, Li Q, Deng J, Mu D, Sun YG (2017) Organization of functional long-range circuits controlling the activity of serotonergic neurons in the dorsal raphe nucleus. Cell Rep 18:3018–3032

Chapter 8
Advance in Stress for Depressive Disorder

Yuqiang Ding and Jinxia Dai

Abstract Stress is an adaptive response to environment aversive stimuli and a common life experience of one's daily life. Chronic or excessive stress especially that happened in early life is found to be deleterious to individual's physical and mental health, which is highly related to depressive disorders onset. Stressful life events are consistently considered to be the high-risk factors of environment for predisposing depressive disorders. In linking stressful life events with depressive disorder onset, dysregulated HPA axis activity is supposed to play an important role in mediating aversive impacts of life stress on brain structure and function. Increasing evidence have indicated the strong association of stress, especially the chronic stress and early life stress, with depressive disorders development, while the association of stress with depression is moderated by genetic risk factors, including polymorphism of *SERT*, *BDNF*, *GR*, *FKBP5*, *MR*, and *CRHR1*. Meanwhile, stressful life experience particularly early life stress will exert epigenetic modification in these risk genes via DNA methylation and miRNA regulation to generate long-lasting effects on these genes expression, which in turn cause brain structural and functional alteration, and finally increase the vulnerability to depressive disorders. Therefore, the interaction of environment with gene, in which stressful life exposure interplay with genetic risk factors and epigenetic modification, is essential in predicting depressive disorders development. As the mediator of environmental risk factors, stress will function together with genetic and epigenetic mechanism to influence brain structure and function, physiology and psychology, and finally the vulnerability to depressive disorders.

Keywords Stress · Depressive disorder · Stressful life events · HPA axis · Epigenetic modification

---

Y. Ding (✉)
State Key Laboratory of Medical Neurobiology, Department of Laboratory Animal Sciences, Institutes of Brain Science, Fudan University, Shanghai, China
e-mail: dingyuqiang@vip.163.com

J. Dai (✉)
Bio-Medical Center and College of Veterinary Medicine, Huazhong Agricultural University, Wuhan, China
e-mail: jxdai@mail.hzau.edu.cn

© Springer Nature Singapore Pte Ltd. 2019

Y. Fang (ed.), *Depressive Disorders: Mechanisms, Measurement and Management*, Advances in Experimental Medicine and Biology 1180,
https://doi.org/10.1007/978-981-32-9271-0_8

8.1 Introduction

Stress, including physiological and psychological stress, is a common experience of individual life. In general, the stress refers to the response to the stressful life events which are challenging, taxing, harmful, and even threating to individual. Disagreeable emotions like anxiety, anger, and bitterness could be evoked, leading to psychological tension. Stressed individuals may become tense, irritable, aggressive, disinterested, anxious, or agitated, being accompanied by disturbed sleep, decreased or increased appetite, and declined sexual desire. In addition, physiological alterations containing activation of hypothalamic–pituitary–adrenal (HPA) axis (Belda et al. 2015; Bonfiglio et al. 2011), and changes in catecholaminergic systems and immune system, will be induced. Does stress cause depression? This essential issue has drawn a great deal of attention but still remained to be fully addressed. A large body of studies from human and animal models focusing on the association between stress and depressive disorders defined the strong impact of stress on risk of developing depressive disorders, supporting the causal association of stress with depressive disorders onset (Hammen 2005; Kendler et al. 1999).

Depressive disorder, also simply known as depression, is a common but serious mood disorder characterized by persistent feelings (at least 2 weeks) of sadness, hopelessness, and worthlessness, which is usually accompanied by loss of interest in normally pleasurable activities, irritability, impaired cognition, low energy, sleeping or eating problem, pain without a clear cause, and even tendency to suicide (Otte et al. 2016; Schulz and Arora 2015). Depression is a highly prevalent psychiatric disorder in current society, and is associated with high levels of morbidity and mortality (Chirita et al. 2015; Kessler and Bromet 2013; Laursen et al. 2016; Otte et al. 2016). Although large attentions and research have been attracted to the exact cause of depressive disorder, the etiology of depression is still not fully understood.

Unlike other diseases determined by specific gene malfunction, no true depression genes are identified to be directly responsible for the onset and cure of the depression due to the heterogeneity and complexity of depressive disorders. Several genetic factors have been recognized to be associated with increased risk for depressive disorders like major depressive disorder based on family, twin, and epidemiologic studies (Clarke et al. 2010; Dunn et al. 2015; Flint and Kendler 2014; Gao et al. 2012; Gatt et al. 2015; Kishi et al.2013; Lee et al. 2012; Lopez-Leon et al. 2008; Shadrina et al. 2018; Smoller 2016; Wray et al. 2012; Zhao et al. 2014). Each susceptibility gene contributes to a small fraction of the total genetic risk. It was estimated that genetic contribution to the risk of depression is probably ~40% and can be increased to 75% in recurrent depression (McGuffin et al. 1996; Sullivan et al. 2000), the remaining main risk for depression development was attributed to environmental factors (Klengel and Binder 2013; Lopizzo et al. 2015; Richter-Levin and Xu 2018). The role of environmental factors in predisposing depressive disorders essentially reflects the crucial effects of stress on etiology of depressive disorders, since environmental risking factors actually trigger stress and dysregulated stress response are supposed to mediate environmental impact on depression onset. These environmental risk factors, which have been found to be correlated with depressive disorders vulnerability, include prenatal infection, maternal stress, child abuse and neglect, social stress, traumatic events, cancers, endocrine abnormalities, and so on (Glover 2014; Heim et al. 2010; Hollis and Kabbaj 2014; Horowitz and Zunszain 2015; Larrieu and Sandi 2018; Laugharne et al. 2010; Lin and Wang 2014; Lindert et al. 2014; Nemeroff 2016; Sotelo et al. 2014; Sperner-Unterweger 2015; Takahashi et al. 2018; Verdolini et al. 2015; Weinstock 2017). It is noticeable that the effect of environmental risk factors in pathology of depressive disorder can happen as early as in embryonic stage, since prenatal maternal stress (e.g., mother's anxiety or depression) has been shown to be correlated with increased risk of psychological disorders including depression in offspring (Babenko et al. 2015; Barker et al. 2011; Braithwaite et al. 2014; Fatima et al. 2017; Weinstock 2017).

Among these environmental risk factors, stressful life events (SLEs) like cancers, losing job or beloved one, which can evoke psychic tension to trigger a series of stress response including physiological, psychological, and behavioral changes, have attracted much attention and been demonstrated to exert important effect on etiology of depression (Chirita et al. 2015; Hammen 2005; Kendler et al. 1999; Kessing and Bukh 2013; Palazidou 2012; Park et al. 2015; Richter-Levin and Xu 2018; Yang et al. 2015). Stress induced by stressful life events, especially traumatic and chronic life stress, is considered to be a crucial linker of stressful life events with etiology of depressive disorder, and the increasing attention has been focused on the possible mechanisms of life stress in depression development. Stress exposure across the life span can cumulatively increase the risk for the development of depressive disorder (Abravanel and Sinha 2015; Agorastos et al. 2014; Steine et al. 2017; Vinkers et al. 2014). The severity and number of stressful life events are shown to be positively correlated with depressive disorders (Chapman et al. 2004; Kendler et al. 1998; Lueboonthavatchai 2009; Roca et al. 2013; You and Conner 2009). Many changes in the brain happened during depression are found to resemble the effects of severe and prolonged stress, suggesting a strong association of stress with depression. More appropriate animal models, which are easy and faithful methods to induce physical and psychological stress, are developed for addressing pathological mechanism of stress in depressive disorders.

Although stress is a major risk factor for depression, most people do not develop a depression with stress exposure. This is supposed to be due to the moderating effect of genetic factors on the role of stress in depression development (Kessing and Bukh 2013; Klengel and Binder 2013; Lopizzo et al. 2015). Although no true depression genes are identified to be responsible for the onset and cure of depression, genetic background is considered to be an important risk factor in predisposing depressive disorder by interacting with environment factors (Kessing and Bukh 2013; Klengel and Binder 2013; Lopizzo et al. 2015). Many studies have focused on identifying genetic factors that are interacted with stressful life exposure in modulating vulnerability and severity of depressive disorders. Several genes' polymorphisms including serotonin transporter (5-HTT) gene-linked polymorphic region (5-HTTLPR), brain-derived neurotrophic factor (BDNF) Val66Met polymorphism, and SNP polymorphism of key factors involved in stress response [glucocorticoid receptor (GR) gene, mineralocorticoid receptor (MR) gene, FK506 binding protein 51 gene (*FKBP5*), and corticotropin-releasing hormone receptor 1 gene (*CRHR1*)] are shown to contribute to the predictive role of stressful life events in depressive disorders (Kessing and Bukh 2013; Klengel and Binder 2013; Lopizzo et al. 2015).

Now, it is believed that genetic vulnerable genes and environmental factors (mainly stressful life events such as early life stress, traumatic events) are combinatorially and cumulatively involved in the onset of depressive disorder (Chirita et al. 2015; Januar et al. 2015; Klengel and Binder 2013; Lopizzo et al. 2015; Mullins et al. 2016; Northoff 2013; Uher 2014), in which multiple and partial overlapping susceptible genes interact with each other and with environment to predispose individuals to depressive disorders. The gene–environment interaction is considered to be accountable for the etiopathogenesis of depression, implying that stress and gene function together in predicting depression development. Meanwhile, environment-risk factor can induce epigenetic modification to cause brain structural and functional alterations, finally increasing the vulnerability to mental disorders including depressive disorders.

In this chapter, based on human and animal model studies, we will summarize and discuss the effects of stress caused by stressful life events in the onset of depressive disorders, the genetic contributions to the vulnerability of depression by stressful life events, and epigenetic modification in linking stress with depression, to elucidate the correlation of stress with depressive disorders. More recent findings and proposals dedicated to uncovering the interconnections between life stress and depression will be included.

8.2 Biology of Stress

Human being and other living organisms need to keep homeostasis, which can be threatened by external and internal, or physiological (e.g., injury, pain, infection) and psychological (can be real or perceived) stimuli, namely, stressors. In response to stressors, the organism will generate a series of physiological, psychological, and behavioral changes to maintain or re-establish its homeostasis. All of these adaptive responses are collectively called stress response, in short, namely, stress. Thus, stress is the physical, psychological, and behavioral responses evoked by the stressor to cope with it, which is a fundamental requirement for survival and well-being. Stress is an adaptive process in coping with stressors during evolution and a common experience of daily life.

During the process of stress response, autonomic nervous system and HPA axis are two primary systems being activated and functioning (Belda et al. 2015; Ulrich-Lai and Herman 2009). The activation of autonomic nervous system evokes the most immediate physiological response to stressor via its sympathetic (sympathoadrenal medullary axis) and parasympathetic branches with opposing activity, which play an important role in translating stress into a response. Excitation of sympathoadrenal medullary axis increases adrenaline and noradrenaline level in blood, heart rate, vasoconstriction, and energy dedication, representing classical fight-or-flight response, while the parasympathetic arm activation attenuates sympathetic excitation and returns the body to homeostasis. The second major physiological response under stress, HPA axis activation, causes the release of corticotropin-releasing factor (CRF) from the parvocellular part of paraventricular neurons (PVN) of hypothalamus to trigger the secretion of adrenocorticotrophic hormone (ACTH) from the pituitary, and finally lead to the liberation of glucocorticoids (GCs; named cortisol in human and corticosterone in rodent, respectively) into blood, which influence many bodily functions such as metabolic, psychological, and immunological functions. The GCs can also exert self-limiting effect through negative feedback on HPA axis to prevent the rise of GCs level, which otherwise fail in some cases like chronic stress. Normally, the activation of HPA axis and autonomic nervous system is tightly regulated, ensuring the body quickly respond to stressful events and return to homeostasis rapidly. There are much complementary actions between HPA axis and sympathetic system in bodily responses such as energy mobilization and blood pressure maintenance, as circulating GCs can potentiate sympathetically mediated effects; conversely, sympathetic nervous system can facilitate GCs release via its innervation of adrenal cortex.

Besides the activation of physiological function for survival under stress (e.g., increasing energy availability and accelerating oxygen supply), other physical functions which are not for immediate survival, including appetite, digestion, and immune, will be downregulated. Meanwhile, stressed subjects will also exhibit increased alertness, vigilance, and attention, which are thought to be good for responding to threatening situation. Moreover, stress may also affect memory, reward, and susceptibility to diseases (Aich et al. 2009; Ragen et al. 2016; Gilpin 2014; Becker 2010; Laudenslager 1987; Musić and Rossell 2016; Myruski et al. 2018; Wolf 2017). In emotion, based on the appraisal of stressor by exposed individuals as routine or challenging, gratifying or taxing, benign or harmful if taxing, manageable or overpowering in the case of negative appraisal, a diversity of emotions will be generated, ranging from joy to despondency, from tranquility to anxiety, and from self-confidence to sham. Under psychic tension evoked by aggravating and disagreeable emotions, stressed subjects may become irritable, aggressive, anxious, agitated, and distracted. Meanwhile, behavior changes including disturbed sleep, decreased sex, declined, or ravenous appetite will be caused.

These physiological, psychological, and behavioral responses under stress are mediated and regulated by multiple brain regions including brain stem, hypothalamus, amygdala, prefrontal cortex, as well as hippocampus (Ulrich-Lai and Herman 2009). It is noteworthy that the effect of stress on an individual is not a uniform syndrome, but is strongly dependent on not only the stressor characteristics including nature, number, or persistence (acute or chronic) but also individual factors such as age, physical and psychological well-being, genetic vulnerability, past stressful experiences, especially coping ability and personality characteristics (Schneiderman et al. 2005).

As mentioned that these stress responses are indispensable for individual homeostasis, the stress should be properly initiated when needed and terminated when homeostasis is re-established. In general, short-term, slight, and mild stress can be beneficial and healthy, which can improve individual motivation, adaptation, and reaction to the environment. However, in some circumstances like intense stressor, chronic stressor, or acute stressor with long-lasting effect, stress response may be improper, excessive, even prolonged, which is thought to be harmful, and also increase the risk of illness including psychological disease such as anxiety and depression (Jeon and Kim 2018; Lucassen et al. 2014). Chronic stressors (e.g., divorce, unemployment, living in a dangerous neighborhood) may be not intense like a natural disaster or a severe accident, but they last for longer periods of time and may be hard to be avoided or coped with, which tend to have more negative effects on individual health and provoke more frequent and excessive stress response (Jeon and Kim 2018). Too much or chronic stress will over-activate HPA axis to cause prolonged increase of stress hormone glucocorticoid levels, which may induce inertia, fatigue, amotivational syndrome, loss of bone mass, hippocampal atrophy, and acceleration of aging (Myers et al. 2014; van Praag et al. 2004). Many studies showed that human exposed to chronic stressor (e.g., caregivers of patients, job stress, chronic illness) and animal model with persistent stress (e.g., unpredictable chronic mild stress, long-term corticosterone exposure) represent slightly worse physical health, higher levels of depression, and defective cognition (Benson 2018; Jeon and Kim 2018; Ngoupaye et al. 2018; Pinquart and Sorensen 2003; van Donkelaar et al. 2014; Yin et al. 2018; Zhu et al. 2014). In addition, chronic stresses are found to be correlated with high rates of several diseases (e.g., cardiovascular disease, diabetes) and mortality, precipitate a serious or relapse into alcohol abuse (Eisenmann et al. 2016; Golbidi et al. 2015; Kivimaki and Steptoe 2018; Ohlin et al. 2004; Pashkow 1999; Rutters et al. 2014; Schneiderman et al. 2005; Sinha 2012; Spanagel et al. 2014; Steptoe and Kivimaki 2012). Thus, excessive or chronic stresses are harmful to not only physical health but also psychological health.

8.3 Association of Stress with Depressive Disorders

The etiology of depressive disorders is heterogeneous and multifactorial, which is considered to depend on the interaction of environmental and multi-genetic risking factors. Environment-risk factors have been shown to exert strong impact on the vulnerability to depressive disorders, of which stressful life events (maternal stress, social defeat and maltreatment, traumatic events, childhood abuse and neglect, cancers, and so on) are widely studied in predisposing depression (Braithwaite et al. 2014; Hollis and Kabbaj 2014; Lindert et al. 2014; Nemeroff 2016; Park et al. 2015; Sperner-Unterweger 2015; Takahashi et al. 2018; Verdolini et al. 2015). Exposure to stressful live events (SLEs) has been consistently implicated in the pathophysiology of depression disorders like major depressive disorder, especially those life stress exposure in early life (childhood and adolescence). Although depressive disorders can arise without any stressful life events exposure and most people do not develop depression even if they experience life stress, most depressive episodes often develop after stressful life events. So SLEs are considered to be high-risk factors for predicting depressive disorder. Stressful life events such as divorce, losing job, or beloved one, which exceed one's ability to effectively cope with, could evoke a series of psychological and physiological changes including activation of HPA axis and autonomous system, which can be referred to as psychological stress. The psychological stress is a perspective of stress from psychology, which is a feeling of strain and pressure, and a kind of psychological pain. The effects of SLEs in depression development actually reflect a crucial role of stress in etiology of depression. A vast majority of research supports the causal association of stressful life events, essentially stress, with depressive disorder onset (Cohen et al. 2007; Colman and Ataullahjan 2010; Hammen 2005; Kendler et al. 1999; Mothersill and Donohoe 2016; Paykel 2003; Richter-Levin and Xu 2018; Tafet and Nemeroff 2016; Tennant 2002). Besides, most animal models of depressive behaviors are established on chronic and inescapable stress paradigms such as social defeat stress and early life stress, further supporting the causal association of stress with depressive disorder development.

Stress could be evoked when an individual perceives that a situation (e.g., stressful life event) exceed his or her adaptive capacity to handle, including a feature of uncontrollability and (or) inescapability (Richter-Levin and Xu 2018). According to the duration, stress can be acute (e.g., under traumatic events, surgical operation) or chronic (e.g., under chronic illness, marital conflicts), of which chronic stress could be disconnected or persistent. The level of stress is not only affected by the intensity, duration, and frequency of triggering events but also by individual genetic predisposition, personality characteristics, coping abilities, as well as subjective perception on how a traumatic event is. Chronic stresses are more strongly related to depressive symptom than acute stresses (Avison and Turner 1988; Eckenrode 1984). Under chronic or traumatic stress, more or sustained glucocorticoids (GCs) will be released due to dysregulation of HPA axis activation, exerting deleterious impacts on multiple brain functions including neurogenesis, synaptic plasticity, learning and memory, hippocampal size, emotional appraisal of events, as well as periphery functions such as metabolism and immunity (Murray et al. 2008; Palazidou 2012; Pariante and Lightman 2008; Teicher et al. 2012). Hyperactivity of HPA axis is one of the commonest neurobiological changes in depressive patients (Pariante and Lightman 2008). And reduced hippocampal size, decreased neurotrophic factors, and neurogenesis is also characteristic feature of depressed patients and depressive-like animal models (Luo et al. 2014; Sheline et al. 1996; Treadway et al. 2015). Meanwhile, sustained overproduction of cortisol induced by chronic stress leads to reduced activity of dopaminergic (DAergic) system, noradrenergic (NAergic) system, and serotonergic (5-HTergic) system, of which changes are shared by some depressed patients presenting indiscriminate NAergic activity, and diminished DAergic and 5-HTergic activity (van Praag et al. 2004; Yang et al. 2015). Thus, stress response and depressive disorder shared many mediators, circuitries, and phenomenology, in which depression represents a dysregulation of stress response, indicating a strong association between stress and depressive disorder onset. Besides, the severity and number of SLEs are positively correlated with probability of depression onset

(Chapman et al. 2004; Kendler et al. 1998; Lueboonthavatchai 2009; Roca et al. 2013; You and Conner 2009), in which moderate to severe stress predisposes higher risk of depressive disorder than mild stress in subjected people, meanwhile the depressed subjects also experienced more stressful life events than the nondepressed subjects. More importantly, cumulative stress exposures across the life span are related with increased risk for the development of depressive disorder (Abravanel and Sinha 2015; Agorastos et al. 2014; Steine et al. 2017; Vinkers et al. 2014).

The effect of stress on the vulnerability of depressive disorder can happen as early as in utero. Epidemiological studies of human populations and studies on animal models indicate that prenatal maternal stress is linked with adverse health outcomes in the offspring (Beydoun and Saftlas 2008). More and more studies show that prenatal maternal psychological stress (e.g., mother's anxiety and depression) is correlated with increased risk of psychological disorders including depression onset in offspring (Babenko et al. 2015; Barker et al. 2011; Braithwaite et al. 2014; Fatima et al. 2017; Monti and Rudolph 2017; Slykerman et al. 2015; Weinstock 2017), reflecting an intergenerational effect of stress. During pregnancy, maternal stress can exert a major impact on brain development and thereby contribute to the pathogenesis of neuropsychiatric illnesses including depression in offspring. In this process, cortisol which is highly lipophilic to pass placenta has generally been identified as the major mediator in transferring maternal stress to the fetus (Osborne et al. 2018; Rakers et al. 2017; Van den Bergh et al. 2005). Excessive maternal cortisol could persistently impair the development of the fetal HPA axis and crucial brain areas including amygdala, hippocampus, and frontal cortex (Nemoda and Szyf 2017; Van den Bergh 2011; Van den Bergh et al. 2005). Importantly, exogenous glucocorticoids treatment of pregnant women can result in similar deficits in offspring development to those related to postpartum depression, supporting the view that the stress response may mediate the adverse effects of maternal depression on offspring (Conti et al. 2017; Owen et al. 2005). Depressive symptoms during pregnancy are very common, and about 10% of pregnant women have major depressive disorder, which can be increased to higher frequency (about 17%) in low- and middle-income countries (Bennett et al. 2004; Fellmeth et al. 2017; Gelaye et al. 2016). So the impact of maternal depression on vulnerability of depressive disorders in offspring has drawn a lot of attention. It is considered that the prenatal maternal depression could also have a long-term indirect effect on offspring depression via poor physical health in early childhood and its psychosocial consequences, in which the prenatal maternal depressive symptoms predict worse physical health of offspring during early childhood to further predict increased health-related stress and poor social functioning, finally predicting increased risk of depressive symptoms later in young adulthood (Raposa et al. 2014).

Childhood and adolescence period are particularly sensitive to stressful life events, which increase the vulnerability to depressive disorder (Andersen and Teicher 2008; Fuhrmann et al. 2015; Laceulle et al. 2014). During these periods, neural plasticity, brain regions involved in regulating emotion and mediating the stress response appear to be particularly sensitive to the effects of stressful events. Excessive or chronic stress evoked by intense or chronic stressful life events (such as child abuse and neglect, parental loss, marital conflict of parents) during childhood and adolescence can have long-lasting neurobiological effects and increase the risk for psychological disorders including depression (Heim and Binder 2012; Infurna et al. 2016; Lopizzo et al. 2015; Mandelli et al. 2015; Oldehinkel et al. 2014). It is suggested that early programming of neurobiological systems in regulating emotion and stress responses can mediate the increased stress vulnerability and depression risk in later life (Chen and Baram 2016; Heim and Binder 2012; Lopizzo et al. 2015; van Bodegom et al. 2017).

During early childhood, postpartum maternal depression, which is more common than prenatal depression, can also negatively impact offspring mental health. So the intergenerational effects of maternal stress also happen at early childhood. Research focused on the impact of maternal postnatal depression on offspring depressive psychopathology has documented a link between them, in which offspring of postnatally depressed mothers are at increased vulnerability for depression (Halligan et al. 2007b; Hammen and Brennan 2003; Murray et al. 2011; Pearson et al. 2013; Sanger et al. 2015). Elevated basal levels of the cortisol have been found in offspring of mothers with postnatal depression and been associated with the presence of maternal postnatal depression (Brennan et al. 2008; Essex et al. 2002; Halligan et al. 2004; Murray et al. 2010). And biological sensitivity to social stress in adulthood is also increased in the offspring of mothers with postnatal depression, as indicated by greater cortisol reactivity to the stress test (Barry et al. 2015). The over-activated HPA axis is supposed to mediate the deleterious impact of maternal depression on offspring in early childhood (Halligan et al. 2007a; Murray et al. 2010). Studies in animal models also suggest that hyperactivity of the HPA axis may mediate the adverse effects of maternal depression on offspring behavior (Maguire and Mody 2016).

As humans are social beings, keeping good social relationship is fundamentally important and beneficial for one's life. Thus, any stimulant disrupting or threatening one's social relationship, esteem, or sense of belonging in a social group can lead to stress. Social stress is the most frequent and intense stress experienced by us in daily lives, including life events characterized by abrupt and severe incidents (e.g., death of loved one, sexual assault), chronic strains (e.g., marital conflict, unemployment, bullying), and daily hassles (e.g., traffic jams, argument). Persistent social stress such as workplace bullying and adolescent bullying has been shown to increase the risk of developing mental disorders such as anxiety and depression in stressed subjects (Pitney et al. 2016; Ttofi 2015; Williams et al. 2017). A lot of supportive views on the association of social stress with depressive disorders are from animal model studies. Two common social stress paradigms, social defeat, and social isolation are introduced in rodent models to explore the neurological mechanism of social stress and the correlation with mental disorders (Chaouloff 2013; Hollis and Kabbaj 2014; Zanier-Gomes et al. 2015). In social defeat paradigm, social stress can be induced in the nonaggressive male rodent when being attacked by the aggressive intruder. Singly housed animals for 21 days are usually introduced to establish social isolation paradigm. All these social-stressed models can induce an array of behavioral and physiological changes in susceptible rodents, which are reminiscent of depression- and anxiety-related symptoms in human (Chaouloff 2013; Liu et al.

2017b; Zanier-Gomes et al. 2015), suggesting social stress is an important risk factor in pathogenesis of psychiatric disorders including depressive disorders.

As mentioned above, more vulnerability of depressive disorders is observed during childhood and adolescent stages under stressful life events, implicating the age of subjected individual will influence the final aversive outcome of life stress. Besides age sensitivity, as well as event severity and number, the impact of stress on the onset of depressive disorders is also influenced by gender, history of depression (family and individual), and personality characteristics of subjected individuals (Assari and Lankarani 2016; Mazure et al. 2000; Monroe et al. 2014; Morse and Robins 2005; You and Conner 2009). Overall, in response to stressful life events with some exceptions of specific event types, women were found to be more likely than men to develop major depression (Maciejewski et al. 2001). No gender difference was detected in risk for depression associated with loss of beloved ones, marital conflicts, or events corresponding to acute financial or legal difficulties, while increased risk for depression associated with more distant interpersonal losses (death of a close friend or relative) and other types of events (change of residence, physical attack, or life-threatening illness/injury) were found in woman as compared with man (Maciejewski et al. 2001). Elevated risk in women for major depression was found to be related with low but not high level of stress exposure (Kendler et al. 2004). A longitudinal study on the long-term predictive role of stressful life events (SLEs) on the subsequent risk of major depressive disorder found a stronger predictive role of SLEs for subsequent clinical depression for men as compared with women (Assari and Lankarani 2016). In rat model of prenatal stress exposure, increased vulnerability for depression-like behaviors was observed in females (Sickmann et al. 2015). So, gender is an important element to consider when analyzing predictive role of SLE on the risk of depressive disorders. The congruence of cognitive-personality characteristics (sociotropy and autonomy) and life stress can confer susceptibility to depressive disorders, in which congruent interaction of sociotropy with negative interpersonal events and autonomy with negative autonomy events significantly predicted depression (Mazure et al. 2000; Morse and Robins 2005). Moreover, individuals with high neuroticism were at increased risk for major depression and more sensitive to the depressogenic effects of stressful life events (Kendler et al. 2004). Severe life events were significantly associated with current depressive symptoms among women without depression history as compared with women depressed before (You and Conner 2009). Individuals with family history of depression would have more lifetime episodes of depression under a major life event than those without family history of depression (Monroe et al. 2014). And the risk of depression relapse in one's life is higher when the first episode occurs at earlier age or when there is family history of depression (Palazidou 2012). Thus, personality characteristics and depression history are also potential elements to impacting depression onset under stressful life events.

8.4 Genetic Contributions in the Association of Stress with Depressive Disorders

On the etiology of depressive disorders, the interaction of gene–environment is considered to be essential in the development of depressive disorders although environment may have more intense impact (~60%) than genetic factors (~40%). Besides the important role of stressful life experience in predisposing depressive disorders, individual genetic background which can in turn regulate and affect coping mechanism in response to stressful life stimuli is another important effector involved in the final depressive outcome of stressed subjects. So not all people with adverse life exposure develop depression despite the intensity of stressful life events, which is supposed to be due to the moderating effects of genetic factors on individual sensitivity to life stress (Klengel and Binder 2013; Lopizzo et al. 2015; Uher 2014). A large body of research has focused on identifying genetic factors that are interacted with stressful life exposure in modulating vulnerability and severity of depressive disorders (Bleys et al. 2018; Caspi et al. 2003; Dalton et al. 2014; Kessing and Bukh 2013; Klengel and Binder 2013; Lopizzo et al. 2015; Risch et al. 2009; Tafet and Nemeroff 2016).

The first supportive evidence of gene–stressful life experience interaction, or gene–stress interaction, in modulating the risk for depressive disorder development is from the study of Caspi et al. (2003). They found 5-HTTLPR polymorphism involving short or long alleles in the promoter region of serotonin transporter gene (*SERT* or *SLC6A4*) can modulate the influence of stressful life experience on depression, in which subjects with short allele of 5-HTTLPR (homozygotes or heterozygotes) exhibited more depressive symptoms, diagnosable depression, and suicidality than individuals with long allele homozygotes of 5-HTTLPR when exposed to life stress (Caspi et al. 2003). Short allele variant of 5-HTTLPR is associated with lower promoter transcriptional efficiency of *5-HTT* gene as compared with long allele (Heils et al. 1996; Hranilovic et al. 2004). The association of 5-HTTLPR short variant with increased vulnerability to depressive disorders under stressful life exposure is supported by a lot of studies replicating the results of Caspi et al., but still controversial as some others failed to repeat it (Bleys et al. 2018; Caspi et al. 2010; Chipman et al. 2007; Dalton et al. 2014; Karg et al. 2011; Risch et al. 2009; Saul et al. 2018; Sharpley et al. 2014). This disagreement could be mainly due to the methodological differences in the assessment of stress and depression between studies (Wankerl et al. 2010).

Another gene modulating stress impact on the vulnerability of depressive disorders is brain-derived neurotrophic factor (BDNF), an important factor involved in normal brain development and function, as well as pathological changes of brain structure and function in brain disorders including depression (Begni et al. 2017; Brunoni et al. 2008; Bus and Molendijk 2016; Lu et al. 2014). Stress is hypothesized to decrease BDNF activity to result in reduced function of brain regions particularly linked to emotion processing and cognition (e.g., hippocampus, amygdala, neocortex), which finally alters mood and may cause depression (Martinowich et al. 2007; Molendijk et al. 2014; Stein et al. 2008). The *BDNF* gene has been implicated in stress and depression vulnerability in human, even animal model (Blugeot et al. 2011; Groves 2007; Homberg et al. 2014; La Greca et al. 2013; Seo et al. 2016). Many studies on the effect of *BDNF* gene in predictive role of life stress to depression are focusing on BDNF Val66Met polymorphism, which is valine to methionine substitution at codon 66 (Val66Met) of *BDNF* gene due to a functional SNP (rs6265) in the promoter region, and found a significant interaction between BDNF Val66Met polymorphism and stressful life experience in depression(Gatt et al. 2009; He et al. 2018; Hosang et al. 2014; Jiang et al. 2013; Lopizzo et al. 2015; Zhao et al. 2018). Such Val66Met polymorphism affects BDNF activity including intracellular distribution, packaging, and release of BDNF protein, in which Met allele leads to adverse effect of BDNF activity (Egan et al. 2003). The Met allele of BDNF Val66Met polymorphism is also linked with impaired memory, reduced brain volume, and harm avoidance (Jiang et al. 2005; Lamb et al. 2015; Toh et al. 2018). Interestingly, individuals with Met allele (heterozygotes or homozygotes) show elevated evening cortisol levels and a significantly attenuated HPA axis response to acute psychosocial stressor as compared with Val/Val genotype (Alexander et al. 2010; Shalev et al. 2009; Vinberg et al. 2009). Transgenic mice with Met allele of BDNF Val66Met polymorphism exhibited increased anxiety under stress conditions (Chen et al. 2006). More prominent effects of Met allele are indicated in the impact of early life stress on depression, while the impact of stress in adulthood on depressive symptoms seems more significant in Val/Val genotype individuals, which might due to the different functions of BDNF at different life stages (Aguilera et al. 2009; Gatt et al. 2009; He et al. 2018; Jiang et al. 2013). Meta-analyses consistently indicated that the Met allele of BDNF Val66Met polymorphism significantly moderates the relationship between life stress and depression (Hosang et al. 2014; Zhao et al. 2018). Furthermore, a combinatory effect of 5-HTTLPR and BDNF Val66Met is implicated to interact with early life stress (childhood adversity) in relation to depression, as indicated by more common depressive symptoms among carriers of either the ss/sl +Val/Val or the ll+Met genotypes in the presence of early life adversities (Comasco et al. 2013).

Key factors involved in stress response have also been studied as possible genetic risking factors for depression vulnerability under life stress exposure. Polymorphisms within four critical factors involved in stress including glucocorticoid receptor (GR) gene, mineralocorticoid receptor (MR) gene, FK506 binding protein 51 gene (*FKBP5*), and corticotropin-releasing hormone receptor 1 gene (*CRHR1*) are suggested to be risk factors in predicting depression (Claes 2009; de Kloet et al. 2016; Grabe et al. 2010; Grimm et al. 2017; Szczepankiewicz et al. 2014). During stress response, stress hormone cortisol will be released to bind with GR, while more and more binding of cortisol with GR provides signal for reduction of corticotropin-releasing hormone (CRH) and adrenocorticotropic hormone release. GR is kept in an inactive state by binding with FKBP5 without cortisol, which will be activated by binding with cortisol to trigger GR-dependent gene expression. FKBP5 is a significant player in stress response through regulating the GR receptor sensitivity to cortisol, and chronic stress was found to increase FKBP5 expression to generate greater inhibition of GR sensitivity (Guidotti et al. 2013; O'Leary et al. 2013). A dysregulation of GR function, called GR resistance, is a common feature of depressed subjects and is related with HPA axis hyperactivity (Juruena et al. 2004; Rodriguez et al. 2016). Binding of cortisol with MR receptor will exert tonic inhibitory effects on basal HPA axis activity, and MR is found to be an important stress modulator and influences basal as well as stress-induced HPA axis activity.

Several GR gene polymorphisms (BclI RFLP, N363S, ER22/23EK, 9β, TthI-III, NR3C1-1) and MR gene polymorphisms (-2C/G and I180V) are implicated to affect HPA axis activity to psychosocial stress, and interact with stressful life experience in predicting the development of stress-related disorders like depression (Bet et al. 2009; de Kloet et al. 2016; Derijk and de Kloet 2008; Hardeveld et al. 2015; Juruena et al. 2015; Kumsta et al. 2007; ter Heegde et al. 2015; Vinkers et al. 2015; Wust et al. 2004). Similarly, single nucleotide polymorphisms in *FKBP5* gene (rs1360780, rs3800373, rs9296158, rs4713916, rs9470080), which will affect FKBP5 basal level, are also shown to modulate the risk of depression to stressful life experience (Appel et al. 2011; Binder 2009; Kohrt et al. 2015; Lavebratt et al. 2010; Menke et al. 2013; Szczepankiewicz et al. 2014; Zimmermann et al. 2011). Besides, single nucleotide polymorphisms within *CRHR1* gene (rs4792887, rs12936511, rs4792887, rs17689882, rs7209436, rs110402, and rs242924) can modulate predictive role of early life stress in depression (Grabe et al. 2010; Kranzler et al. 2011; Polanczyk et al. 2009; Rogers et al. 2013; Wasserman et al. 2009). Interaction of 5-HTTLPR S allele with CRHR1 haplotypes moderates effect of child abuse on predicting depressive symptoms (Ressler et al. 2010).

Taken together, in the association of stress with depressive disorders development, genetic element is an important risking factor in increasing vulnerability of depression under environmental life stress. It should be kept in mind that different risking genes will interact with each other and with stressful life events to exert cumulative role in predisposing depressive disorders.

8.5 Epigenetic Modification in Linking Stress with Depression

Epigenetic regulation means stable change of gene expression and function via chromatin structure alteration or noncoding RNA regulation without change of DNA sequence, which is heritable and can be modified by environmental stimuli. Epigenetic modification caused by environment factors, such as stress especially early life stress, is considered as a key mechanism through which environmental risk factors can modify gene function without affecting DNA sequence to cause brain structural and functional alteration, finally increasing the vulnerability to mental disorders including depressive disorders (Bagot et al. 2014; Bakusic et al. 2017; Roy et al. 2017b; Sun et al. 2013; Vialou et al. 2013). And epigenetic mechanism also underlies extraordinary interindividual variability in sensitivity to stressful life exposure and variability in depressive symptoms (Bagot et al. 2014; Sun et al. 2013).

Among epigenetic modification, DNA methylation and histone acetylation are extensively studied in the context of long-term effects of stressful life events on gene transcription and on mental disorders etiology especially depression (Bagot et al. 2014; Bakusic et al. 2017; Deussing and Jakovcevski 2017; Fatima et al. 2017; Klengel et al. 2014; Li et al. 2017; Liu et al. 2014; Nemoda and Szyf 2017; Weder et al. 2014; Zheng et al. 2016). DNA methylation changes within promoter and enhancer regions of genes are closely related to the alteration of gene expression, as they reduce the entry of transcription factors into regulatory elements to promote silencing or downregulation of gene expression. The role of DNA methylation in several genetic risk genes, such as *BDNF*, *5-HTT*, *GR*, and *FKBP5*, has been extensively investigated in the pathology of stress-related depressive disorders. Acetylation of histone is associated with greater level of gene expression, and its alteration in mental disorders and the role in vulnerability to depressive disorder under stressful life exposure are also widely studied. For example, in depressed rat model with chronically unpredicted stress treatment, increased depression-like behaviors were associated with decreased acetylated histone (H3K14, H3K23, H4K16), suggesting the decrease of histone acetylation modification level may contribute to the mechanism of depression-like behaviors (Li et al. 2017).

Environmental factors may also modify gene expression via changing microRNAs (miRNAs) synthesis (Carthew and Sontheimer 2009), a type of small noncoding RNAs (20–22 nt), which play a major role in posttranscriptional regulation of gene expression. Individual miRNAs are able to target hundreds of different mRNAs, and a single gene can be modulated by several different miRNAs. Many studies have observed expression change of miRNAs targeting to important genes for brain development and function in rodent brain with life stress exposure, indicating the contribution of miRNAs in epigenetic modulation by stress to affect the vulnerability of depressive disorders (Babenko et al. 2015; Dirven et al. 2017; Higuchi et al. 2016; Liu et al. 2017a; Lopizzo et al. 2015; Ma et al. 2016). With early life stress exposure, the expression level of two miRNAs which are closely related to 5-HT neurotransmitter system was found to be altered in rat brain, with a reduction of miRNA-135a in the prefrontal cortex and increase of miRNA-16 in the hippocampus (Liu et al. 2017a). In depression-like mice induced by chronic unpredictable mild stress, upregulation of miRNAs responsible for downregulating genes functioning in GABAergic synapses, dopaminergic synapses, myelination, synaptic vesicle cycle, and neuronal growth were detected in the medial prefrontal cortex (Ma et al. 2016). Interestingly, expression change of several miRNAs (such as miR-709, miR-132, miR-124) under life stress can persist for a long time (Babenko et al. 2012; Liu et al. 2017a; Uchida et al. 2010), which might be due to the epigenetic changes of miRNAs, as hypermethylation in 31 miRNAs from adult males exposed to childhood abuse were detected (Suderman et al. 2014).

Due to the high sensitivity of depression risk gene *BDNF* to stress and its decreased level in depressed patients and animal models, epigenetic modification of *BDNF* gene attracted much attention. In rodents studies, stress exposure at gestation, early life and adult stages can induce epigenetic change of *BDNF* gene to decrease its expression level in several brain regions including hippocampus and prefrontal cortex, which could increase the vulnerability to stress or mental disorder including depression (Roth et al. 2009; Seo et al. 2016; Weinstock 2017; Xu et al. 2018; Zheng et al. 2016). In the hippocampus of rodents with stress exposure, increased methylation within *BDNF* promoter, elevated level of DNA (cytosine-5)-methyltransferase and histone deacetylase (such as HDAC1, HDAC2), as well as decreased expression of acetylated histone and its binding on specific *BDNF* promoters were observed (Seo et al. 2016; Zheng et al. 2016). Similarly, an increase of methylation at CpG sites within the promoter region of the *BDNF* exon IV was detected in the prefrontal cortex of mice with early maltreatment from adolescence till adult stages, which account for the reduced *BDNF* mRNA level in this brain region (Roth et al. 2009). And such increased methylation of *BDNF* gene promoter in the frontal cortex can even be transmitted from one generation to the next generation, as this epigenetic modification was also observed in the offspring of females with early maltreatment (Roth et al. 2009). Changed epigenetic modification of *BDNF* has also been detected in human brain or peripheral blood or saliva from depressive subjects. Fuchikami et al. reported a significant difference of CpG methylation within *BDNF* gene-specific promoter of exon I but not IV, in peripheral blood between depressed patients and controls (Fuchikami et al. 2011). Roy et al. also observed that *BDNF* gene CpG methylation was significantly increased in peripheral blood of depressed patients (Roy et al. 2017b). An increased methylation in *BDNF* gene promoter of exon IV was found to be associated with reduced *BDNF* expression in Wernicke's area of postmortem brain from suicide victims with major depression (Keller et al. 2010). Significant difference of *BDNF* gene methylation from saliva sample between maltreatment children and control was observed, which may confer risk for depression in children (Weder et al. 2014). In contrast, prenatal maternal stress (maternal depression) significantly predicted decreased DNA methylation in *BDNF* IV in both male and female infants, which was proposed to reflect a molecular basis for the rapid maturation induced by adverse prenatal events (Braithwaite et al. 2015). It is noteworthy that BDNF dysregulation may also be contributed by alterations of miRNAs level. In depressed patients and depression-like animal models with early life stress exposure, upregulation of several miRNAs (miR-132, miR-182, miR-16) which can regulate *BDNF* translation are correlated with decreased *BDNF* expression in brain or serum (Bai et al. 2012; Li et al. 2013, 2016; Su et al. 2015).

Another genetic risk gene for depressive disorders, 5-HTT (*SLC6A4* or *SERT*), has also been investigated on its epigenetic modification correlated with life stress exposure and depression pathology. Since *SLC6A4* gene plays an important role in the normal development and function of brain regions, and its malfunctioning is supposed to be involved in the pathology of mental disorders including depressive disorders, the alteration of 5-HTT expression will be closely related to brain structure and function, behavior change, as well as vulnerability of mental disorders. The expression of 5-HTT is dependent not only on genetic variation (5-HTTLPR polymorphism) but also epigenetic modification in response to environmental stress exposure (such as hypermethylation with early life exposure or chronic stress) (Palma-Gudiel and Fananas 2017; Vijayendran et al. 2012). Besides that short(s) 5HTTLPR allele is associated with lower amounts of *5-HTT* mRNA transcription, several in vitro and in vivo studies consistently observed higher level of specific-site methylation in *SLC6A4* promoter which is associated with its lower mRNA expression (Abdolmaleky et al. 2014; Palma-Gudiel and Fananas 2017; Philibert et al. 2008; Wang et al. 2012). Methylation level of *SLC6A4* has been associated with a number of environmental stresses originated from maternal stress, childhood trauma and abuse, and other early life stress, which is confirmed as an epigenetic biomarker of early adversity exposures in human (Palma-Gudiel and Fananas 2017; Provenzi et al. 2016). In patients with major depression, higher *5-HTT* promoter methylation status was found to be significantly associated with childhood adversities and worse clinical presentation (family history of depression, higher perceived stress, and more severe psychopathology) (Kang et al. 2013). And the increased methylation within *SLC6A4* promoter can affect brain responses to negative stimuli, as indicated by the results that twins with higher peripheral SLC6A4 methylation levels showed greater orbitofrontal cortical (OFC) activity, greater connectivity of left amygdala–anterior cingulate cortex (ACC) and left amygdala–right OFC in response to sadness, as well as greater ACC–left amygdala and ACC–left insula connectivity in response to fearful stimuli (Ismaylova et al. 2018).

Further, miRNAs also contribute to the epigenetic regulation of 5-HTT expression in response to life stress exposure. In depression-like rat model induced by chronic mild stress, the expression of seven miRNAs targeting 5-HTT (miR-18a-5p, miR-34a-5p, miR-135a-5p, miR-195-5p, miR-320-3p, miR-674-3p, miR-872-5p) was increased in ventral tegmental area (VTA) but decreased in prefrontal cortex, and more profound increase of these miRNA in VTA was detected in resilient rat accompanied by lower 5-HTT level (Zurawek et al. 2017). Cell line and animal studies demonstrated that miR-16 (also targeting on BDNF) and miR-15a can also negatively regulate 5-HTT expression, which might be associated with major depressive disorder, as alteration of these two miRNAs level can be detected under life stress and in depression patients and animal models (Baudry et al. 2010; Launay et al. 2011; Moya et al. 2013; Shao et al. 2018; Song et al. 2015). In the same paradigm of chronic mild stress-induced depression, miR-16 level in cerebrospinal fluid (CSF) and raphe nuclei was significantly decreased with obviously increase of 5-HTT protein in raphe (Shao et al. 2018). And with anti-miR-16 treatment, rats exhibited depression-like behaviors, extremely lower CSF miR-16, and obviously higher raphe 5-HTT protein than control (Song et al. 2015). Similarly, CSF miR-16 is also decreased in major depression patients (Song et al. 2015). Interestingly, antidepressants drugs like selective serotonin reuptake inhibitors (SSRI) can increase miR-16 levels in serotonergic raphe nuclei, which was accompanied by reduction of 5-HTT expression in raphe (Baudry et al. 2010; Yang et al. 2017), supporting the role of miR-16 in depressive disorders pathology.

Due to the essential role of HPA axis in stress response under aversive environment stimuli, the involved genes are also investigated on their epigenetic modulation in stress exposure and depressive disorders onset. Numerous studies have focused on the relationship of methylation in glucocorticoid receptor (GR) gene, also known as NR3C1 (nuclear receptor subfamily 3, group C, member 1), with early life stress. And increased methylation of GR promoter in brain or periphery samples from rodent or human subjected with early life stress (e.g., prenatal or postnatal maternal stress, paternal stress, child abuse, low maternal care) are consistently reported among these studies (Bockmuhl et al. 2015; Braithwaite et al. 2015; Efstathopoulos et al. 2018; Farrell et al. 2018; Smart et al. 2015; Turecki and Meaney 2016). Mainly increased methylation of rat NR3C1 exon 17 and human homolog NR3C1 exon 1F are found to be related with early life stress. Prenatal depressive symptoms were shown to significantly predict increased NR3C1 exon 1F DNA methylation in male infants (Braithwaite et al. 2015). Mean NR3C1 exon 1F DNA methylation levels were significantly increased in depressed patients, and the degree of methylation was positively associated with morning cortisol concentrations (Farrell et al. 2018). Meanwhile, DNA methylation level at specific CG sites in the NR3C1 exon 1F was related to childhood emotional abuse severity (Farrell et al. 2018). The methylation level of NR3C1 exon 1F was also significantly increased in the hippocampus from suicide victims with a history of childhood abuse, as compared with those from suicide victims without childhood abuse or with control samples (McGowan et al. 2009). Furthermore, the severity, number, and types of childhood maltreatments positively correlated with NR3C1 exon 1F methylation (Perroud et al. 2011).

The GR co-chaperone, *FKBP5*, which is also one of the genetic risk factors in depressive disorders, has likewise been widely investigated on its epigenetic changes induced by life stress especially early life stress. The methylation of the *FKBP5* gene is associated with reductions in transcription; conversely, *FKBP5* demethylation will increase FKBP5 level to interfere GR sensitivity to cortisol. In human, childhood maltreatment has been associated with demethylation in *FKBP5* gene, as indicated by lower DNA methylation at *FKBP5* intron 7 in the blood or saliva of maltreated subjects as compared with those unexposed (Klengel et al. 2013; Non et al. 2016; Tyrka et al. 2015). Since polymorphisms of *FKBP5* gene have been shown to influence glucocorticoid receptor sensitivity, stress response regulation, and depression risk in traumatized subjects, the epigenetic modification in these functional variants especially rs1360780 also attracts much attentions (Han et al. 2017; Mulder et al. 2017; Tyrka et al. 2015). For example, *FKBP5* methylation was positively correlated with the thickness of the right transverse frontopolar gyrus in the C allele (FKBP5 rs1360780 polymorphism) homozygote group, suggesting that the FKBP5 genetic and epigenetic changes may affect morphology of emotion-related brain regions, which may be involved in depression development (Han et al. 2017).

Besides DNA methylation affecting HPA axis responsiveness, miRNAs which target key factors in HPA axis have also been shown to be changed under life stress. The miR-124, an important effector in brain development, can also target GR. Chronic stress led to the upregulation of miR-124 in the hippocampus and basolateral amygdala of rats (Bahi et al. 2014; Xu et al. 2017), and a higher level of miR-124 was detected in prefrontal cortex of depressive-like rats and in serum of major depressive patients (Roy et al. 2017a). In rat model, depressive-like behaviors induced by chronic unpredictable mild stress were positively correlated with the level of miR-124a, whereas GR levels were negatively correlated with miR-124a level in both adolescent and adult brain (Xu et al. 2017). Inhibition of miR-124 by its antagomir was shown to reverse the depressive-like behaviors in mice exposed to chronic corticosterone (Wang et al. 2017). The exact role of increased mir-124 in regulating GR expression and HPA axis response activity under life stress, and in linking stress with depressive disorder need to be investigated. Upregulation of miR-34c in mice exposed to acute and chronic stress was also observed (Haramati et al. 2011), which may regulate stress response by targeting stress-related corticotropin-releasing factor receptor type 1 (CRFR1) mRNA. Under chronic unpredictable mild stress, the expression of another miRNA targeting GR, miR-18a, was found increased in the basolateral amygdala only in adolescent rats but not adult rat (Xu et al. 2017).

Overall, in this part, epigenetic modification including DNA methylation and miRNA on key depression risk genes (*BDNF*, *SERT*, *GR*, and *FKBP5*) are demonstrated. The changes in DNA methylation of these genes or expression of miRNAs targeting on these genes are sensitive to life stress and can generate long-lasting effects on brain structure and function, behavior, and most important the vulnerability to mental disorders including depressive disorders. These epigenetic modifications provide a potential mechanism in linking environmental risk factors such as early life stress with depression disorders onset. It is of noteworthy that these epigenetic regulations will function together with genetic background to affect individual vulnerability to depressive disorders under stressful life events.

8.6 Conclusion

Stress is an adaptive mechanism of individuals to cope with environment aversive stimuli during evolution. However, excessive or prolonged stress triggered by stressful life events will cause deleterious impacts on individuals' both physiological and psychological health. Dysregulated HPA axis activity under overloaded stress is considered to play important role in depression development. More and more evidence have been supporting the causal association of stress, especially the chronic stress and early life stress, with depressive disorders development. Thus, stressful life events are high-risk factors of environment in predisposing depressive disorders.

While the effects of stressful life exposure on the vulnerability to depressive disorders are moderated by individual's genetic background, polymorphism of several high-risk genes including *SERT*, *BDNF*, as well as key genes involved in stress response (*GR*, *FKBP5*, *MR*, and *CRHR1*) have been found to affect aversive outcome of life stress in depression onset. The interaction of environment–gene, meaning the combinatorial function of stressful life exposure with genetic background, actually plays the essential roles in pathology of depressive disorders. Now it is widely accepted that depressive disorders are multifactorial and polygenic mental disorders.

The close association of stressful life exposure to depressive disorders development is also dependent on the epigenetic mechanism. Environmental risk factors will lead to epigenetic modification in depression risk genes (*BDNF*, *SERT*, *GR*, and *FKBP5*) through DNA methylation and miRNA regulation, to generate long-lasting even heritable effects on risk genes expression. Such epigenetic mechanism can cause brain structural and functional alteration, and finally increasing the vulnerability to depressive disorders. Meanwhile, epigenetic regulation will interact with genetic background to affect the predicting role of stressful life events in vulnerability to depressive disorders.

Therefore, as the mediator of environmental risk factors, stress will function together with genetic risk factors and epigenetic modification to generate impacts on brain structure and function, as well as physiology and psychology, to predispose depressive disorders.

References

Abdolmaleky HM, Nohesara S, Ghadirivasfi M, Lambert AW, Ahmadkhaniha H, Ozturk S, Wong CK, Shafa R, Mostafavi A, Thiagalingam S (2014) DNA hypermethylation of serotonin transporter gene promoter in drug naive patients with schizophrenia. Schizophr Res 152:373–380

Abravanel BT, Sinha R (2015) Emotion dysregulation mediates the relationship between lifetime cumulative adversity and depressive symptomatology. J Psychiatr Res 61:89–96

Agorastos A, Pittman, JO, Angkaw AC, Nievergelt CM, Hansen CJ, Aversa LH, Parisi SA, Barkauskas DA, Marine Resiliency Study T, Baker DG (2014) The cumulative effect of different childhood trauma types on self-reported symptoms of adult male depression and PTSD, substance abuse and health-related quality of life in a large active-duty military cohort. J Psychiatr Res 58:46–54

Aguilera M, Arias B, Wichers M, Barrantes-Vidal N, Moya J, Villa H, van Os J, Ibanez MI, Ruiperez MA, Ortet G et al (2009) Early adversity and 5-HTT/BDNF genes: new evidence of gene-environment interactions on depressive symptoms in a general population. Psychol Med 39:1425–1432

Aich P, Potter AA, Griebel PJ (2009) Modern approaches to understanding stress and disease susceptibility: a review with special emphasis on respiratory disease. Int J Gen Med 2:19–32

Alexander N, Osinsky R, Schmitz A, Mueller E, Kuepper Y, Hennig J (2010) The BDNF Val66Met polymorphism affects HPA-axis reactivity to acute stress. Psychoneuroendocrinology 35:949–953

Andersen SL, Teicher MH (2008) Stress, sensitive periods and maturational events in adolescent depression. Trends Neurosci 31:183–191

Appel K, Schwahn C, Mahler J, Schulz A, Spitzer C, Fenske K, Stender J, Barnow S, John U, Teumer A et al (2011) Moderation of adult depression by a polymorphism in the FKBP5 gene and childhood physical abuse in the general population. Neuropsychopharmacology 36:1982–1991

Assari S, Lankarani MM (2016) Stressful life events and risk of depression 25 years later: race and gender differences. Front Public Health 4:49

Avison WR, Turner RJ (1988) Stressful life events and depressive symptoms: disaggregating the effects of acute stressors and chronic strains. J Health Soc Behav 29:253–264

Babenko O, Kovalchuk I, Metz GA (2015) Stress-induced perinatal and transgenerational epigenetic programming of brain development and mental health. Neurosci Biobehav Rev 48:70–91

Babenko O, Golubov A, Ilnytskyy Y, Kovalchuk I, Metz GA (2012) Genomic and epigenomic responses to chronic stress involve miRNA-mediated programming. PLoS ONE 7:e29441

Bagot RC, Labonte B, Pena CJ, Nestler EJ (2014) Epigenetic signaling in psychiatric disorders: stress and depression. Dialogues Clin Neurosci 16:281–295

Bahi A, Chandrasekar V, Dreyer JL (2014) Selective lentiviral-mediated suppression of microRNA124a in the hippocampus evokes antidepressants-like effects in rats. Psychoneuroendocrinology 46:78–87

Bai M, Zhu X, Zhang Y, Zhang S, Zhang L, Xue L, Yi J, Yao S, Zhang X (2012) Abnormal hippocampal BDNF and miR-16 expression is associated with depression-like behaviors induced by stress during early life. PLoS ONE 7:e46921

Bakusic J, Schaufeli W, Claes S, Godderis L (2017) Stress, burnout and depression: a systematic review on DNA methylation mechanisms. J Psychosom Res 92:34–44

Barker ED, Jaffee SR, Uher R, Maughan B (2011) The contribution of prenatal and postnatal maternal anxiety and depression to child maladjustment. Depress Anxiety 28:696–702

Barry TJ, Murray L, Fearon RM, Moutsiana C, Cooper P, Goodyer IM, Herbert J, Halligan SL (2015) Maternal postnatal depression predicts altered offspring biological stress reactivity in adulthood. Psychoneuroendocrinology 52:251–260

Baudry A, Mouillet-Richard S, Schneider B, Launay JM, Kellermann O (2010) miR-16 targets the serotonin transporter: a new facet for adaptive responses to antidepressants. Science 329:1537–1541

Becker JB (2010). Stress and reward. In: Koob GF, Moal ML, Thompson RF (eds) Encyclopedia of behavioral neuroscience. Academic Press, pp 328–333

Begni V, Riva MA, Cattaneo A (2017) Cellular and molecular mechanisms of the brain-derived neurotrophic factor in physiological and pathological conditions. Clin Sci (Lond) 131:123–138

Belda X, Fuentes S, Daviu N, Nadal R, Armario A (2015) Stress-induced sensitization: the hypothalamic-pituitary-adrenal axis and beyond. Stress 18:269–279

Bennett HA, Einarson A, Taddio A, Koren G, Einarson TR (2004) Prevalence of depression during pregnancy: systematic review. Obstet Gynecol 103:698–709

Benson PR (2018) The impact of child and family stressors on the self-rated health of mothers of children with autism spectrum disorder: associations with depressed mood over a 12-year period. Autism 22:489–501

Bet PM, Penninx BW, Bochdanovits Z, Uitterlinden AG, Beekman AT, van Schoor NM, Deeg DJ, Hoogendijk WJ (2009) Glucocorticoid receptor gene polymorphisms and childhood adversity are associated with depression: New evidence for a gene-environment interaction. Am J Med Genet B Neuropsychiatr Genet 150B:660–669

Beydoun H, Saftlas AF (2008) Physical and mental health outcomes of prenatal maternal stress in human and animal studies: a review of recent evidence. Paediatr Perinat Epidemiol 22:438–466

Binder EB (2009) The role of FKBP5, a co-chaperone of the glucocorticoid receptor in the pathogenesis and therapy of affective and anxiety disorders. Psychoneuroendocrinology 34(Suppl 1):S186–S195

Bleys D, Luyten P, Soenens B, Claes S (2018) Gene-environment interactions between stress and 5-HTTLPR in depression: a meta-analytic update. J Affect Disord 226:339–345

Blugeot A, Rivat C, Bouvier E, Molet J, Mouchard A, Zeau B, Bernard C, Benoliel JJ, Becker C (2011) Vulnerability to depression: from brain neuroplasticity to identification of biomarkers. J Neurosci 31:12889–12899

Bockmuhl Y, Patchev AV, Madejska A, Hoffmann A, Sousa JC, Sousa N, Holsboer F, Almeida OF, Spengler D (2015) Methylation at the CpG island shore region upregulates Nr3c1 promoter activity after early-life stress. Epigenetics 10:247–257

Bonfiglio JJ, Inda C, Refojo D, Holsboer F, Arzt E, Silberstein S (2011) The corticotropin-releasing hormone network and the hypothalamic-pituitary-adrenal axis: molecular and cellular mechanisms involved. Neuroendocrinology 94:12–20

Braithwaite EC, Murphy SE, Ramchandani PG (2014) Prenatal risk factors for depression: a critical review of the evidence and potential mechanisms. J Dev Orig Health Dis 5:339–350

Braithwaite EC, Kundakovic M, Ramchandani PG, Murphy SE, Champagne FA (2015) Maternal prenatal depressive symptoms predict infant NR3C1 1F and BDNF IV DNA methylation. Epigenetics 10:408–417

Brennan PA, Pargas R, Walker EF, Green P, Newport DJ, Stowe Z (2008) Maternal depression and infant cortisol: influences of timing, comorbidity and treatment. J Child Psychol Psychiatry 49:1099–1107

Brunoni AR, Lopes M, Fregni F (2008) A systematic review and meta-analysis of clinical studies on major depression and BDNF levels: implications for the role of neuroplasticity in depression. Int J Neuropsychopharmacol 11:1169–1180

Bus BA, Molendijk ML (2016) The neurotrophic hypothesis of depression. Tijdschr Psychiatr 58:215–222

Carthew RW, Sontheimer EJ (2009) Origins and mechanisms of miRNAs and siRNAs. Cell 136:642–655

Caspi A, Hariri AR, Holmes A, Uher R, Moffitt TE (2010) Genetic sensitivity to the environment: the case of the serotonin transporter gene and its implications for studying complex diseases and traits. Am J Psychiatry 167:509–527

Caspi A, Sugden K, Moffitt TE, Taylor A, Craig IW, Harrington H, McClay J, Mill J, Martin J, Braithwaite A et al (2003) Influence of life stress on depression: moderation by a polymorphism in the 5-HTT gene. Science 301:386–389

Chaouloff F (2013) Social stress models in depression research: what do they tell us? Cell Tissue Res 354:179–190

Chapman DP, Whitfield CL, Felitti VJ, Dube SR, Edwards VJ, Anda RF (2004) Adverse childhood experiences and the risk of depressive disorders in adulthood. J Affect Disord 82:217–225

Chen Y, Baram TZ (2016) Toward understanding how early-life stress reprograms cognitive and emotional brain networks. Neuropsychopharmacology 41:197–206

Chen ZY, Jing D, Bath KG, Ieraci A, Khan T, Siao CJ, Herrera DG, Toth M, Yang C, McEwen BS et al (2006) Genetic variant BDNF (Val66Met) polymorphism alters anxiety-related behavior. Science 314:140–143

Chipman P, Jorm AF, Prior M, Sanson A, Smart D, Tan X, Easteal S (2007) No interaction between the serotonin transporter polymorphism (5-HTTLPR) and childhood adversity or recent stressful life events on symptoms of depression: results from two community surveys. Am J Med Genet B Neuropsychiatr Genet 144B:561–565

Chirita AL, Gheorman V, Bondari D, Rogoveanu I (2015) Current understanding of the neurobiology of major depressive disorder. Rom J Morphol Embryol 56:651–658

Claes S (2009) Glucocorticoid receptor polymorphisms in major depression. Ann NY Acad Sci 1179:216–228

Clarke H, Flint J, Attwood AS, Munafo MR (2010) Association of the 5-HTTLPR genotype and unipolar depression: a meta-analysis. Psychol Med 40:1767–1778

Cohen S, Janicki-Deverts D, Miller GE (2007) Psychological stress and disease. JAMA 298:1685–1687

Colman I, Ataullahjan A (2010) Life course perspectives on the epidemiology of depression. Can J Psychiatry 55:622–632

Comasco E, Aslund C, Oreland L, Nilsson KW (2013) Three-way interaction effect of 5-HTTLPR, BDNF Val66Met, and childhood adversity on depression: a replication study. Eur Neuropsychopharmacol 23:1300–1306

Conti M, Spulber S, Raciti M, Ceccatelli S (2017) Depressive-like phenotype induced by prenatal dexamethasone in mice is reversed by desipramine. Neuropharmacology 126:242–249

Dalton ED, Hammen CL, Najman JM, Brennan PA (2014) Genetic susceptibility to family environment: BDNF Val66met and 5-HTTLPR influence depressive symptoms. J Fam Psychol 28:947–956

de Kloet ER, Otte C, Kumsta R, Kok L, Hillegers MH, Hasselmann H, Kliegel D, Joels M (2016) Stress and depression: a crucial role of the mineralocorticoid receptor. J Neuroendocrinol 28

Derijk RH, de Kloet ER (2008) Corticosteroid receptor polymorphisms: determinants of vulnerability and resilience. Eur J Pharmacol 583:303–311

Deussing JM, Jakovcevski M (2017) Histone modifications in major depressive disorder and related rodent models. Adv Exp Med Biol 978:169–183

Dirven BCJ, Homberg JR, Kozicz T, Henckens M (2017) Epigenetic programming of the neuroendocrine stress response by adult life stress. J Mol Endocrinol 59:R11–R31

Dunn EC, Brown RC, Dai Y, Rosand J, Nugent NR, Amstadter AB, Smoller JW (2015) Genetic determinants of depression: recent findings and future directions. Harv Rev Psychiatry 23:1–18

Farrell C, Doolin K, O' Leary N, Jairaj C, Roddy D, Tozzi L, Morris D, Harkin A, Frodl T, Nemoda Z et al (2018) DNA methylation differences at the glucocorticoid receptor gene in depression are related to functional alterations in hypothalamic-pituitary-adrenal axis activity and to early life emotional abuse. Psychiatry Res 265:341–348

Eckenrode J (1984) Impact of chronic and acute stressors on daily reports of mood. J Pers Soc Psychol 46:907–918

Efstathopoulos P, Andersson F, Melas PA, Yang LL, Villaescusa JC, Ruegg J, Ekstrom TJ, Forsell Y, Galanti MR, Lavebratt C (2018) NR3C1 hypermethylation in depressed and bullied adolescents. Transl Psychiatry 8:121

Egan MF, Kojima M, Callicott JH, Goldberg TE, Kolachana BS, Bertolino A, Zaitsev E, Gold B, Goldman D, Dean M et al (2003) The BDNF val66met polymorphism affects activity-dependent secretion of BDNF and human memory and hippocampal function. Cell 112:257–269

Eisenmann ED, Rorabaugh BR, Zoladz PR (2016) Acute stress decreases but chronic stress increases myocardial sensitivity to ischemic injury in rodents. Front Psychiatry 7:71

Essex MJ, Klein MH, Cho E, Kalin NH (2002) Maternal stress beginning in infancy may sensitize children to later stress exposure: effects on cortisol and behavior. Biol Psychiatry 52:776–784

Fatima M, Srivastav S, Mondal AC (2017) Prenatal stress and depression associated neuronal development in neonates. Int J Dev Neurosci 60:1–7

Fellmeth G, Fazel M, Plugge E (2017) Migration and perinatal mental health in women from low- and middle-income countries: a systematic review and meta-analysis. BJOG 124:742–752

Flint J, Kendler KS (2014) The genetics of major depression. Neuron 81:1214

Fuchikami M, Morinobu S, Segawa M, Okamoto Y, Yamawaki S, Ozaki N, Inoue T, Kusumi I, Koyama T, Tsuchiyama K et al (2011) DNA methylation profiles of the brain-derived neurotrophic factor (BDNF) gene as a potent diagnostic biomarker in major depression. PLoS ONE 6:e23881

Fuhrmann D, Knoll LJ, Blakemore SJ (2015) Adolescence as a sensitive period of brain development. Trends Cogn Sci 19:558–566

Gao J, Pan Z, Jiao Z, Li F, Zhao G, Wei Q, Pan F, Evangelou E (2012) TPH2 gene polymorphisms and major depression—a meta-analysis. PLoS ONE 7:e36721

Gatt JM, Burton KL, Williams LM, Schofield PR (2015) Specific and common genes implicated across major mental disorders: a review of meta-analysis studies. J Psychiatr Res 60:1–13

Gatt JM, Nemeroff CB, Dobson-Stone C, Paul RH, Bryant RA, Schofield PR, Gordon E, Kemp AH, Williams LM (2009) Interactions between BDNF Val66Met polymorphism and early life stress predict brain and arousal pathways to syndromal depression and anxiety. Mol Psychiatry 14:681–695

Gelaye B, Rondon MB, Araya R, Williams MA (2016) Epidemiology of maternal depression, risk factors, and child outcomes in low-income and middle-income countries. Lancet Psychiatry 3:973–982

Gilpin NW (2014) Brain reward and stress systems in addiction. Front Psychiatry 5:79

Glover V (2014) Maternal depression, anxiety and stress during pregnancy and child outcome; what needs to be done. Best Pract Res Clin Obstet Gynaecol 28:25–35

Golbidi S, Frisbee JC, Laher I (2015) Chronic stress impacts the cardiovascular system: animal models and clinical outcomes. Am J Physiol Heart Circ Physiol 308:H1476–H1498

Grabe HJ, Schwahn C, Appel K, Mahler J, Schulz A, Spitzer C, Fenske K, Barnow S, Lucht M, Freyberger HJ et al (2010) Childhood maltreatment, the corticotropin-releasing hormone receptor gene and adult depression in the general population. Am J Med Genet B Neuropsychiatr Genet 153B:1483–1493

Grimm S, Wirth K, Fan Y, Weigand A, Gartner M, Feeser M, Dziobek I, Bajbouj M, Aust S (2017) The interaction of corticotropin-releasing hormone receptor gene and early life stress on emotional empathy. Behav Brain Res 329:180–185

Groves JO (2007) Is it time to reassess the BDNF hypothesis of depression? Mol Psychiatry 12:1079–1088

Guidotti G, Calabrese F, Anacker C, Racagni G, Pariante CM, Riva MA (2013) Glucocorticoid receptor and FKBP5 expression is altered following exposure to chronic stress: modulation by antidepressant treatment. Neuropsychopharmacology 38:616–627

Halligan SL, Herbert J, Goodyer IM, Murray L (2004) Exposure to postnatal depression predicts elevated cortisol in adolescent offspring. Biol Psychiatry 55:376–381

Halligan SL, Herbert J, Goodyer I, Murray L (2007a) Disturbances in morning cortisol secretion in association with maternal postnatal depression predict subsequent depressive symptomatology in adolescents. Biol Psychiatry 62:40–46

Halligan SL, Murray L, Martins C, Cooper PJ (2007b) Maternal depression and psychiatric outcomes in adolescent offspring: a 13-year longitudinal study. J Affect Disord 97:145–154

Hammen C (2005) Stress and depression. Annu Rev Clin Psychol 1:293–319

Hammen C, Brennan PA (2003) Severity, chronicity, and timing of maternal depression and risk for adolescent offspring diagnoses in a community sample. Arch Gen Psychiatry 60:253–258

Han KM, Won E, Sim Y, Kang J, Han C, Kim YK, Kim SH, Joe SH, Lee MS, Tae WS et al (2017) Influence of FKBP5 polymorphism and DNA methylation on structural changes of the brain in major depressive disorder. Sci Rep 7:42621

Haramati S, Navon I, Issler O, Ezra-Nevo G, Gil S, Zwang R, Hornstein E, Chen A (2011) MicroRNA as repressors of stress-induced anxiety: the case of amygdalar miR-34. J Neurosci 31:14191–14203

Hardeveld F, Spijker J, Peyrot WJ, de Graaf R, Hendriks SM, Nolen WA, Penninx BW, Beekman AT (2015) Glucocorticoid and mineralocorticoid receptor polymorphisms and recurrence of major depressive disorder. Psychoneuroendocrinology 55:154–163

He SC, Wu S, Wang C, Du XD, Yin G, Jia Q, Zhang Y, Wang L, Soares JC, Zhang XY (2018) Interaction between job stress and the BDNF Val66Met polymorphism affects depressive symptoms in Chinese healthcare workers. J Affect Disord 236:157–163

Heils A, Teufel A, Petri S, Stober G, Riederer P, Bengel D, Lesch KP (1996) Allelic variation of human serotonin transporter gene expression. J Neurochem 66:2621–2624

Heim C, Binder EB (2012) Current research trends in early life stress and depression: review of human studies on sensitive periods, gene-environment interactions, and epigenetics. Exp Neurol 233:102–111

Heim C, Shugart M, Craighead WE, Nemeroff CB (2010) Neurobiological and psychiatric consequences of child abuse and neglect. Dev Psychobiol 52:671–690

Higuchi F, Uchida S, Yamagata H, Abe-Higuchi N, Hobara T, Hara K, Kobayashi A, Shintaku T, Itoh Y, Suzuki T et al (2016) Hippocampal MicroRNA-124 enhances chronic stress resilience in mice. J Neurosci 36:7253–7267

Hollis F, Kabbaj M (2014) Social defeat as an animal model for depression. ILAR J 55:221–232

Homberg JR, Molteni R, Calabrese F, Riva MA (2014) The serotonin-BDNF duo: developmental implications for the vulnerability to psychopathology. Neurosci Biobehav Rev 43:35–47

Horowitz MA, Zunszain PA (2015) Neuroimmune and neuroendocrine abnormalities in depression: two sides of the same coin. Ann NY Acad Sci 1351:68–79

Hosang GM, Shiles C, Tansey KE, McGuffin P, Uher R (2014) Interaction between stress and the BDNF Val66Met polymorphism in depression: a systematic review and meta-analysis. BMC Med 12:7

Hranilovic D, Stefulj J, Schwab S, Borrmann-Hassenbach M, Albus M, Jernej B, Wildenauer D (2004) Serotonin transporter promoter and intron 2 polymorphisms: relationship between allelic variants and gene expression. Biol Psychiatry 55:1090–1094

Infurna MR, Reichl C, Parzer P, Schimmenti A, Bifulco A, Kaess M (2016) Associations between depression and specific childhood experiences of abuse and neglect: a meta-analysis. J Affect Disord 190:47–55

Ismaylova E, Levesque ML, Pomares FB, Szyf M, Nemoda Z, Fahim C, Vitaro F, Brendgen M, Dionne G, Boivin M et al (2018) Serotonin transporter promoter methylation in peripheral cells and neural responses to negative stimuli: a study of adolescent monozygotic twins. Transl Psychiatry 8:147

Januar V, Saffery R, Ryan J (2015) Epigenetics and depressive disorders: a review of current progress and future directions. Int J Epidemiol 44:1364–1387

Jeon SW, Kim YK (2018) Application of assessment tools to examine mental health in workplaces: job stress and depression. Psychiatry Investig 15:553–560

Jiang R, Brummett BH, Babyak MA, Siegler IC, Williams RB (2013) Brain-derived neurotrophic factor (BDNF) Val66Met and adulthood chronic stress interact to affect depressive symptoms. J Psychiatr Res 47:233–239

Jiang X, Xu K, Hoberman J, Tian F, Marko AJ, Waheed JF, Harris CR, Marini AM, Enoch MA, Lipsky RH (2005) BDNF variation and mood disorders: a novel functional promoter polymorphism and Val66Met are associated with anxiety but have opposing effects. Neuropsychopharmacology 30:1353–1361

Juruena MF, Cleare AJ, Pariante CM (2004) The hypothalamic pituitary adrenal axis, glucocorticoid receptor function and relevance to depression. Rev Bras Psiquiatr 26:189–201

Juruena MF, Werne Baes CV, Menezes IC, Graeff FG (2015) Early life stress in depressive patients: role of glucocorticoid and mineralocorticoid receptors and of hypothalamic-pituitary-adrenal axis activity. Curr Pharm Des 21:1369–1378

Kang HJ, Kim JM, Stewart R, Kim SY, Bae KY, Kim SW, Shin IS, Shin MG, Yoon JS (2013) Association of SLC6A4 methylation with early adversity, characteristics and outcomes in depression. Prog Neuropsychopharmacol Biol Psychiatry 44:23–28

Karg K, Burmeister M, Shedden K, Sen S (2011) The serotonin transporter promoter variant (5-HTTLPR), stress, and depression meta-analysis revisited: evidence of genetic moderation. Arch Gen Psychiatry 68:444–454

Keller S, Sarchiapone M, Zarrilli F, Videtic A, Ferraro A, Carli V, Sacchetti S, Lembo F, Angiolillo A, Jovanovic N et al (2010) Increased BDNF promoter methylation in the Wernicke area of suicide subjects. Arch Gen Psychiatry 67:258–267

Kendler KS, Karkowski LM, Prescott CA (1998) Stressful life events and major depression: risk period, long-term contextual threat, and diagnostic specificity. J Nerv Ment Dis 186:661–669

Kendler KS, Karkowski LM, Prescott CA (1999) Causal relationship between stressful life events and the onset of major depression. Am J Psychiatry 156:837–841

Kendler KS, Kuhn J, Prescott CA (2004) The interrelationship of neuroticism, sex, and stressful life events in the prediction of episodes of major depression. Am J Psychiatry 161:631–636

Kessing LV, Bukh JD (2013) Genetics and stressful life events interact in depression. Ugeskr Laeger 175:948–951

Kessler RC, Bromet EJ (2013) The epidemiology of depression across cultures. Annu Rev Public Health 34:119–138

Kishi T, Yoshimura R, Fukuo Y, Okochi T, Matsunaga S, Umene-Nakano W, Nakamura J, Serretti A, Correll CU, Kane JM et al (2013) The serotonin 1A receptor gene confer susceptibility to mood disorders: results from an extended meta-analysis of patients with major depression and bipolar disorder. Eur Arch Psychiatry Clin Neurosci 263:105–118

Kivimaki M, Steptoe A (2018) Effects of stress on the development and progression of cardiovascular disease. Nat Rev Cardiol 15:215–229

Klengel T, Binder EB (2013) Gene-environment interactions in major depressive disorder. Can J Psychiatry 58:76–83

Klengel T, Pape J, Binder EB, Mehta D (2014) The role of DNA methylation in stress-related psychiatric disorders. Neuropharmacology 80:115–132

Klengel T, Mehta D, Anacker C, Rex-Haffner M, Pruessner JC, Pariante CM, Pace TW, Mercer KB, Mayberg HS, Bradley B et al (2013) Allele-specific FKBP5 DNA demethylation mediates gene-childhood trauma interactions. Nat Neurosci 16:33–41

Kohrt BA, Worthman CM, Ressler KJ, Mercer KB, Upadhaya N, Koirala S, Nepal MK, Sharma VD, Binder EB (2015) Cross-cultural gene- environment interactions in depression, post-traumatic stress disorder, and the cortisol awakening response: FKBP5 polymorphisms and childhood trauma in South Asia. Int Rev Psychiatry 27:180–196

Kranzler HR, Feinn R, Nelson EC, Covault J, Anton RF, Farrer L, Gelernter J (2011) A CRHR1 haplotype moderates the effect of adverse childhood experiences on lifetime risk of major depressive episode in African–American women. Am J Med Genet B Neuropsychiatr Genet 156B:960–968

Kumsta R, Entringer S, Koper JW, van Rossum EF, Hellhammer DH, Wust S (2007) Sex specific associations between common glucocorticoid receptor gene variants and hypothalamus-pituitary-adrenal axis responses to psychosocial stress. Biol Psychiatry 62:863–869

Laceulle OM, O'Donnell K, Glover V, O'Connor TG, Ormel J, van Aken MAG, Nederhof E (2014) Stressful events and psychological difficulties: testing alternative candidates for sensitivity. Eur Child Adoles Psy 23:103–113

La Greca AM, Lai BS, Joormann J, Auslander BB, Short MA (2013) Children's risk and resilience following a natural disaster: genetic vulnerability, posttraumatic stress, and depression. J Affect Disord 151:860–867

Lamb YN, McKay NS, Thompson CS, Hamm JP, Waldie KE, Kirk IJ (2015) Brain-derived neurotrophic factor Val66Met polymorphism, human memory, and synaptic neuroplasticity. Wiley Interdiscip Rev Cogn Sci 6:97–108

Larrieu T, Sandi C (2018) Stress-induced depression: is social rank a predictive risk factor? BioEssays 40:e1800012

Laudenslager ML (1987) Psychosocial stress and susceptibility to infectious disease. In: Kurstak E, Lipowski ZJ, Morozov PV (eds) Viruses, immunity, and mental disorders. Springer, US, pp 391–402

Laugharne J, Lillee A, Janca A (2010) Role of psychological trauma in the cause and treatment of anxiety and depressive disorders. Curr Opin Psychiatry 23:25–29

Launay JM, Mouillet-Richard S, Baudry A, Pietri M, Kellermann O (2011) Raphe-mediated signals control the hippocampal response to SRI antidepressants via miR-16. Transl Psychiatry 1:e56

Laursen TM, Musliner KL, Benros ME, Vestergaard M, Munk-Olsen T (2016) Mortality and life expectancy in persons with severe unipolar depression. J Affect Disord 193:203–207

Lavebratt C, Aberg E, Sjoholm LK, Forsell Y (2010) Variations in FKBP5 and BDNF genes are suggestively associated with depression in a Swedish population-based cohort. J Affect Disord 125:249–255

Lee PH, Perlis RH, Jung JY, Byrne EM, Rueckert E, Siburian R, Haddad S, Mayerfeld CE, Heath AC, Pergadia ML et al (2012) Multi-locus genome-wide association analysis supports the role of glutamatergic synaptic transmission in the etiology of major depressive disorder. Transl Psychiatry 2:e184

Li Y, Li S, Yan J, Wang D, Yin R, Zhao L, Zhu Y, Zhu X (2016) miR-182 (microRNA-182) suppression in the hippocampus evokes antidepressant-like effects in rats. Prog Neuropsychopharmacol Biol Psychiatry 65:96–103

Li HY, Jiang QS, Fu XY, Jiang XH, Zhou QX, Qiu HM (2017) Abnormal modification of histone acetylation involved in depression-like behaviors of rats induced by chronically unpredicted stress. NeuroReport 28:1054–1060

Li YJ, Xu M, Gao ZH, Wang YQ, Yue Z, Zhang YX, Li XX, Zhang C, Xie SY, Wang PY (2013) Alterations of serum levels of BDNF-related miRNAs in patients with depression. PLoS ONE 8:e63648

Lin YL, Wang S (2014) Prenatal lipopolysaccharide exposure increases depression-like behaviors and reduces hippocampal neurogenesis in adult rats. Behav Brain Res 259:24–34

Lindert J, von Ehrenstein OS, Grashow R, Gal G, Braehler E, Weisskopf MG (2014) Sexual and physical abuse in childhood is associated with depression and anxiety over the life course: systematic review and meta-analysis. Int J Public Health 59:359–372

Liu Y, Liu D, Xu J, Jiang H, Pan F (2017a) Early adolescent stress-induced changes in prefrontal cortex miRNA-135a and hippocampal miRNA-16 in male rats. Dev Psychobiol 59:958–969

Liu YY, Zhou XY, Yang LN, Wang HY, Zhang YQ, Pu JC, Liu LX, Gui SW, Zeng L, Chen JJ et al (2017b) Social defeat stress causes depression-like behavior with metabolite changes in the prefrontal cortex of rats. PLoS ONE 12:e0176725

Liu D, Qiu HM, Fei HZ, Hu XY, Xia HJ, Wang LJ, Qin LJ, Jiang XH, Zhou QX (2014) Histone acetylation and expression of mono-aminergic transmitters synthetases involved in CUS-induced depressive rats. Exp Biol Med (Maywood) 239:330–336

Lopez-Leon S, Janssens AC, Gonzalez-Zuloeta Ladd AM, Del-Favero J, Claes SJ, Oostra BA, van Duijn CM (2008) Meta-analyses of genetic studies on major depressive disorder. Mol Psychiatry 13:772–785

Lopizzo N, Bocchio Chiavetto L, Cattane N, Plazzotta G, Tarazi FI, Pariante CM, Riva MA, Cattaneo A (2015) Gene-environment interaction in major depression: focus on experience-dependent biological systems. Front Psychiatry 6:68

Lu B, Nagappan G, Lu Y (2014) BDNF and synaptic plasticity, cognitive function, and dysfunction. Handb Exp Pharmacol 220:223–250

Lucassen PJ, Pruessner J, Sousa N, Almeida OF, Van Dam AM, Rajkowska G, Swaab DF, Czeh B (2014) Neuropathology of stress. Acta Neuropathol 127:109–135

Lueboonthavatchai P (2009) Role of stress areas, stress severity, and stressful life events on the onset of depressive disorder: a case-control study. J Med Assoc Thai 92:1240–1249

Luo Y, Cao Z, Wang D, Wu L, Li Y, Sun W, Zhu Y (2014) Dynamic study of the hippocampal volume by structural MRI in a rat model of depression. Neurol Sci 35:1777–1783

Ma K, Guo L, Xu A, Cui S, Wang JH (2016) Molecular mechanism for stress-induced depression assessed by sequencing miRNA and mRNA in medial prefrontal Cortex. PLoS ONE 11:e0159093

Maciejewski PK, Prigerson HG, Mazure CM (2001) Sex differences in event-related risk for major depression. Psychol Med 31:593–604

Maguire J, Mody I (2016) Behavioral deficits in Juveniles mediated by maternal stress hormones in mice. Neural Plast 2016:2762518

Mandelli L, Petrelli C, Serretti A (2015) The role of specific early trauma in adult depression: A meta-analysis of published literature: childhood trauma and adult depression. Eur Psychiatry 30:665–680

Martinowich K, Manji H, Lu B (2007) New insights into BDNF function in depression and anxiety. Nat Neurosci 10:1089–1093

Mazure CM, Bruce ML, Maciejewski PK, Jacobs SC (2000) Adverse life events and cognitive-personality characteristics in the prediction of major depression and antidepressant response. Am J Psychiatry 157:896–903

McGowan PO, Sasaki A, D'Alessio AC, Dymov S, Labonte B, Szyf M, Turecki G, Meaney MJ (2009) Epigenetic regulation of the glucocorticoid receptor in human brain associates with childhood abuse. Nat Neurosci 12:342–348

McGuffin P, Katz R, Watkins S, Rutherford J (1996) A hospital-based twin register of the heritability of DSM-IV unipolar depression. Arch Gen Psychiatry 53:129–136

Menke A, Klengel T, Rubel J, Bruckl T, Pfister H, Lucae S, Uhr M, Holsboer F, Binder EB (2013) Genetic variation in FKBP5 associated with the extent of stress hormone dysregulation in major depression. Genes Brain Behav 12:289–296

Molendijk ML, Spinhoven P, Polak M, Bus BA, Penninx BW, Elzinga BM (2014) Serum BDNF concentrations as peripheral manifestations of depression: evidence from a systematic review and meta-analyses on 179 associations (N = 9484). Mol Psychiatry 19:791–800

Monroe SM, Slavich GM, Gotlib IH (2014) Life stress and family history for depression: the moderating role of past depressive episodes. J Psychiatr Res 49:90–95

Monti JD, Rudolph KD (2017) Maternal depression and trajectories of adolescent depression: the role of stress responses in youth risk and resilience. Dev Psychopathol 29:1413–1429

Morse JQ, Robins CJ (2005) Personality–life event congruence effects in late-life depression. J Affect Disord 84:25–31

Mothersill O, Donohoe G (2016) Neural effects of social environmental stress—an activation likelihood estimation meta-analysis. Psychol Med 46:2015–2023

Moya PR, Wendland JR, Salemme J, Fried RL, Murphy DL (2013) miR-15a and miR-16 regulate serotonin transporter expression in human placental and rat brain raphe cells. Int J Neuropsychopharmacol 16:621–629

Mulder RH, Rijlaarsdam J, Luijk MP, Verhulst FC, Felix JF, Tiemeier H, Bakermans-Kranenburg MJ, Van Ijzendoorn MH (2017) Methylation matters: FK506 binding protein 51 (FKBP5) methylation moderates the associations of FKBP5 genotype and resistant attachment with stress regulation. Dev Psychopathol 29:491–503

Mullins N, Power RA, Fisher HL, Hanscombe KB, Euesden J, Iniesta R, Levinson DF, Weissman MM, Potash JB, Shi J et al (2016) Polygenic interactions with environmental adversity in the aetiology of major depressive disorder. Psychol Med 46:759–770

Murray F, Smith DW, Hutson PH (2008) Chronic low dose corticosterone exposure decreased hippocampal cell proliferation, volume and induced anxiety and depression like behaviours in mice. Eur J Pharmacol 583:115–127

Murray L, Halligan SL, Goodyer I, Herbert J (2010) Disturbances in early parenting of depressed mothers and cortisol secretion in offspring: a preliminary study. J Affect Disord 122:218–223

Murray L, Arteche A, Fearon P, Halligan S, Goodyer I, Cooper P (2011) Maternal postnatal depression and the development of depression in offspring up to 16 years of age. J Am Acad Child Adolesc Psychiatry 50:460–470

Music S, Rossell SL (2016) Stress, memory, and memory impairment. In: Fink G (ed) Stress: concepts, cognition, emotion, and behavior. Academic Press, pp 145–152

Myers B, McKlveen JM, Herman JP (2014) Glucocorticoid actions on synapses, circuits, and behavior: implications for the energetics of stress. Front Neuroendocrinol 35:180–196

Myruski S, Denefrio S, Dennis-Tiwary TA (2018) Stress and emotion regulation: the dynamic fit model. In: Harkness K, Hayden EP (eds) The Oxford handbook of stress and mental health. Oxford University Press

Nemeroff CB (2016) Paradise lost: the neurobiological and clinical consequences of child Abuse and Neglect. Neuron 89:892–909

Nemoda Z, Szyf M (2017) Epigenetic alterations and prenatal maternal depression. Birth Defects Res 109:888–897

Ngoupaye GT, Yassi FB, Bahane DAN, Bum EN (2018) Combined corticosterone treatment and chronic restraint stress lead to depression associated with early cognitive deficits in mice. Metab Brain Dis 33:421–431

Non AL, Hollister BM, Humphreys KL, Childebayeva A, Esteves K, Zeanah CH, Fox NA, Nelson CA, Drury SS (2016) DNA methylation at stress-related genes is associated with exposure to early life institutionalization. Am J Phys Anthropol 161:84–93

Northoff G (2013) Gene, brains, and environment-genetic neuroimaging of depression. Curr Opin Neurobiol 23:133–142

Ohlin B, Nilsson PM, Nilsson JA, Berglund G (2004) Chronic psychosocial stress predicts long-term cardiovascular morbidity and mortality in middle-aged men. Eur Heart J 25:867–873

Oldehinkel AJ, Ormel J, Verhulst FC, Nederhof E (2014) Childhood adversities and adolescent depression: a matter of both risk and resilience. Dev Psychopathol 26:1067–1075

O'Leary JC 3rd, Zhang B, Koren J 3rd, Blair L, Dickey CA (2013) The role of FKBP5 in mood disorders: action of FKBP5 on steroid hormone receptors leads to questions about its evolutionary importance. CNS Neurol Disord: Drug Targets 12:1157–1162

Osborne S, Biaggi A, Chua TE, Du Preez A, Hazelgrove K, Nikkheslat N, Previti G, Zunszain PA, Conroy S, Pariante CM (2018) Antenatal depression programs cortisol stress reactivity in offspring through increased maternal inflammation and cortisol in pregnancy: the psychiatry research and motherhood—depression (PRAM-D) study. Psychoneuroendocrinology

Otte C, Gold SM, Penninx BW, Pariante CM, Etkin A, Fava M, Mohr DC, Schatzberg AF (2016) Major depressive disorder. Nat Rev Dis Primers 2:16065

Owen D, Andrews MH, Matthews SG (2005) Maternal adversity, glucocorticoids and programming of neuroendocrine function and behaviour. Neurosci Biobehav Rev 29:209–226

Palazidou E (2012) The neurobiology of depression. Br Med Bull 101:127–145

Palma-Gudiel H, Fananas L (2017) An integrative review of methylation at the serotonin transporter gene and its dialogue with environmental risk factors, psychopathology and 5-HTTLPR. Neurosci Biobehav Rev 72:190–209

Pariante CM, Lightman SL (2008) The HPA axis in major depression: classical theories and new developments. Trends Neurosci 31:464–468

Park S, Hatim A, Si TM, Jeon HJ, Srisurapanont M, Bautista D, Liu SI, Chua HC, Hong JP (2015) Stressful life events preceding the onset of depression in Asian patients with major depressive disorder. Int J Soc Psychiatry 61:735–742

Pashkow FJ (1999) Is stress linked to heart disease? The evidence grows stronger. Cleve Clin J Med 66:75–77

Paykel ES (2003) Life events and affective disorders. Acta Psychiatr Scand Suppl 61–66

Pearson RM, Evans J, Kounali D, Lewis G, Heron J, Ramchandani PG, O'Connor TG, Stein A (2013) Maternal depression during pregnancy and the postnatal period: risks and possible mechanisms for offspring depression at age 18 years. JAMA Psychiatry 70:1312–1319

Perroud N, Paoloni-Giacobino A, Prada P, Olie E, Salzmann A, Nicastro R, Guillaume S, Mouthon D, Stouder C, Dieben K et al (2011) Increased methylation of glucocorticoid receptor gene (NR3C1) in adults with a history of childhood maltreatment: a link with the severity and type of trauma. Transl Psychiatry 1:e59

Philibert RA, Sandhu H, Hollenbeck N, Gunter T, Adams W, Madan A (2008) The relationship of 5HTT (SLC6A4) methylation and genotype on mRNA expression and liability to major depression and alcohol dependence in subjects from the Iowa Adoption Studies. Am J Med Genet B Neuropsychiatr Genet 147B:543–549

Pinquart M, Sorensen S (2003) Differences between caregivers and noncaregivers in psychological health and physical health: a meta-analysis. Psychol Aging 18:250–267

Pitney WA, Weuve C, Mazerolle SM (2016) Experiences with and perceptions of workplace bullying among athletic trainers in the secondary school setting. J Athl Train 51:709–716

Polanczyk G, Caspi A, Williams B, Price TS, Danese A, Sugden K, Uher R, Poulton R, Moffitt TE (2009) Protective effect of CRHR1 gene variants on the development of adult depression following childhood maltreatment: replication and extension. Arch Gen Psychiatry 66:978–985

Provenzi L, Giorda R, Beri S, Montirosso R (2016) SLC6A4 methylation as an epigenetic marker of life adversity exposures in humans: a systematic review of literature. Neurosci Biobehav Rev 71:7–20

Ragen BJ, Roach AE, Chollak CL (2016) Chronic stress, regulation of emotion, and functional activity of the brain. In: Fink G (ed) Stress: concepts, cognition, emotion, and behavior. Academic Press, pp 241–249

Rakers F, Rupprecht S, Dreiling M, Bergmeier C, Witte OW, Schwab M (2017) Transfer of maternal psychosocial stress to the fetus. Neurosci Biobehav Rev pii: S0149-7634(16)30719-9

Raposa E, Hammen C, Brennan P, Najman J (2014) The long-term effects of maternal depression: early childhood physical health as a pathway to offspring depression. J Adolesc Health 54:88–93

Ressler KJ, Bradley B, Mercer KB, Deveau TC, Smith AK, Gillespie CF, Nemeroff CB, Cubells JF, Binder EB (2010) Polymorphisms in CRHR1 and the serotonin transporter loci: gene x gene x environment interactions on depressive symptoms. Am J Med Genet B Neuropsychiatr Genet 153B:812–824

Richter-Levin G, Xu L (2018) How could stress lead to major depressive disorder? IBRO Reports 4:38–43

Risch N, Herrell R, Lehner T, Liang KY, Eaves L, Hoh J, Griem A, Kovacs M, Ott J, Merikangas KR (2009) Interaction between the serotonin transporter gene (5-HTTLPR), stressful life events, and risk of depression: a meta-analysis. JAMA 301:2462–2471

Roca M, Gili M, Garcia-Campayo J, Armengol S, Bauza N, Garcia-Toro M (2013) Stressful life events severity in patients with first and recurrent depressive episodes. Soc Psychiatry Psychiatr Epidemiol 48:1963–1969

Rodriguez JM, Monsalves-Alvarez M, Henriquez S, Llanos MN, Troncoso R (2016) Glucocorticoid resistance in chronic diseases. Steroids 115:182–192

Rogers J, Raveendran M, Fawcett GL, Fox AS, Shelton SE, Oler JA, Cheverud J, Muzny DM, Gibbs RA, Davidson RJ et al (2013) CRHR1 genotypes, neural circuits and the diathesis for anxiety and depression. Mol Psychiatry 18:700–707

Roth TL, Lubin FD, Funk AJ, Sweatt JD (2009) Lasting epigenetic influence of early-life adversity on the BDNF gene. Biol Psychiatry 65:760–769

Roy B, Dunbar M, Shelton RC, Dwivedi Y (2017a) Identification of MicroRNA-124-3p as a putative epigenetic signature of major depressive disorder. Neuropsychopharmacology 42:864–875

Roy B, Shelton RC, Dwivedi Y (2017b) DNA methylation and expression of stress related genes in PBMC of MDD patients with and without serious suicidal ideation. J Psychiatr Res 89:115–124

Rutters F, Pilz S, Koopman AD, Rauh SP, Te Velde SJ, Stehouwer CD, Elders PJ, Nijpels G, Dekker JM (2014) The association between psychosocial stress and mortality is mediated by lifestyle and chronic diseases: the Hoorn Study. Soc Sci Med 118:166–172

Sanger C, Iles JE, Andrew CS, Ramchandani PG (2015) Associations between postnatal maternal depression and psychological outcomes in adolescent offspring: a systematic review. Arch Womens Ment Health 18:147–162

Saul A, Taylor B, Simpson S Jr, Ponsonby AL, Blizzard L, Dwyer T, McMorran B, Wood B, van der Mei IA (2018) Polymorphism in the serotonin transporter gene polymorphisms (5-HTTLPR) modifies the association between significant life events and depression in people with multiple sclerosis. Mult Scler 25(6):848–855

Schneiderman N, Ironson G, Siegel SD (2005) Stress and health: psychological, behavioral, and biological determinants. Annu Rev Clin Psychol 1:607–628

Schulz PE, Arora G (2015) Depression. Continuum (Minneap Minn) 21:756–771

Seo MK, Ly NN, Lee CH, Cho HY, Choi CM, Nhu LH, Lee JG, Lee BJ, Kim GM, Yoon BJ et al (2016) Early life stress increases stress vulnerability through BDNF gene epigenetic changes in the rat hippocampus. Neuropharmacology 105:388–397

Shadrina M, Bondarenko EA, Slominsky PA (2018) Genetics factors in major depression disease. Front Psychiatry 9:334

Shalev I, Lerer E, Israel S, Uzefovsky F, Gritsenko I, Mankuta D, Ebstein RP, Kaitz M (2009) BDNF Val66Met polymorphism is associated with HPA axis reactivity to psychological stress characterized by genotype and gender interactions. Psychoneuroendocrinology 34:382–388

Shao QY, You F, Zhang YH, Hu LL, Liu WJ, Liu Y, Li J, Wang SD, Song MF (2018) CSF miR-16 expression and its association with miR-16 and serotonin transporter in the raphe of a rat model of depression. J Affect Disord 238:609–614

Sharpley CF, Palanisamy SK, Glyde NS, Dillingham PW, Agnew LL (2014) An update on the interaction between the serotonin transporter promoter variant (5-HTTLPR), stress and depression, plus an exploration of non-confirming findings. Behav Brain Res 273:89–105

Sheline YI, Wang PW, Gado MH, Csernansky JG, Vannier MW (1996) Hippocampal atrophy in recurrent major depression. Proc Natl Acad Sci USA 93:3908–3913

Sickmann HM, Arentzen TS, Dyrby TB, Plath N, Kristensen MP (2015) Prenatal stress produces sex-specific changes in depression-like behavior in rats: implications for increased vulnerability in females. J Dev Orig Health Dis 6:462–474

Sinha R (2012) How Does stress lead to risk of alcohol relapse? Alcohol Res-Curr Rev 34:432–440

Slykerman RF, Thompson J, Waldie K, Murphy R, Wall C, Mitchell EA (2015) Maternal stress during pregnancy is associated with moderate to severe depression in 11-year-old children. Acta Paediatr 104:68–74

Smart C, Strathdee G, Watson S, Murgatroyd C, McAllister-Williams RH (2015) Early life trauma, depression and the glucocorticoid receptor gene–an epigenetic perspective. Psychol Med 45:3393–3410

Smoller JW (2016) The genetics of stress-related disorders: PTSD, depression, and anxiety disorders. Neuropsychopharmacology 41:297–319

Song MF, Dong JZ, Wang YW, He J, Ju X, Zhang L, Zhang YH, Shi JF, Lv YY (2015) CSF miR-16 is decreased in major depression patients and its neutralization in rats induces depression-like behaviors via a serotonin transmitter system. J Affect Disord 178:25–31

Sotelo JL, Musselman D, Nemeroff C (2014) The biology of depression in cancer and the relationship between depression and cancer progression. Int Rev Psychiatry 26:16–30

Spanagel R, Noori HR, Heilig M (2014) Stress and alcohol interactions: animal studies and clinical significance. Trends Neurosci 37:219–227

Sperner-Unterweger B (2015) Depression in cancer patients. Wien Med Wochenschr 165:297–303

Stein DJ, Daniels WM, Savitz J, Harvey BH (2008) Brain-derived neurotrophic factor: the neurotrophin hypothesis of psychopathology. CNS Spectr 13:945–949

Steine IM, Winje D, Krystal JH, Bjorvatn B, Milde AM, Gronli J, Nordhus IH, Pallesen S (2017) Cumulative childhood maltreatment and its dose-response relation with adult symptomatology: Findings in a sample of adult survivors of sexual abuse. Child Abuse Negl 65:99–111

Steptoe A, Kivimaki M (2012) Stress and cardiovascular disease. Nat Rev Cardiol 9:360–370

Su M, Hong J, Zhao Y, Liu S, Xue X (2015) MeCP2 controls hippocampal brain-derived neurotrophic factor expression via homeostatic interactions with microRNA132 in rats with depression. Mol Med Rep 12:5399–5406

Suderman M, Borghol N, Pappas JJ, Pinto Pereira SM, Pembrey M, Hertzman C, Power C, Szyf M (2014) Childhood abuse is associated with methylation of multiple loci in adult DNA. BMC Med Genomics 7:13

Sullivan PF, Neale MC, Kendler KS (2000) Genetic epidemiology of major depression: review and meta-analysis. Am J Psychiatry 157:1552–1562

Sun H, Kennedy PJ, Nestler EJ (2013) Epigenetics of the depressed brain: role of histone acetylation and methylation. Neuropsychopharmacology 38:124–137

Szczepankiewicz A, Leszczynska-Rodziewicz A, Pawlak J, Narozna B, Rajewska-Rager A, Wilkosc M, Zaremba D, Maciukiewicz M, Twarowska-Hauser J (2014) FKBP5 polymorphism is associated with major depression but not with bipolar disorder. J Affect Disord 164:33–37

Tafet GE, Nemeroff CB (2016) The links between stress and depression: psychoneuroendocrinological, genetic, and environmental interactions. J Neuropsychiatry Clin Neurosci 28:77–88

Takahashi A, Flanigan ME, McEwen BS, Russo SJ (2018) Aggression, social stress, and the immune system in humans and animal models. Front Behav Neurosci 12:56

Teicher MH, Anderson CM, Polcari A (2012) Childhood maltreatment is associated with reduced volume in the hippocampal subfields CA3, dentate gyrus, and subiculum. Proc Natl Acad Sci U S A 109:E563–E572

Tennant C (2002) Life events, stress and depression: a review of recent findings. Aust N Z J Psychiatry 36:173–182

ter Heegde F, De Rijk RH, Vinkers CH (2015) The brain mineralocorticoid receptor and stress resilience. Psychoneuroendocrinology 52:92–110

Toh YL, Ng T, Tan M, Tan A, Chan A (2018) Impact of brain-derived neurotrophic factor genetic polymorphism on cognition: A systematic review. Brain Behav 8:e01009

Treadway MT, Waskom ML, Dillon DG, Holmes AJ, Park MTM, Chakravarty MM, Dutra SJ, Polli FE, Iosifescu DV, Fava M et al (2015) Illness progression, recent stress, and morphometry of hippocampal subfields and medial prefrontal cortex in major depression. Biol Psychiatry 77:285–294

Ttofi MM (2015) Adolescent bullying linked to depression in early adulthood. BMJ 350:h2694

Turecki G, Meaney MJ (2016) Effects of the social environment and stress on glucocorticoid receptor gene methylation: a systematic review. Biol Psychiatry 79:87–96

Tyrka AR, Ridout KK, Parade SH, Paquette A, Marsit CJ, Seifer R (2015) Childhood maltreatment and methylation of FK506 binding protein 5 gene (FKBP5). Dev Psychopathol 27:1637–1645

Uchida S, Hara K, Kobayashi A, Funato H, Hobara T, Otsuki K, Yamagata H, McEwen BS, Watanabe Y (2010) Early life stress enhances behavioral vulnerability to stress through the activation of REST4-mediated gene transcription in the medial prefrontal cortex of rodents. J Neurosci 30:15007–15018

Uher R (2014) Gene-environment interactions in common mental disorders: an update and strategy for a genome-wide search. Soc Psychiatry Psychiatr Epidemiol 49:3–14

Ulrich-Lai YM, Herman JP (2009) Neural regulation of endocrine and autonomic stress responses. Nat Rev Neurosci 10:397–409

van Bodegom M, Homberg JR, Henckens M (2017) Modulation of the hypothalamic-pituitary-adrenal axis by early life stress exposure. Front Cell Neurosci 11:87

Van den Bergh BR (2011) Developmental programming of early brain and behaviour development and mental health: a conceptual framework. Dev Med Child Neurol 53(Suppl 4):19–23

Van den Bergh BR, Mulder EJ, Mennes M, Glover V (2005) Antenatal maternal anxiety and stress and the neuro behavioural development of the fetus and child: links and possible mechanisms: a review. Neurosci Biobehav Rev 29:237–258

van Donkelaar EL, Vaessen KR, Pawluski JL, Sierksma AS, Blokland A, Canete R, Steinbusch HW (2014) Long-term corticosterone exposure decreases insulin sensitivity and induces depressive-like behaviour in the C57BL/6NCrl mouse. PLoS ONE 9:e106960

van Praag HM, de Kloet R, van Os J (2004) Stress, the brain and depression. Cambridge University Press

Verdolini N, Attademo L, Agius M, Ferranti L, Moretti P, Quartesan R (2015) Traumatic events in childhood and their association with psychiatric illness in the adult. Psychiatr Danub 27(Suppl 1):S60–S70

Vialou V, Feng J, Robison AJ, Nestler EJ (2013) Epigenetic mechanisms of depression and antidepressant action. Annu Rev Pharmacol Toxicol 53:59–87

Vijayendran M, Beach SR, Plume JM, Brody GH, Philibert RA (2012) Effects of genotype and child abuse on DNA methylation and gene expression at the serotonin transporter. Front Psychiatry 3:55

Vinberg M, Trajkovska V, Bennike B, Knorr U, Knudsen GM, Kessing LV (2009) The BDNF Val66Met polymorphism: relation to familiar risk of affective disorder, BDNF levels and salivary cortisol. Psychoneuroendocrinology 34:1380–1389

Vinkers CH, Joels M, Milaneschi Y, Kahn RS, Penninx BW, Boks MP (2014) Stress exposure across the life span cumulatively increases depression risk and is moderated by neuroticism. Depress Anxiety 31:737–745

Vinkers CH, Joels M, Milaneschi Y, Gerritsen L, Kahn RS, Penninx BW, Boks MP (2015) Mineralocorticoid receptor haplotypes sex-dependently moderate depression susceptibility following childhood maltreatment. Psychoneuroendocrinology 54:90–102

Wang SS, Mu RH, Li CF, Dong SQ, Geng D, Liu Q, Yi LT (2017) microRNA-124 targets glucocorticoid receptor and is involved in depression-like behaviors. Prog Neuropsychopharmacol Biol Psychiatry 79:417–425

Wang D, Szyf M, Benkelfat C, Provencal N, Turecki G, Caramaschi D, Cote SM, Vitaro F, Tremblay RE, Booij L (2012) Peripheral SLC6A4 DNA methylation is associated with in vivo measures of human brain serotonin synthesis and childhood physical aggression. PLoS ONE 7:e39501

Wankerl M, Wust S, Otte C (2010) Current developments and controversies: does the serotonin transporter gene-linked polymorphic region (5-HTTLPR) modulate the association between stress and depression? Curr Opin Psychiatry 23:582–587

Wasserman D, Wasserman J, Rozanov V, Sokolowski M (2009) Depression in suicidal males: genetic risk variants in the CRHR1 gene. Genes Brain Behav 8:72–79

Weder N, Zhang H, Jensen K, Yang BZ, Simen A, Jackowski A, Lipschitz D, Douglas-Palumberi H, Ge M, Perepletchikova F et al (2014) Child abuse, depression, and methylation in genes involved with stress, neural plasticity, and brain circuitry. J Am Acad Child Adolesc Psychiatry 53(417–424):e415

Weinstock M (2017) Prenatal stressors in rodents: effects on behavior. Neurobiol Stress 6:3–13

Williams SG, Langhinrichsen-Rohling J, Wornell C, Finnegan H (2017) Adolescents transitioning to high school: sex differences in bullying victimization associated with depressive symptoms, suicide ideation, and suicide Attempts. J Sch Nurs 33:467–479

Wolf OT (2017). Stress effects on learning and memory in humans. In: Fink G (ed) Stress: neuroendocrinology and neurobiology. Academic Press, pp 309–315

Wray NR, Pergadia ML, Blackwood DH, Penninx BW, Gordon SD, Nyholt DR, Ripke S, MacIntyre DJ, McGhee KA, Maclean AW et al (2012) Genome-wide association study of major depressive disorder: new results, meta-analysis, and lessons learned. Mol Psychiatry 17:36–48

Wust S, Van Rossum EF, Federenko IS, Koper JW, Kumsta R, Hellhammer DH (2004) Common polymorphisms in the glucocorticoid receptor gene are associated with adrenocortical responses to psychosocial stress. J Clin Endocrinol Metab 89:565–573

Xu J, Wang R, Liu Y, Liu D, Jiang H, Pan F (2017) FKBP5 and specific microRNAs via glucocorticoid receptor in the basolateral amygdala involved in the susceptibility to depressive disorder in early adolescent stressed rats. J Psychiatr Res 95:102–113

Xu H, Wang J, Zhang K, Zhao M, Ellenbroek B, Shao F, Wang W (2018) Effects of adolescent social stress and antidepressant treatment on cognitive inflexibility and Bdnf epigenetic modifications in the mPFC of adult mice. Psychoneuroendocrinology 88:92–101

Yang L, Zhao Y, Wang Y, Liu L, Zhang X, Li B, Cui R (2015) The effects of psychological stress on depression. Curr Neuropharmacol 13:494–504

Yang Y, Hu Z, Du X, Davies H, Huo X, Fang M (2017) miR-16 and fluoxetine both reverse autophagic and apoptotic change in chronic unpredictable mild stress model rats. Front Neurosci 11:428

Yin Q, Shang Z, Zhou N, Wu L, Liu G, Yu X, Zhang H, Xue H, Liu W (2018) An investigation of physical and mental health consequences among Chinese parents who lost their only child. BMC Psychiatry 18:45

You S, Conner KR (2009) Stressful life events and depressive symptoms: influences of gender, event severity, and depression history. J Nerv Ment Dis 197:829–833

Zanier-Gomes PH, de Abreu Silva TE, Zanetti GC, Benati ER, Pinheiro NM, Murta BM, Crema VO (2015) Depressive behavior induced by social isolation of predisposed female rats. Physiol Behav 151:292–297

Zhao X, Sun L, Sun YH, Ren C, Chen J, Wu ZQ, Jiang YH, Lv XL (2014) Association of HTR2A T102C and A-1438G polymorphisms with susceptibility to major depressive disorder: a meta-analysis. Neurol Sci 35:1857–1866

Zhao M, Chen L, Yang J, Han D, Fang D, Qiu X, Yang X, Qiao Z, Ma J, Wang L et al (2018) BDNF Val66Met polymorphism, life stress and depression: a meta-analysis of gene-environment interaction. J Affect Disord 227:226–235

Zheng Y, Fan W, Zhang X, Dong E (2016) Gestational stress induces depressive-like and anxiety-like phenotypes through epigenetic regulation of BDNF expression in offspring hippocampus. Epigenetics 11:150–162

Zhu S, Shi R, Wang J, Wang JF, Li XM (2014) Unpredictable chronic mild stress not chronic restraint stress induces depressive behaviours in mice. NeuroReport 25:1151–1155

Zimmermann P, Bruckl T, Nocon A, Pfister H, Binder EB, Uhr M, Lieb R, Moffitt TE, Caspi A, Holsboer F et al (2011) Interaction of FKBP5 gene variants and adverse life events in predicting depression onset: results from a 10-year prospective community study. Am J Psychiatry 168:1107–1116

Zurawek D, Kusmider M, Faron-Gorecka A, Gruca P, Pabian P, Solich J, Kolasa M, Papp M, Dziedzicka-Wasylewska M (2017) Reciprocal MicroRNA expression in mesocortical circuit and its interplay with serotonin transporter define resilient rats in the chronic mild stress. Mol Neurobiol 54:5741–5751

Chapter 9
Advance in Diagnosis of Depressive Disorder

Yiru Fang and Zhiguo Wu

Abstract Diagnosis for MDD in modern psychiatry has developed for decades based on long traceable historic efforts on conceptualizing the illness. This article reviews the historical background of current diagnostic framework for MDD, diagnostic criteria and two newly added specifiers ("with anxious distress" and "with mixed features" specifiers) of MDD in the DSM-5, the most influential diagnostic instrument in the world, as well as problems and limitations of symptom-based diagnosis for sake of better understanding about the inter-relationship between diagnostic criteria and MDD.

Keywords Major depressive disorder · Diagnosis · Criterion symptoms · Specifiers

9.1 Introduction

Large cross-national variations in the prevalence of major depressive disorder (MDD) have been reported (Smith 2014). For example, the prevalence of MDD in the Chinese population was consecutively reported to be in the low-average range around the world (Lee et al. 2009). Findings from the latest epidemiological data in China

---

Y. Fang (✉)
Clinical Research Center and Division of Mood Disorders, Shanghai Mental Health Center, Shanghai Jiao Tong University School of Medicine, Shanghai, China
e-mail: yirufang@aliyun.com

CAS Center for Excellence in Brain Science and Intelligence Technology, Shanghai, China

Shanghai Key Laboratory of Psychotic Disorders, Shanghai, China

Z. Wu (✉)
Division of Mood Disorders, Shanghai Mental Health Center, Shanghai Jiao Tong University School of Medicine, Shanghai, China
e-mail: zhiguo_wu@yeah.net

Department of Psychiatry and Neuropsychology, Shanghai Deji Hospital, Qingdao University, Shanghai, China

© Springer Nature Singapore Pte Ltd. 2019
Y. Fang (ed.), *Depressive Disorders: Mechanisms, Measurement and Management*, Advances in Experimental Medicine and Biology 1180,
https://doi.org/10.1007/978-981-32-9271-0_9

showed the 12-month and lifetime prevalence of MDD was 2.1 and 3.4% (Huang et al. 2019) based on DSM-IV criteria. By contrast, in the USA, the 12-month and lifetime prevalence of MDD were 6.6% and 16.2%, respectively (Kessler et al. 2007). When answering why this is the case, what cannot be ignored include two key aspects: how MDD is diagnosed and how it is experienced.

There is a "rich tradition" of historical descriptions in psychopathology of depression/melancholia in psychiatry (Kendler 2016, 2017b). Major depressive disorder (MDD) has been conceptualized as a complicated illness entity encompassing a variety of symptoms across emotional, cognitive, somatic, and interpersonal domains for centuries (Davison 2009; Kendler 2016, 2017b). Therefore, the current conceptualization and diagnosis of MDD (as well as other mental disorders) are deeply based on its historical roots. Diagnosis of MDD is largely based on certain number of key clinical symptoms (diagnostic criteria). To qualify for a diagnosis of MDD, the one has to exhibit at least 5-criterion symptoms, at least one of which has to be one of the two "core" criteria "depressed mood" or "loss of interest/pleasure" during the same period.

The latest edition of the fifth version of the Diagnostic and Statistical Manual of Mental Disorders (DSM-5) provides an opportunity for clinicians to use a common language for describing psychopathology of MDD and other mental disorders (American Psychiatric Association 2013). Two of the new specifiers of MDD "with anxious distress" and "with mixed features" in DSM-5 allow characterization of additional symptoms in the clinic. However, DSM-5 (as well as other diagnostic frameworks), which should be viewed as a theoretical construct, has limitations. For example, the simplified method of permutation and combination of diagnostic symptoms for diagnosing MDD, used as equivalent illness indicators, generates obvious heterogeneity in constructing the diagnosis and depicting the spectrum of psychopathology of MDD (Fried and Nesse 2014; Zimmerman et al. 2015). It would be useful for recognizing the limitations to improve the development of diagnostic systems and advance knowledge in psychiatry.

This review examines the historical background of current diagnostic framework for MDD, its diagnostic criteria in the DSM-5, the most influential diagnostic instrument in the world, and potential problems and limitations of current symptom-based diagnosis for sake of better understanding about the inter-relationship between diagnostic criteria and MDD.

9.2 Historical Background of Diagnosis for Major Depressive Disorder

There is a long traceable history of conceptualization of depressive disorder or depression from antiquity to modern era either in Western (Davison 2009) or in Eastern (Xiu et al. 2008) societies. Kindler et al. (Kendler 2016, 2017b) reviewed monographs, textbooks, and reviews about melancholia/depression published through 1880–1960 and found the concept of depression as an independent form of mood disorders had stabilized as early as 1880. At that time, melancholia evolved as a diagnostic concept with a variety of characteristic symptoms of changes in mood, cognition and psychomotor behavior, loss of interest, and many other signs not covered by current diagnostic framework (Kendler 2017b). In particular, Kraepelin incorporated melancholia into his diagnostic concept of manic-depressive illness (MDI) and renamed this syndrome as "depressive states" in 1883.

The setup of the operationalized diagnostic criteria for the depressive disorder can be traced back to the middle twentieth century. In 1957, Cassidy et al. (1957) proposed a symptom list for diagnosing manic-depressive disorders based on the findings resulted from their descriptive study. The diagnostic requirement for depressive disorder in Cassidy's report included the presence of low mood along with six out of ten secondary symptoms, which were slow thinking, poor appetite, constipation, insomnia, fatigue, loss of concentration, suicidal ideas, weight loss, decreased sex interest, and agitation (Cassidy et al. 1957). This list was reworked in 1972 by Feighner et al. (1972) by deleting constipation and adding other symptoms like guilt, worthlessness, indecisiveness, hypersomnia, and anhedonia in the inclusion criteria for depression. However, the basis for changing the diagnostic criteria developed by Cassidy et al. was not interpreted and no empirical data to support the selection of the criterion symptoms were published (Zimmerman et al. 2006). Though Feighner et al. claimed the criteria should not be considered the final, closed system, and encouraged further test for better changing (Feighner et al. 1972), the symptom inclusion criteria have remained largely unchanged during the past four decades.

The current diagnostic criteria for depressive disorder are largely based on the Diagnostic and Statistical Manual of Mental Disorders, 3rd edition (DSM-III), produced by the American Psychiatric Association (APA). It endorsed revolutionary symptom-based descriptive approaches to make diagnoses in response to growing criticism about the blurred boundary between the normal and the sick and low reliability of diagnosis across assessors raised by the two prior editions of the DSM, which were barely based on assumptions of the psychosocial model, in particular, psychodynamic theory, and focused more on "*What is behind the symptom*?" than the symptom itself (Wilson 1993). The symptom inclusion criteria of the DSM-III and the later DSM-IV for major depressive disorder both included nine characteristic criterion A symptoms, i.e., low mood, anhedonia, changes in appetite or weight, sleeping problems, psychomotor changes, loss of energy or fatigue, worthlessness or guilt, poor concentration or indecisiveness, and suicidality. It requires at least five or more out of the nine symptoms, at least one of which is low mood or anhedonia (loss of interest or pleasure) to qualify a diagnosis of major depressive disorder. Requirements for duration of the current episode (lasting at least 2-week period), significant functioning impairment and exclusion criteria (e.g., not meeting the diagnostic criteria for a mixed episode, not secondary to substance use or physical conditions) were also needed when making such a diagnosis. There was bereavement exclusion in the diagnostic criteria of DSM-III, DSM-III R, and DSM-IV given the fact that depression symptoms could be involved in normal grief reaction. In DSM-IV, a diagnosis of depressive disorder could be made only if the bereaved individual met the full criteria for a major depressive episode, lasted longer (more than 2 months), or had depression symptoms not to be characteristic of normal grief.

The construction of a depression diagnosis using a method of permutation and combination of symptoms in the DSM is also used in another currently accepted major diagnostic system, the International Classification of Diseases, 10th version (ICD-10) produced by the World Health Organization. The symptom inclusion criteria in the ICD-10 are much similar to those in the DSM-IV.

9.3 Criterion Symptoms for Major Depressive Disorder in DSM-5

The DSM system has prevailing influence on psychiatry in the world. Some changes on diagnostic criteria for major depressive disorder have been made in the newly updated DSM-5 (American Psychiatric Association 2013). This is expected to significantly influence how the diagnosis for depressive disorder is used in both clinical settings and research fields. For example, "hopelessness" has been added to the subjective descriptors of depressed mood (e.g., sadness, emptiness, hopelessness) in DSM-5. This means a patient who endorses feeling hopeless but not sad or low-mood fulfills the mood change criterion in DSM-5 but not in DSM-IV. The change obviously broadens the context of this core criterion and then broadens the diagnosis of MDD in clinical settings and epidemiological researches since hopelessness occurs in the absence of depressed mood (Marsiglia et al. 2011; Uher et al. 2008). Such a change may be of potential importance in Eastern Psychiatry, because both of empirical evidence and clinical experience have shown that Chinese (Kleinman 2004; Lee et al. 2007; Parker et al. 2001) and Korean (Chang et al. 2008) patients with MDD may be reluctant to express sadness/low mood directly, which would increase difficulty in valid assessment and recognition of "depressed mood" and then possibly contribute to higher diagnostic threshold of the illness (Chang et al. 2008).

Another important change in the DSM-5 is the removal of the "bereavement exclusion." The DSM-5 task force decided to eliminate the exclusion based on evidence that bereavement-related depressive episodes did not differ from non bereavement-related episodes in symptom profile (Karam et al. 2009), family history (Zisook and Kendler 2007), response to treatment and short-term prognosis (Corruble et al. 2011), as well as consensus on findings of no meaningful difference between the depressive episodes starting after the death of a loved person and other severe stressful life events (Kendler et al. 2008; Wakefield et al. 2007). Instead, the DSM-5 added "note" and "footnote" about the meaningfulness of a significant loss and grief. However, this change has provoked tremendous criticism (Sabin and Daniels 2017; Wakefield and First 2012a; Wakefield 2012), including lack of enough evidence to support the elimination of bereavement exclusion (Wakefield and First 2012b) and medicalization of normal grief including overdiagnosis and overtreatment (Bandini 2015). Besides, operationalized criteria for "significant loss" note and "grief" footnote are not provided in DSM-5; instead, it calls for "clinical judgment" when making distinction between normal grief and a major depressive episode, which may reduce the reliability of the diagnosis (Uher et al. 2014).

9.4 Specifiers of Major Depressive Disorder in DSM-5

In DSM-5, a list of uncoded specifiers was provided in the depressive disorders section, which can be added to diagnoses of coded subcategories. Those included "with anxious distress," "with mixed features," "with melancholic features," "with atypical features," "with mood-congruent psychotic features," "with mood-incongruent psychotic features," "with catatonia", with "peripartum onset," and "with seasonal pattern." Here we mainly discuss the two new specifiers "with anxious distress" and "with mixed features".

"With anxious distress" specifier is defined as the presence of at least two of the following five anxiety symptoms during the most of days of a major depressive episode.

1. Feeling keyed up or tense.
2. Feeling unusually restless.
3. Difficulty concentrating because of worry.
4. Fear that something awful may happen.
5. Feeling that the individual might lose control of himself or herself.

The specifier can be further ranked as mild (with two of the above symptoms), moderate (with three symptoms), moderate-severe (with four or five symptoms), and severe (with four or five symptoms and psychomotor agitation) to specify current severity. Co-occurring anxiety either at symptom (Fava et al. 2004; Wu et al. 2013, 2018) or at syndromal level (Kessler et al. 2007; Phillips et al. 2009; Wu et al. 2018) of MDD is common and negatively impacts its treatment outcomes (Fava et al. 2008; Howland et al. 2009; Wu et al. 2013). This raised the clinical importance of assessment on co-occurring anxiety symptoms and diagnoses of anxiety disorders in patients with MDD. However, making comorbid diagnoses in real-world settings is far from simple and comorbidity of anxiety disorders with MDD may be the most frequently undetected diagnosis in routine clinical practice (Wu and Fang 2014; Zimmerman and Mattia 1999). Underdiagnosis of comorbidity is exacerbated in areas of the world where psychiatric care is not readily available, or in regions where the mental health practitioner is inexperienced or the primary care physician has limited psychiatric knowledge. In China, for example, there is not only a paucity of research on the prevalence of comorbid anxiety disorders in patients with MDD but also widespread neglect of anxiety disorders in clinical settings. In the busy environment of clinical settings in China, psychiatrists usually do not ask patients about anxiety symptoms or make diagnoses of comorbid anxiety disorders when making a diagnosis of depression (Wu and Fang 2014). In comparison, identification and assessment of anxiety at symptom (Wu et al. 2018) or dimensional level (Fava et al. 2004, 2006; Wu et al. 2013; Wu and Fang 2014) is much easier, less time-consuming, and more feasible than making comorbidity diagnoses. The introduction of anxious distress specifier in DSM-5 is useful to routinely record anxiety information during a major depressive episode and then may help ameliorate inadequate assessment of anxiety and underdiagnosis of comorbid anxiety disorders in patients with MDD.

Studies (Hasin et al. 2018) have reported that up to 56–74.6% (Hasin et al. 2018; McIntyre et al. 2016; Rosellini et al. 2018) of the participants with major depressive disorder endorsed anxious distress specifier, which are close to the proportions of MDD patients with concurrent anxiety defined by other dimensional approaches (e.g., with a HRSD-17anxiety/somatization factor score $\geq$7) (Fava et al. 2004, 2006; Wu et al. 2018, 2013). The anxious distress specifier for depression has been reliably used in assessing anxiety in depressed patients with or without anxiety disorders (Zimmerman et al. 2017a, b) and was more likely to be associated with higher prevalence of comorbid generalized anxiety disorder (GAD), panic disorders (PD), agoraphobia, and social phobia (Gaspersz et al. 2017b; Rosellini et al. 2018). Compared to the patients without anxious distress specifier, those with anxious distress specifier showed differential clinical presentations like higher number of hospitalizations, higher suicidality, greater depression severity, great functioning impairment, and decreased quality of life (Hasin et al. 2018; McIntyre et al. 2016) and worse treatment outcomes (Gaspersz et al. 2017a, b). Emerging evidence have shown that routine assessment on anxious distress specifier and utilizing it to develop treatment strategy for patients with MDD is merited and more studies focusing on culture-distinctive populations to test its reliability and validity of assessing anxiety during a major depressive episode are needed.

It is of note that another newly designated specifier "with mixed features" should be understood in the context of parallel changes in criteria for bipolar disorders (Uher et al. 2014). In DSM-IV, the coded "mixed episodes" was limited to cases who met full criteria for a major depressive episode and manic episode. In DSM-5, the "mixed episodes" has been replaced with "mixed features" for manic, hypomanic, and major depressive episodes. The "with mixed features" specifier can only be used for cases that have subthreshold bipolar features during a predominantly depressive episode but do not meet the criteria for full-fledged bipolar disorders. In other words, if criteria for a hypomanic or manic episode are met during the current depressive episode, diagnosis of bipolar disorder should be made. The specifier is defined as the presence of at least three out of the following seven symptoms nearly every day during the same period of a major depressive episode. Psychomotor agitation, irritability, and distractibility that were thought as core symptoms of mixed depression were excluded from the definition considering those might overlap with a major depressive episode. This has raised criticism for researchers that found higher prevalence of distractibility (59.2%), irritability (57.3%), and psychomotor agitation (36.5%) than euphoria (18%) and grandiosity (6.6%) in MDD with mixed features (Targum et al. 2016).

1. Elevated, expansive mood.
2. Inflated self-esteem or grandiosity.
3. More talkative than usual or pressure to keep talking.
4. Flight of ideas or subjective experience about racing thoughts.
5. Increased energy or goal-directed activity.
6. Increase or excessive involvement in activities that have a high potential for painful consequences.
7. Decreased need for sleep.

The introduction of the "with mixed features" specifier was based on evidence that have shown high prevalence of subthreshold manic symptoms, which are associated with higher risk of suicide (Olgiati et al. 2006), increased familial risk of bipolar disorder (Benazzi 2004), and poor antidepressant treatment outcome (Smith et al. 2009) during a major depressive episode. Findings of studies have shown mixed features specifier is commonly presented in a major depressive episode. McIntyre et al. (2016) reported that 26% of patients with MDD endorsed "with mixed features" specifier in a clinical sample and exhibited greater illness severity and higher proportion of cardiovascular comorbidity. Efforts have been made in developing measures for assessing mixed features during a major depressive episode. For instance, Zimmerman et al. (2014) have developed a 13-item self-rating questionnaire assessing both symptoms of depression and symptoms of the DSM-5 mixed specifier, *the Clinically Useful Depression Outcome Scale (CUDOS) supplemented with questions for the DSM-5 mixed features specifier (CUDOS-M)*, which has demonstrated good internal consistency (Cronbach's $\alpha = 0.84$) and high test–retest reliability ($r = 0.72$) (Zimmerman et al. 2014).

Although there are preliminary evidence that supports the effective use of some antipsychotics like lurasidone for the treatment of MDD with mixed features (Suppes et al. 2016), the "with mixed features" specifier during major depressive episodes still poses challenge in terms of pharmacological management, in particular, use of antidepressants (Faedda and Marangoni 2017) given that no controlled studies have been conducted. More studies on this field are needed to determine how this specifier can affect treatment choice.

9.5 How to Understand Inter-relationship Between Diagnostic Criteria and MDD?

The strength of DSM and other diagnostic frameworks has been supposed to be "reliability"—each has ensured that clinicians use the same terms in the same ways. However, the disappointing findings of DSM-5 field trials that MDD was among the least reliable diagnoses (with an interrater reliability kappa = 0.28) (Regier et al. 2013) suggest the question of how to reliably measure MDD that has not been well resolved till now. This problem of low agreement between clinicians may further weaken the validity of MDD diagnosis and even the credibility of psychiatry (Uher et al. 2014).

There are some questions about how to better understand the inter-relationship between diagnostic criteria and MDD that should be discussed.

First, diagnostic criteria have limitations in capturing the spectrum of MDD psychopathology. There is evidence that clinical experience and presentation of depression symptoms often go beyond current diagnostic criteria (Fried et al. 2015; Haroz et al. 2017; Kendler 2016; Kennedy 2008). Haroz et al. (2017) reported in a review that some most frequently complained symptoms across studies focusing on culture-distinctive populations were not captured by DSM criteria. Those included social isolation/loneliness, anger, and general pain. While some DSM criteria poor concentration and psychomotor changes were not among the frequently reported symptoms. Patients with MDD in Korea had a higher diagnostic threshold for symptom complaints than those in the USA by being more likely to express "low energy" and less likely to express to "depressed mood" (Chang et al. 2008). Results of a qualitative study in China indicated some important culture-specific symptoms including preverbal pain, distress of social disharmony, and centrality of sleeplessness. These are not reflected in the current DSM and have even not been regarded as core symptoms of MDD by psychiatrists. However, these were regarded as the most distressing experiences of Chinese patients with MDD (Lee et al. 2007). Thus, whether this would cause missing of key clinical information of MDD has never been taken seriously in clinical practice. This may not only influence reliable diagnosis and assessment of depression but possibly contribute to prevalence difference across nations and lower detection of MDD in some regions in the world (Parker et al. 2001; Smith 2014).

Second, diagnostic criteria are not superior to nondiagnostic symptoms in manifesting MDD. It is not surprising that criterion symptoms like "depressed mood" and "loss of interest or pleasure" are naturally thought to be the "core" of major depressive disorder (MDD) under the prevailing influence of DSM diagnostic criteria. Then it is common to focus mainly, if not solely, on those core symptoms and other criterion symptoms in our training, clinical practice, and research (Kendler 2016). This parallels the tendency that describes DSM as a "Bible" for psychiatry. However, this trend risks producing an impoverished view of psychopathology (Andreasen 2007), to which our knowledge of MDD are largely confined. This further poses challenge in understanding of the construct and measurement of MDD.

The recently developed network approach to psychopathology uses pairwise interactions among symptoms to represent a disorder as a web of symptoms with causal connections, which is helpful to explore pathogenesis and maintenance of mental disorders and has been successfully applied to MDD (Fried et al. 2015; Kendler et al. 2017). The network perspective provides new insights that may be obfuscated by conventional analytic methods regarding the importance of specific symptoms and answer questions like what real *core* symptoms are. Findings from network analytical studies (Fried et al. 2015; Kendler et al. 2017) have shown that DSM criteria and non-DSM symptoms of MDD had equivalent clinical relevance as reflected by centrality measures, which represent the connectedness of a given symptom with all other symptoms and thus the clinical relevance of the symptom in the network. And the two sets of symptoms were interacted within a single psychopathological network with no significant qualitative differences (Fried et al. 2015; Kendler et al. 2017). This suggests that it is of clinical importance to comprehensively assess both of diagnostic criteria and symptoms not included in the diagnostic system when making diagnosis, which would be helpful to valid use of criteria and may offer new insights.

Finally, the fact that determination of criteria for MDD was based more on historical consensus than empirical evidence and findings that both diagnostic criteria and other symptoms outside the diagnostic system are equivalent illness indicators support the notion that criteria chosen for MDD just reflect one subset of a broader number of plausible criteria that could have been chosen (Kendler 2016, 2017b; Kendler et al. 2017). Kindler (Kendler 2016, 2017a; Kendler et al. 2017) proposed two distinct approaches (constitutive vs. indexical) to understand the inter-relationship between mental disorders and diagnostic criteria:

> In a constitutive relationship, criteria definitively define the disorder so that having a disorder is nothing more than meeting the criteria. In an indexical relationship, by contrast, the criteria are fallible indices of a disorder understood as a tentative or hypothetical diagnostic construct. (Kendler et al. 2017).

According to constitutive perspectives, a patient with depression is just a person who meets the criteria for depression. This poses us at risk of mistaking the criteria with the illness itself they were developed to measure, while the indexical perspectives tell that MDD is more than just the criteria. This is suggestive to better understanding and use of diagnostic criteria for MDD and reduction in diagnostic literalism and reification in psychiatry (Kendler 2016, 2017a).

9.6 Conclusion

Current diagnostic systems have improved standardization and reliability of the diagnostic process for MDD and measurement of the phenomenology of MDD after decades of development and revisions. Historical inquiries and empirical evidence suggest that there remain unresolved problems of current symptom-based diagnostic framework and the current diagnostic criteria should be better understood as an indexical position with respect to the illness.

References

American Psychiatric Association (2013) Diagnostic and statistical manual of mental disorders, 5th edn. American Psychiatric Association, Washington, DC

Andreasen NC (2007) DSM and the death of phenomenology in America: an example of unintended consequences. Schizophr Bull 33:108–112

Bandini J (2015) The medicalization of bereavement: (ab)normal grief in the DSM-5. Death Stud 39:347–352

Benazzi F (2004) Intra-episode hypomanic symptoms during major depression and their correlates. Psychiatry Clin Neurosci 58:289–294

Cassidy WL, Flanagan NB, Spellman M, Cohen ME (1957) Clinical observations in manic-depressive disease: a quantitative study of one hundred manic-depressive patients and fifty medically sick controls. J Am Med Assoc 164:1535–1546

Chang SM, Hahm BJ, Lee JY, Shin MS, Jeon HJ, Hong JP, Lee HB, Lee DW, Cho MJ (2008) Cross-national difference in the prevalence of depression caused by the diagnostic threshold. J Affect Disord 106:159–167

Corruble E, Falissard B, Gorwood P (2011) Is DSM-IV bereavement exclusion for major depression relevant to treatment response? A case-control, prospective study. J Clin Psychiatry 72:898–902

Davison K (2009) Historical aspects of mood disorders. Psychiatry 8:47–51

Faedda GL, Marangoni C (2017) What is the role of conventional antidepressants in the treatment of major depressive episodes with mixed features specifier? CNS Spectr 22:120–125

Fava M, Alpert JE, Carmin CN, Wisniewski SR, Trivedi MH, Biggs MM, Shores-Wilson K, Morgan D, Schwartz T, Balasubramani GK, Rush AJ (2004) Clinical correlates and symptom patterns of anxious depression among patients with major depressive disorder in STAR*D. Psychol Med 34:1299–1308

Fava M, Rush AJ, Alpert JE, Carmin CN, Balasubramani GK, Wisniewski SR, Trivedi MH, Biggs MM, Shores-Wilson K (2006) What clinical and symptom features and comorbid disorders characterize outpatients with anxious major depressive disorder: a replication and extension. Can J Psychiatry 51:823–835

Fava M, Rush AJ, Alpert JE, Balasubramani GK, Wisniewski SR, Carmin CN, Biggs MM, Zisook S, Leuchter A, Howland R, Warden D, Trivedi MH (2008) Difference in treatment outcome in outpatients with anxious versus nonanxious depression: a STAR*D report. Am J Psychiatry 165:342–351

Feighner JP, Robins E, Guze SB, Woodruff RA Jr, Winokur G, Munoz R (1972) Diagnostic criteria for use in psychiatric research. Arch Gen Psychiatry 26:57–63

Fried EI, Nesse RM (2014) Depression is not a consistent syndrome: an investigation of unique symptom patterns in the STAR*D study. J Affect Disord 172C:96–102

Fried EI, Epskamp S, Nesse RM, Tuerlinckx F, Borsboom D (2015) What are 'good' depression symptoms? Comparing the centrality of DSM and non-DSM symptoms of depression in a network analysis. J Affect Disord 189:314–320

Gaspersz R, Lamers F, Kent JM, Beekman AT, Smit JH, van Hemert AM, Schoevers RA, Penninx BW (2017a) Longitudinal predictive validity of the DSM-5 anxious distress specifier for clinical outcomes in a large cohort of patients with major depressive disorder. J Clin Psychiatry 78:207–213

Gaspersz R, Lamers F, Kent JM, Beekman ATF, Smit JH, van Hemert AM, Schoevers RA, Penninx B (2017b) Anxious distress predicts subsequent treatment outcome and side effects in depressed patients starting antidepressant treatment. J Psychiatr Res 84:41–48

Haroz EE, Ritchey M, Bass JK, Kohrt BA, Augustinavicius J, Michalopoulos L, Burkey MD, Bolton P (2017) How is depression experienced around the world? A systematic review of qualitative literature. Soc Sci Med 183:151–162

Hasin DS, Sarvet AL, Meyers JL, Saha TD, Ruan WJ, Stohl M, Grant BF (2018) Epidemiology of adult DSM-5 major depressive disorder and its specifiers in the United States. JAMA Psychiatry 75:336–346

Howland RH, Rush AJ, Wisniewski SR, Trivedi MH, Warden D, Fava M, Davis LL, Balasubramani GK, McGrath PJ, Berman SR (2009) Concurrent anxiety and substance use disorders among outpatients with major depression: clinical features and effect on treatment outcome. Drug Alcohol Depend 99:248–260

Huang Y, Wang Y, Wang H, Liu Z, Yu X, Yan J, Yu Y, Kou C, Xu X, Lu J, Wang Z, He S, Xu Y, He Y, Li T, Guo W, Tian H, Xu G, Xu X, Ma Y, Wang L, Wang L, Yan Y, Wang B, Xiao S, Zhou L, Li L, Tan L, Zhang T, Ma C, Li Q, Ding H, Geng H, Jia F, Shi J, Wang S, Zhang N, Du X, Du X, Wu Y (2019) Prevalence of mental disorders in China: a cross-sectional epidemiological study. Lancet Psychiatry 6:211–224

Karam EG, Tabet CC, Alam D, Shamseddeen W, Chatila Y, Mneimneh Z, Salamoun MM, Hamalian M (2009) Bereavement related and non-bereavement related depressions: a comparative field study. J Affect Disord 112:102–110

Kendler KS (2016) The phenomenology of major depression and the representativeness and nature of DSM criteria. Am J Psychiatry 173:771–780

Kendler KS (2017a) DSM disorders and their criteria: how should they inter-relate? Psychol Med 47:2054–2060

Kendler KS (2017b) The genealogy of major depression: symptoms and signs of melancholia from 1880 to 1900. Mol Psychiatry 22:1539–1553

Kendler KS, Myers J, Zisook S (2008) Does bereavement-related major depression differ from major depression associated with other stressful life events? Am J Psychiatry 165:1449–1455

Kendler KS, Aggen SH, Flint J, Borsboom D, Fried EI (2017) The centrality of DSM and non-DSM depressive symptoms in Han Chinese women with major depression. J Affect Disord 227:739–744

Kennedy SH (2008) Core symptoms of major depressive disorder: relevance to diagnosis and treatment. Dialogues Clin Neurosci 10:271–277

Kessler RC, Merikangas KR, Wang PS (2007) Prevalence, comorbidity, and service utilization for mood disorders in the United States at the beginning of the twenty-first century. Annu Rev Clin Psychol 3:137–158

Kleinman A (2004) Culture and depression. N Engl J Med 351:951–953

Lee DT, Kleinman J, Kleinman A (2007) Rethinking depression: an ethnographic study of the experiences of depression among Chinese. Harv Rev Psychiatry 15:1–8

Lee S, Tsang A, Huang YQ, He YL, Liu ZR, Zhang MY, Shen YC, Kessler RC (2009) The epidemiology of depression in metropolitan China. Psychol Med 39:735–747

Marsiglia FF, Kulis S, Garcia Perez H, Bermudez-Parsai M (2011) Hopelessness, family stress, and depression among Mexican-heritage mothers in the southwest. Health Soc Work 36:7–18

McIntyre RS, Woldeyohannes HO, Soczynska JK, Vinberg M, Cha DS, Lee Y, Gallaugher LA, Dale RS, Alsuwaidan MT, Mansur RB, Muzina DJ, Carvalho A, Kennedy S (2016) The prevalence and clinical characteristics associated with diagnostic and statistical manual version-5-defined anxious distress specifier in adults with major depressive disorder: results from the international mood disorders collaborative project. Ther Adv Chronic Dis 7:153–159

Olgiati P, Serretti A, Colombo C (2006) Retrospective analysis of psychomotor agitation, hypomanic symptoms, and suicidal ideation in unipolar depression. Depress Anxiety 23:389–397

Parker G, Gladstone G, Chee KT (2001) Depression in the planet's largest ethnic group: the Chinese. Am J Psychiatry 158:857–864

Phillips MR, Zhang J, Shi Q, Song Z, Ding Z, Pang S, Li X, Zhang Y, Wang Z (2009) Prevalence, treatment, and associated disability of mental disorders in four provinces in China during 2001–05: an epidemiological survey. Lancet 373:2041–2053

Regier DA, Narrow WE, Clarke DE, Kraemer HC, Kuramoto SJ, Kuhl EA, Kupfer DJ (2013) DSM-5 field trials in the United States and Canada, Part II: test-retest reliability of selected categorical diagnoses. Am J Psychiatry 170:59–70

Rosellini AJ, Bourgeois ML, Correa J, Tung ES, Goncharenko S, Brown TA (2018) Anxious distress in depressed outpatients: prevalence, comorbidity, and incremental validity. J Psychiatr Res 103:54–60

Sabin JE, Daniels N (2017) Seeking legitimacy for DSM-5: the bereavement exception as an example of failed process. AMA J Ethics 19:192–198

Smith K (2014) Mental health: a world of depression. Nature 515:181

Smith DJ, Forty L, Russell E, Caesar S, Walters J, Cooper C, Jones I, Jones L, Craddock N (2009) Sub-threshold manic symptoms in recurrent major depressive disorder are a marker for poor outcome. Acta Psychiatr Scand 119:325–329

Suppes T, Silva R, Cucchiaro J, Mao Y, Targum S, Streicher C, Pikalov A, Loebel A (2016) Lurasidone for the treatment of major depressive disorder with mixed features: a randomized, double-blind, placebo-controlled study. Am J Psychiatry 173:400–407

Targum SD, Suppes T, Pendergrass JC, Lee S, Silva R, Cucchiaro J, Loebel A (2016) Major depressive disorder with subthreshold hypomania (mixed features): clinical characteristics of patients entered in a multiregional, placebo-controlled study. Prog Neuropsychopharmacol Biol Psychiatry 68:9–14

Uher R, Farmer A, Maier W, Rietschel M, Hauser J, Marusic A, Mors O, Elkin A, Williamson RJ, Schmael C, Henigsberg N, Perez J, Mendlewicz J, Janzing JG, Zobel A, Skibinska M, Kozel D, Stamp AS, Bajs M, Placentino A, Barreto M, McGuffin P, Aitchison KJ (2008) Measuring depression: comparison and integration of three scales in the GENDEP study. Psychol Med 38:289–300

Uher R, Payne JL, Pavlova B, Perlis RH (2014) Major depressive disorder in DSM-5: implications for clinical practice and research of changes from DSM-IV. Depress Anxiety 31:459–471

Wakefield JC (2012) DSM-5: proposed changes to depressive disorders. Curr Med Res Opin 28:335–343

Wakefield J, First M (2012a) Fallacious reasoning in the argument to eliminate the major depression bereavement exclusion in DSM-5. World Psychiatry 11:203

Wakefield JC, First MB (2012b) Validity of the bereavement exclusion to major depression: does the empirical evidence support the proposal to eliminate the exclusion in DSM-5? World Psychiatry 11:3–10

Wakefield JC, Schmitz MF, First MB, Horwitz AV (2007) Extending the bereavement exclusion for major depression to other losses: evidence from the National Comorbidity Survey. Arch Gen Psychiatry 64:433–440

Wilson M (1993) DSM-III and the transformation of American psychiatry: a history. Am J Psychiatry 150:399–410

Wu Z, Fang Y (2014) Comorbidity of depressive and anxiety disorders: challenges in diagnosis and assessment. Shanghai Arch Psychiatry 26:227–231

Wu Z, Chen J, Yuan C, Hong W, Peng D, Zhang C, Cao L, Fang Y (2013) Difference in remission in a Chinese population with anxious versus nonanxious treatment-resistant depression: a report of operation study. J Affect Disord 150:834–839

Wu Z, Cao L, Peng D, Mellor D, Zhang C, Li H, Wang Z, Song Y, Li C, Fang Y (2018) The clinical correlates of comorbid anxiety symptoms and syndromal anxiety in patients with major depressive disorder. Psychiatry Res 269:251–257

Xiu LJ, Yang YX, Yu S, Sun DZ, Wei PK (2008) Research of depression in traditional Chinese medicine: review and prospect. J Chin Integr Med 6, 416–421

Zimmerman M, Mattia JI (1999) Psychiatric diagnosis in clinical practice: is comorbidity being missed? Compr Psychiatry 40:182–191

Zimmerman M, McGlinchey JB, Young D, Chelminski I (2006) Diagnosing major depressive disorder I: A psychometric evaluation of the DSM-IV symptom criteria. J Nerv Ment Dis 194:158–163

Zimmerman M, Chelminski I, Young D, Dalrymple K, Martinez JH (2014) A clinically useful self-report measure of the DSM-5 mixed features specifier of major depressive disorder. J Affect Disord 168:357–362

Zimmerman M, Ellison W, Young D, Chelminski I, Dalrymple K (2015) How many different ways do patients meet the diagnostic criteria for major depressive disorder? Compr Psychiatry 56:29–34

Zimmerman M, Clark H, McGonigal P, Harris L, Holst CG, Martin J (2017a) Reliability and validity of the DSM-5 anxious distress specifier interview. Compr Psychiatry 76:11–17

Zimmerman M, Martin J, Clark H, McGonigal P, Harris L, Holst CG (2017b) Measuring anxiety in depressed patients: a comparison of the Hamilton anxiety rating scale and the DSM-5 anxious distress specifier interview. J Psychiatr Res 93:59–63

Zisook S, Kendler KS (2007) Is bereavement-related depression different than non-bereavement-related depression? Psychol Med 37:779–794

Chapter 10
Standardized Treatment Strategy for Depressive Disorder

Zuowei Wang, Xiancang Ma and Chunlan Xiao

Abstract The treatment strategies of depressive disorder include pharmacological treatment, psychotherapy, and physical therapy (electroconvulsive therapy [ECT], transcranial magnetic stimulation [TMS], etc.). The updated CANMAT guidelines recommended the most second-generation antidepressants as first-line treatments for patients with a major depressive disorder (MDD) of moderate or greater severity. Before antidepressant treatment, comprehensive assessment and safety monitoring are necessary. The application of measurement-based care in the diagnosis and treatment of depression would better ensure that enough dosage and response of antidepressant is achieved at each key point, and the final outcome of disease is improved. It is recommended that antidepressant is used with monotherapy in patients with depression. Antidepressants of different types and different mechanisms could be combined to improve the efficacy for patients with treatment-resistant depression (TRD). To prevent the relapse and recurrence of disease, the long-term treatment comprised of acute treatment, consolidation treatment, and maintenance treatment must be considered for all patients.

Keywords Depressive disorder · Treatment · Pharmacology · Standardized

10.1 Introduction

The target of treatment for depression is early diagnosis, standardized treatment to control the symptoms, achieve the clinical remission, reduce the disability rate, prevent the recurrence, improve the social function, and the quality of life. The treatment strategies of depressive disorder include pharmacological treatment, psychotherapy, and physical therapy (electroconvulsive therapy [ECT], transcranial magnetic stimulation [TMS], etc.). Other methods of treating depression (such as vagal nerve

Z. Wang (✉) · C. Xiao
Hongkou District Mental Health Center, Shanghai, China
e-mail: wzwhk@163.com

X. Ma
The First Affiliated Hospital of Xi'an Jiaotong University, Xi'an, China

© Springer Nature Singapore Pte Ltd. 2019

Y. Fang (ed.), *Depressive Disorders: Mechanisms, Measurement and Management*, Advances in Experimental Medicine and Biology 1180,
https://doi.org/10.1007/978-981-32-9271-0_10

stimulation [VNS]) have also emerged but remain somewhat experimental and are not widely available. This chapter will mainly discuss the standardized strategy of pharmacological treatment for depressive disorder.

10.2 Who Should be Treated with Antidepressant?

The mainstay of treatment is the prescription of antidepressants although psychotherapy has found a place as an alternative to antidepressants in mild depression. More and more studies support that the use of antidepressants for depressive patients is effective (Hieronymus et al. 2016). The updated CANMAT guidelines recommended the most second-generation antidepressants as first-line treatments for patients with a major depressive disorder (MDD) of moderate or greater severity (as determined by symptom scales and/or functional impairment) (Kennedy et al. 2016).

10.3 The Strategy of Antidepressant Therapy

As we all know, MDD has a varied longitudinal course due to its heterogeneity of pathogenesis. More than 50% patients will have a chronic or recurrent course of depressive episode. Thus, following a successful treatment of acute phase, whether patients require long-term maintenance treatment must be determined to achieve recovery, improve social function and quality of life, and pretend the relapse and recurrence of disease.

The total duration of treatment for MDD could be divided into three phases that are as follows (Li and Ma 2015):

1. Acute treatment phase (8–12 weeks): Control the depressive and related symptoms to achieve clinical remission as much as possible. The effect of treatment during acute phase determines the outcome and prognosis of the disease.
2. Consolidation treatment phase (4–9 months): In principle, effective drugs during acute treatment phase continue to be used. Researches showed that antidepressants can effectively prevent relapse. Psychotherapy, such as cognitive behavioral therapy, is also recommended to effectively reduce the risk of relapse and recurrence.
3. Maintenance treatment phase (usually 2–3 years): Long-term maintenance treatment is recommended for patients with multiple recurrences or significant residual symptoms. The antidepressants used during consolidation treatment phase should be maintained with a sufficient dosage during the maintenance treatment phase.

The following criteria could be used to evaluate the efficacy of treatment in patients with MDD:

Response: Reduction rate of score ≥50% for 17-items Hamilton Depression Scale (HAMD-17) or Montgomery–Ellsberg Depression Rating Scale (MARDS).
Remission: Depressive symptoms have completely disappeared (HAMD-17 ≤ 7 or MARDS ≤ 10) for ≥2 weeks and <6 months, and the social function recovered well.
Recovery: Patients are fully restored to normal or have stable remission for at least 6 months.
Relapse: Patients have repeated or aggravated symptoms during the phases of acute and consolidation treatment.
Recurrence: A new episode of depression occurs after a recovery.

10.4 The Principle of Antidepressant Therapy (Li and Ma 2015)

1. Comprehensive assessment and monitoring.

The comprehensive assessment includes demographic characteristics, physical condition, depressive and related symptoms, diagnosis, past and current treatment, safety and tolerance to medicine, and comorbidities. The regular laboratory tests and psychometrics assessment are necessary for the efficacy and safety monitoring. In recent years, the application of measurement-based care in the diagnosis and treatment of depression has been paid more and more attention, which would better ensure that enough dosage and response of antidepressant is achieved at each key point, and the final outcome of disease is improved.

2. Determining the timing of drug therapy

For patients with mild depressive disorder, a close observation can be adopted for 2 weeks before determining whether or not to provide antidepressant. For patients with moderate to severe depressive disorder, antidepressant therapy should be initiated as soon as possible.

3. Individualized drug therapy

Individualized treatment should be carried out according to demographic and clinical characteristics, pharmacokinetics, and pharmacodynamics in different populations, previous history of medication and treatment preference for patients should be considered (Detailed suggestions are shown in Chap. 11 Optimized treatment strategy for depressive disorder and Chap. 12 Individualized treatment strategy for depressive disorder). The major factors for decision of antidepressants are safety, efficacy, price, and appropriateness.

4. Antidepressant monotherapy and combination treatment

In general, it is recommended that antidepressant is used with monotherapy in patients with depression. If no response (reduction rate of score <50% for HAMD-17 or MARDS) is achieved with sufficient dosage and duration (more than 4 weeks), the initial treatment could be considered to replace with other antidepressants. Antidepressants of different types and different mechanisms could be combined to improve the efficacy for patients with treatment-resistant depression (TRD). The combination of three or more than antidepressants is usually not recommended. For patients with depression comorbid with psychotic symptoms, antipsychotics can be used as an adjunctive treatment. MECT is also recommended as a combination treatment for patients with serious depression (especially with suicide attempt and behavior), no response to sufficient treatment of antidepressants, psychotic symptoms, and who can not tolerate with antidepressant treatment.

5. Comprehensive and long-term treatment

The etiology and pathogenesis of depression are related to both biological and psychosocial factors. Thus the comprehensive treatment including antidepressant, psychotherapy, and social intervention is necessary for patients with depression. Psychological education should be given to improve treatment compliance. In addition, the comorbidities with other mental and physical diseases must be treated as active as possible. To prevent the relapse and recurrence of disease, the long-term treatment comprised acute treatment, consolidation treatment, and maintenance treatment must be considered for all patients.

10.5 Basic Procedures of Depression Management

The following nine basic procedures were recommended for clinical management of depression referred to the updated Canadian Network for Mood and Anxiety Treatments (CANMAT) Clinical Guidelines for the Management of Adults with Major Depressive Disorder (Lam et al. 2016).

1. Conduct a comprehensive assessment of biopsychosocial factors and comorbidity conditions.
2. Obtain collateral information whenever possible.
3. Formulate a diagnosis and differential diagnosis.
4. Establish and maintain a therapeutic alliance as well as a collaborative care system provided by practitioners.
5. Provide health education and support self-management for patients and their families.
6. Encourage the patients as a partner in determining treatment goals.
7. Make a comprehensive management plan together with the patients and their families (or other supports) if possible.

8. Deliver the evidence-based treatments according to therapeutic goals.
9. Monitor outcomes with measure-based care, including safety, efficacy, social function, and life quality.

10.6 How to Choose an Antidepressant?

The age of patients, depressive and related symptoms, comorbidity diseases, and pharmacological mechanisms of antidepressant should be considered when making a decision of antidepressant therapy. Based on the difference in efficacy of antidepressants, the available evidence is not enough to recommend a specific antidepressant as the best choice. Recently, a meta-analysis included placebo-controlled and head-to-head trials of 21 antidepressants suggested that agomelatine, amitriptyline, escitalopram, mirtazapine, paroxetine, venlafaxine, and vortioxetine were more effective than other antidepressants, and agomelatine, citalopram, escitalopram, fluoxetine, sertraline, and vortioxetine were more tolerable than other antidepressants (Cipriani et al. 2018). The differences in response time and adverse events may be considered when choosing an antidepressant (Gartlehner et al. 2011). The first-line recommendation antidepressants include selective serotonin reuptake inhibitor (SSRI), selective serotonin and norepinephrine reuptake inhibitor (SNRI), norepinephrine and specific serotonin antidepressant (NaSSA), norepinephrine and dopamine reuptake inhibitors (NDRI), selective serotonin reuptake activator (SSRA), selective norepinephrine reuptake inhibitor (NRI), and reversible A monoamine oxidase inhibitor (RIMA). The second-line recommendation antidepressants include serotonin balanced antidepressant (SMA) and tricyclic antidepressants (TCA). For patients with mild or moderate depression, Chinese herbal medicine (e.g., Shuganjieyu capsule, St. John herb preparation) could be prescribed. The antidepressants usually used in clinical practice are listed in Table 10.1.

1. SSRI antidepressant

The SSRI antidepressant includes fluoxetine, paroxetine, sertraline, fluvoxamine, citalopram, and escitalopram. During the acute treatment phase, the efficacy of SSRI antidepressants in the treatment of depression is equivalent to that of TCA and better than placebo. There is no significant difference in the overall efficacy of different SSRI antidepressants.

2. SNRI antidepressant

Meta-analysis showed that the efficacy of venlafaxine, duloxetine, and milnacipran was equivalent to SSRIs. In addition, duloxetine did not seem to provide a significant advantage in efficacy over other antidepressants for the acute treatment of depression, but the improvement of somatic pain symptoms in patients with depression is obviously better than that of SSRIs (Cipriani et al. 2012; Kuga et al. 2017).

Table 10.1 The recommended antidepressants in clinical practice

Antidepressant	Pharmacological mechanism	Daily dose range (China, mg)
First-line recommendation antidepressants		
Fluoxetine	SSRI	20–60
Paroxetine	SSRI	20–50
Fluvoxamine	SSRI	100–300
Sertraline	SSRI	50–200
Citalopram	SSRI	20–60
Escitalopram	SSRI	10–20
Venlafaxine	SNRI	75–225
Duloxetine	SNRI	60–120
Milnacipran	SNRI	100–200
Mitrazapine	NaSSA	15–45
Bupropion	NDRI	150–450
Agomelatine	MRA	25–50
Second-line recommendation antidepressants		
Amitriptyline	TCA	50–250
Clomipramine	TCA	50–250
Doxepin	TCA	50–250
Imipramine	TCA	50–250
Maprotiline	Tetracyclic	50–225
Mianserin	Tetracyclic	30–90
Trazodone	SMA	50–400
Reboxetine	NRI	8–12
Tianeptine	SARI	25–37.5
Third-line recommendation antidepressants		
Moclobemide	RIMA	150–600

SSRI, selective serotonin reuptake inhibitor; SNRI, serotonin and noradrenaline reuptake Inhibitor; NaSSA, noradrenergic and specific serotonergic antidepressant; NDRI, noradrenaline and dopamine reuptake inhibitor; MRA, melatonin receptor agonist; TCA, tricyclic antidepressant; SMA, serotonin balanced antidepressant; NRI, norepinephrine reuptake inhibitor; SARI, serotonin receptor antagonist and reuptake inhibitor; RIMA, reversible monoamine oxidase inhibitor

3. NaSSA antidepressant

Mirtazapine can effectively control depressive symptoms and prevent the recurrence of depression, which is equivalent to that of SSRI. Mirtazapine is likely to have an earlier response than SSRIs during acute treatment phase, and has a good effect on the emotional and physical symptoms of the patients with depression (Watanabe et al. 2011).

4. NDRI antidepressant

The efficacy of bupropion in the treatment of depression is better than placebo and equivalent to SSRI (Nam et al. 2017). Bupropion is superior to SSRI antidepressants in controlling impulsive and attentive symptoms, improving fatigue and sleep, but inferior to SSRI for anxiety symptoms. And bupropion does not have a significant effect on weight gain and sex function.

References

Cipriani A, Koesters M, Furukawa TA, Nosè M, Purgato M, Omori IM, Trespidi C, Barbui C (2012) Duloxetine versus other anti-depressive agents for depression. Cochrane Database Syst Rev 10:CD006533

Cipriani A, Furukawa TA, Salanti G, Chaimani A, Atkinson LZ, Ogawa Y, Leucht S, Ruhe HG, Turner EH, Higgins JPT, Egger M, Takeshima N, Hayasaka Y, Imai H, Shinohara K, Tajika A, Ioannidis JPA, Geddes JR (2018) Comparative efficacy and acceptability of 21 antidepressant drugs for the acute treatment of adults with major depressive disorder: a systematic review and network meta-analysis. Lancet 391(10128):1357–1366

Gartlehner G, Hansen RA, Morgan LC, Thaler K, Lux L, Van Noord M, Mager U, Thieda P, Gaynes BN, Wilkins T, Strobelberger M, Lloyd S, Reichenpfader U, Lohr KN (2011) Comparative benefits and harms of second-generation antidepressants for treating major depressive disorder: an updated meta-analysis. Ann Intern Med 155(11):772–785

Hieronymus F, Emilsson JF, Nilsson S, Eriksson E (2016) Consistent superiority of selective serotonin reuptake inhibitors over placebo in reducing depressed mood in patients with major depression. Mol Psychiatry 21:523–530

Kennedy SH, Lam RW, McIntyre RS, Tourjman SV, Bhat V, Blier P et al (2016) Canadian network for mood and anxiety treatments (CANMAT) 2016 clinical guidelines for the management of adults with major depressive disorder: section 3. Pharmacological Treatments. Can J Psychiatry/La Rev Can Psychiatr 1–21

Kuga A, Tsuji T, Hayashi S, Matsubara M, Fujikoshi S, Tokuoka H, Yoshikawa A, Escobar R, Tanaka K, Azekawa T (2017) An observational study of duloxetine versus SSRI monotherapy for the treatment of painful physical symptoms in Japanese patients with major depressive disorder: primary analysis. Neuropsychiatr Dis Treat 4(13):2105–2114

Lam RW, McIntosh D, Wang J, Enns MW, Kolivakis T, Michalak EE, Sareen J, Song WY, Kennedy SH, MacQueen GM, Milev RV, Parikh SV, Ravindran AV (2016) CANMAT depression work group. Canadian network for mood and anxiety treatments (CANMAT) 2016 clinical guidelines for the management of adults with major depressive disorder: section 1. Disease burden and principles of care. Can J Psychiatry 61(9):510–523

Li LJ, Ma X (2015) Practice guideline for prevention and treatment of depressive disorder (in Chinese). Chin Soc Psychiatry, Beijing

Nam B, Bae S, Kim SM, Hong JS, Han DH (2017) Comparing the effects of bupropion and escitalopram on excessive internet game play in patients with major depressive disorder. Clin Psychopharmacol Neurosci 15(4):361–368

Watanabe N, Omori IM, Nakagawa A, Cipriani A, Barbui C, Churchill R, Furukawa TA (2011) Mirtazapine versus other antidepressive agents for depression. Cochrane Database Syst Rev (12):CD006528. https://doi.org/10.1002/14651858.cd006528

Chapter 11
Optimized Treatment Strategy for Depressive Disorder

Peijun Chen

Abstract Despite many advances in pharmacotherapy over the past half centurye, only a fraction of patients with Major Depressive Disorder (MDD) can achieve remission after the first or second trial of pharmacotherapy. Those who failed standard antidepressant treatment are termed as Treatment-Resistant Depression (TRD). Pharmacotherapy for TRD is more viable over past 15 years in part due to advances in clinical trials such as the Sequenced Treatment Alternatives to Relieve Depression (STAR*D) and the US Department of Veterans Affairs Augmentation and Switching Treatments for Improving Depression Outcomes (VAST-D) study. In general, optimizing pharmacotherapy consists of switching to different agents, combination with different antidepressants, or augmentation with different class of psychotropic medications, and the latter is preferred. Augmenting agents with strong evidence include Bupropion, Lithium, Triiodothyronine (T3), Aripiprazole, Brexpiprazole, Quetiapine, and Olanzapine in combination with Fluoxetine. Many works need to be done to further advance this field. These include (1) Establish agreement on a standardized, systematic, and feasible definition of TRD, (2) Establish safety and tolerability beyond acute treatment phase, (3) Establish individual psychosocial and neurobiological marks such as pharmacogenetic variance, and (4) Utilize multi-treatment modules such as combination of psychotherapy and pharmacotherapy in conjunction with brain stimulation therapy such as electroconvulsive therapy, vagus nerve stimulation and transcranial magnetic stimulation; as well as non-traditional therapy such as nutritional supplements, exercise and light therapy.

Keywords Treatment-resistant depression · Risk factors · Pharmacotherapy · Augmentation

Major Depressive Disorder (MDD) is a chronic, prevalent, and debilitating disorder affecting more than 300 million people worldwide (WHO 2018). Despite many advances in pharmacotherapy over the past half centrury, only a fraction of patients

P. Chen (✉)
VA Northeast Ohio Healthcare System, Cleveland VA Medical Center, Case Western Reserve University School of Medicine, Cleveland, USA
e-mail: Peijun.chen@va.gov; peijun.chen@case.edu

© Springer Nature Singapore Pte Ltd. 2019
Y. Fang (ed.), *Depressive Disorders: Mechanisms, Measurement and Management*, Advances in Experimental Medicine and Biology 1180,
https://doi.org/10.1007/978-981-32-9271-0_11

with MDD are able to achieve remission after the first or second trial of pharmacotherapy.

In a pooled 9 longitudinal studies of 1814 patients with MDD, 34% of patients achieved remission 3 months later (Simon et al. 2006). In a single clinical trial of 2876 with MDD, the Sequenced Treatment Alternatives to Relieve Depression (STAR*D) found similarly 33% remission rate after treatment with citalopram over 12 weeks (Rush et al. 2006a; Trivedi et al. 2006b), Moreover, at least half will not achieve and sustain remission following multiple pharmacological approaches (McIntyre et al. 2014).

Those who failed standard antidepressant treatment are termed as Treatment-Resistant Depression (TRD). Various definitions were used in the literature. The most common definition for TRD requires a minimum of two-prior treatment failures and confirmation of prior adequate dose and duration (generally as a minimum recognized dose over about 6 weeks).

TRD is associated with increased severity and recurrence of disease, negative impacts on prolonged function impairments, increased economic cost, caregiver burden, associated stigma, medical comorbidities, and increased suicidality as well as nonsuicidal mortality (Demyttenaere and Van Duppen 2018; Greden 2001; Simon et al. 2006).

11.1 Risk Factors for TRD

In a systematic review, Bennabi et al. (2014) identified following risk factors: comorbid anxiety disorder, current suicidal risk, non-response to the first antidepressant received in the patient's lifetime, and presence of melancholic features; bipolarity, an early onset of first depressive episode, high rate of depressive recurrences, and lack of full remission after a previous episode; low-reward dependence and low cooperativeness; high neuroticism, low extraversion, low openness, and low conscientiousness (Bennabi et al. 2014).

In the European multicenter study of clinical factors predicting treatment-resistant depression, the following factors were found to be associated with TRD: Symptom severity [odds ratio (OR) = 3.31], psychotic symptoms (OR = 2.52), suicidal risk (OR = 1.74), generalized anxiety disorder (OR = 1.68), inpatient status (OR = 1.65), higher number of antidepressants administered previously (OR = 1.23), and lifetime depressive episodes (OR = 1.15) as well as longer duration of the current episode (OR = 1.022) (Kautzky et al. 2019).

In STAR*D trial stage 2 study, it was found that remission was more likely among participants who were white, employed, cohabiting or married, or privately insured or who had prior intolerance to citalopram or at least a response to citalopram, and no prior suicide attempts. Remission was less likely among participants with concurrent generalized anxiety, obsessive-compulsive, panic, or posttraumatic stress disorders; social phobia; anxious or melancholic features; or more severe depression (Rush et al. 2008).

In the US Department of Veterans Affairs (VA) Augmentation and Switching Treatments for Improving Depression Outcomes (VAST-D) study, 1,522 patients with MDD who did not have an adequate response to at least one course of antidepressant treatment meeting minimal standards for dose and duration, the authors found remission was more likely in those who were employed, less severely and chronically depressed, less anxious, not experiencing complicated grief symptoms, without past childhood adversity, better quality of life and positive mental health (Zisook et al. 2019).

Biological-genetic variation is reported to explain up to 42% of the variance in antidepressant response (Tansey et al. 2013). Pharmacogenetic variance and its testing (PGx) is reported to associate with treatment outcome. Pérez et al. (2017) reported a 12 week, double-blind, parallel, and multi-center randomized controlled trial in 316 adult patients with MDD and found that Pharmacogenetic (PGx)-guided treatment group had a higher responder rate compared to Treatment As Usual group (TAU) at 12 weeks [47.8% vs. 36.1%, $p = 0.0476$; OR = 1.62 (95%CI: 1.00–2.61)], and this difference increased after removing subjects in the PGx-guided group when clinicians explicitly reported not to follow the test recommendations [51.3% vs. 36.1%, $p = 0.0135$; OR = 1.86 (95%CI: 1.13–3.05)]. Effects were more consistent in patients with 1–3 failed drug trials. In subjects reporting side effects burden at baseline, odds of achieving a better tolerability (frequency, intensity, and burden of side ffects effects rating *Burden* subscore ≤ 2) were higher in the PGx-guided group than in controls at 6 weeks and maintained at 12 weeks [68.5% vs. 51.4%, $p = 0.0260$; OR = 2.06 (95%CI: 1.09–3.89)].

In general, these risk or protective factors can be categorized into (1) phenomenal factors: unrecognized bipolarity, mixed hypomanic symptoms, psychotic features, atypical features, severity of symptoms, and duration of disease; (2) Psychosocial and characteristic factors: age, age of onset, early life trauma, personality, and adherence; (3) complexity: comorbid substance use, anxiety disorder, PTSD/panic, medical conditions, and neurological conditions; and (4) biological and genetic factors: neurobiological and genetic variation, including pharmacokinetic and pharmacodynamic variations, and family history.

11.2 General Principle in Managing TRD

In clinical practice, the following steps are commonly adapted in formulating treatment plan to optimize treatment of depression that hasn't responded well to standard treatment.

1. Thorough history taking for previous trials, especially adequacy of medication dose and duration.
2. Detailed personal history of psychosocial risk and positive factors.
3. Accuracy in diagnosis especially unrecognized bipolar spectrum disorder.

4. Comprehensive evaluation of atypical features especially anxious features and mixed feature.
5. Detection of comorbid psychiatric and medical conditions.
6. Utilization of pharmacogenetic testing.

To deal with these risk or protective factors for TRD is beyond the scope of this chapter. Some of those issues will be discussed in the chapter of Individualized treatment strategy for depressive disorder.

Currently, the management of TRD includes the following strategies:

1. Pharmacotherapy through switching to different agents or adding with (augmentation) second agent such as antidepressant and antipsychotic medication.
2. Psychotherapy through adding on or switching to different methods such as CBT and IPT.
3. Other biological therapies such as electroconvulsive therapy, vagus nerve stimulation, and transcranial magnetic stimulation.
4. Psychoeducation and case management to improve rapport, accessibility, acceptance, adherence, and adverse effect monitoring.
5. Non-traditional therapy such as exercise, nutritional supplements, light therapy, and internet driven therapy.

This chapter focuses primarily on pharmacotherapy for TRD, and other types of therapy for TRD will be discussed in the Chaps. 13, 14, and 15.

11.2.1 Pharmacotherapy with Switching to Another Antidepressant

Switching to a different antidepressant after the initially failed trial is commonly used in clinical practice. However, the outcomes from studies are not convincing.

In a meta-analysis of eight studies found no statistically significant difference between switching and continuation ($P = 0.836$) with no superiority of switching in remission rate [OR = 1.05 (95%CI: 0.65–1.69)], and in response rate [OR = 0.97 (95%CI: 0.69–1.36)]. The drawback of this comparison between switching and continuation is that only limited medications were investigated including three studies for nonresponders to Selective Serotonin Reuptake Inhibitors (SSRIs) were switched to duloxetine, and two studies for nonresponders (to venlafaxine and to nortriptyline, respectively) were switched to fluoxetine (Bschor et al. 2018).

Another study pooled data from four clinical trials with 1496 patients, and found that patients randomized to switch to a non-SSRI antidepressant (bupropion, mirtazapine, and venlafaxine) were more likely to experience remission than patients switched to the second SSRI (risk ratio = 1.29, $p = 0.007$). Pooled remission rates were 28% (for non SSRIs) and 23.5% (for SSRIs). Discontinuation rate due to intolerance, thought is high (with risk ratio = 1.23) in switching to non-SSRI group though not statistically significant ($p = 0.1$) (Papakostas et al. 2008).

STAR*D trial found comparable remission rate between SSRI and SNRI. In 727 adult outpatients with a nonpsychotic major depressive disorder who failed an initial trial of citalopram, they were randomized into bupropion sustained-release (up to 400 mg), sertraline (up to 200 mg), or venlafaxine extended-release (up to 375 mg) for up to 14 weeks. The study found that remission rates as assessed by the QIDS-SR-16, were 25.5% for sustained-release bupropion, 26.6% for sertraline, and 25.0% for extended-release venlafaxine. It concluded that after unsuccessful treatment with an SSRI, any one of the medications in the study provided a reasonable second-step choice for patients with depression (Rush et al. 2006b).

VAST-D using strict criteria for remission (QIDS-C16 $\leq$ 5 at two consecutive visits) reported similar remission rate in switching to bupropion sustained-release (22%). VAST-D is a multi-site randomized, single-blind trial including Veterans Health Administration (VHA) patients who failed at least one antidepressant trial meeting minimal standards for dose and duration of treatment, 1,522 patients were randomly assigned to Switch-BUP (up to 400 mg/day), Augment-BUP (up to 400 mg/day), and Augment-ARI (up to 10 mg/day) for 12 weeks. The remission rate was 22% for Switch-BUP, 27% for Augment-BUP, and 29% for Augment-ARI (Mohamed et al. 2017).

STAR-D trial further studied adult outpatients ($N = 235$) with nonpsychotic major depressive disorder after two consecutive unsuccessful antidepressant trials, and found low-remission rates in either treatment with mirtazapine (up to 60 mg/day $\times$ 14 weeks) or nortriptyline (up to 200 mg/day $\times$ 14 weeks); with remission rates were 8.0% in Mirtazapne group per QIDS-SR(16) scores, and 12.4% in nortriptyline group (Fava et al. 2006).

11.2.2 Pharmacotherapy with Adding Another Antidepressant

Combination with another antidepressant to optimize treatment of depression that didn't respond well to initial trial has also been commonly used. Mirtazapine and bupropion are two antidepressants with different pharmacological prosperities than SSRIs or SNRIs, and have been studied in the treatment of refractory depression.

Mirtazapine, an alpha2-adrenoreceptor antagonist, increases central noradrenergic and serotonergic neurotransmission by inhibiting negative feedback from synaptic Noradrenaline (NA) acting on presynaptic alpha2 autoreceptors on noradrenergic neurones and alpha2 heteroreceptors on 5-Hydroxytryptaminergic (5-HT) neurones.

Mirtazapine's benefit as added on agent for depression has been reported in some small size studies, however, a larger study found no benefit from combining antidepressants, including mirtazapine and venlafaxine, over SSRI monotherapy with escitalopram and combined treatment had a higher side-effect burden (Carpenter et al. 2002; Blier et al. 2009, 2010; Rush et al. 2011).

Another study of 480 patients randomized to add-on mirtazapine (up to 30 mg daily for up to 12 months), and add-on placebo for 12 months did not find convincing evidence of a superior efficacy of mirtazapine in addition to an SSRI or SNRI antidepressant in primary care patients with treatment-resistant depression, with remission rates at 12 weeks of 29.4% in mirtazapine group versus 24.4% in placebo group ($p = 0.266$). More participants in the mirtazapine group withdrew from the trial associated with adverse effects (19.1%) than placebo group (3.7%). The most common AEs were drowsiness (41/241 = 17.4%) and weight increase (24/241 = 10.0%) in mirtazapine group, and 2.1% and 3.4%, respectively, in the placebo group (Kessler et al. 2018).

Bupropion, inhibiting the reuptake of both noradrenaline and dopamine, has no effects on serotonin, histamine, acetylcholine, or adrenaline (epinephrine) receptors, therefore, not associated with significant sedation, anticholinergic side effects, hypotension, weight gain, or sexual dysfunction. Bupropion has gained more favorable findings in clinical trials.

A study of 60 patients who had not responded adequately to 4 weeks of monotherapy with a selective serotonin reuptake inhibitor, were randomized into bupropion sustained release or placebo as add-on treatment found remission rate of 60% in bupropion-SR added on group as compared with 24% in placebo added on group (Gulrez et al. 2012).

Another small randomized trial comparing duloxetine plus bupropion with duloxetine plus placebo ($n = 46$, duration = 6 weeks) for treatment of resistant depression, found that response was comparable with adjunctive bupropion or placebo (26% vs. 22%) (Fornaro et al. 2014).

In a large clinical trial of 851 individuals with MDD not remitting on citalopram (mean dose 55 mg/day, mean duration 11.9 weeks) who were randomly assigned to receive either bupropion SR ($n = 565$, dose of up to 400 mg/day) or buspirone ($n = 286$, dose of up to 60 mg/day) for up to 12 weeks (Trivedi et al. 2006a, b), remission rate by QIDS-SR-16 was 39.0% in bupropion-SR group and 32.9% in buspirone group (were 29.7% and 30.1%, respectively, by HRSD-17). Sustained-release bupropion, was associated with a greater reduction (from baseline to the end of this study) in QIDS-SR-16 scores than was buspirone (25.3% vs. 17.1%, $P < 0.04$), a lower QIDS-SR-16 score at the end of this study (8.0% vs. 9.1%, $P < 0.02$), and a lower dropout rate due to intolerance (12.5% vs. 20.6%, $P < 0.009$).

One systematic review of 51 studies on Bupropion's effectiveness as a sole antidepressant in depression, or as added on agent in depression (6 studies), supported the benefit of adding bupropion in the treatment of refractory depression (Patel et al. 2016).

11.2.3 Augmentation with Atypical Antipsychotics

Atypical antipsychotic medications have complex effect at various monoaminergic receptors and transporters, presumed mechanism for treatment of depression such as affinity for the serotonin-1A receptor (Ziprasidone and aripiprazole), affinity for the serotonin-1D receptor (ziprasidone and risperidone), inhibit the reuptake of serotonin and norepinephrine (Ziprasidone), mixed agonist, and antagonist effects at dopamine receptors (aripiprazole and brexpiprazole).

Four antipsychotic medications have received US FDA approval for augmentation treatment for depression, i.e., olanzapine (in combination with fluoxetine), quetiapine XR, aripiparzole, and brexpiprazole. Risperidone is also a commonly used agent, though it is not FDA approved.

A meta-analysis pooled 10 randomized, double-blind, and placebo-controlled clinical trials of atypical antipsychotics in augmentation of antidepressants for TRD found a 47% remission rate for augmentation with atypical antipsychotics versus a 22% remission rate for augmentation with placebo, though discontinuation due to intolerance is higher for augmentation group 37% versus 12% for placebo (Papakostas et al. 2007).

Another meta-analysis pooled 16 studies (1966–2009). Nelson and Papakostas (2009) found the odds ratio for the response with drug versus placebo was 1.69 (95%CI = 1.46–1.95, $z = 7.00$, $N = 16$, $p < 0.00001$). The overall pooled response rate for augmenting treatment with an atypical agent was 44.2%, compared with 29.9% for placebo. The pooled remission rates were 30.7% for atypical antipsychotics, compared with 17.2% for placebo. The odds ratio for remission was 2.00 (95%CI = 1.69–2.37, $z = 8.03$, $N = 16$, $p < 0.00001$). The pooled actual rates of discontinuation for any reason were 19.6% in the atypical treatment group and 15.5% in the placebo group.

The most comprehensive review of second-generation antipsychotics in the treatment of major depressive disorder which summarized eight meta-analyses of SGA in the treatment of MDD and reviewed some new clinical trials for brexpiprazole, aripiprazole, olanzapine/fluoxetine combination, quetiapine, XR, lurasidone, and ziprasidone reaffirmed positive efficacy of SGA as augmentation agents for the treatment of MDD or treatment-refractory depression, with the strongest evidence for these FDA approved agents including olanzapine/Fluoxetine combination, quetiapine XR, aripiparzole, and brexpiprazole (Wang et al. 2016).

While there is convincing efficacy data, SGA's safety and tolerability remain to be of clinically concerned and have to be weighed in treatment of refractory depression. Spielmans et al. reviewed 14 studies including 3549 adjunctive aripiprazole that had a very high rate of akathisia (Number needed to harm NH = 4), linked to sedation (NNH = 14), OF combination had a higher risk of weight gain of ≥ 10% (NNH = 9), sedation (NNH = 5), abnormal metabolic laboratory results (NNH = 10), and elevated prolactin (NNH = 6), adjunctive quetiapine had high rate of sedation (NNH = 3) and metabolic laboratory results (NNH = 6), linked to significant weight gain of ≥ 7% (NNH = 37).

Other studies report discontinuation rate due to adverse events 3.8% in aripiprazole, 11.6% in olanzapine (with fluoxetine), 8.9% in quetiapine ER 150 mg/day, 15.4% in quetiapine er 300 mg/day, 6.5% in risperidone, and in brexpiprazole (Citrome 2010, Nelson and Papakostas 2009; Citrome 2015).

In VAST-D study, 5.3% (27/505) subjects in aripiprazole augmentation group discontinued due to side effects at week 12, and 9% reported weight gain of 7% or greater, 14.9% reported akathisia (Mohamed et al. 2017).

11.2.4 Augmentation with Lithium

Lithium is a mood stabilizer, the first-line therapy for bipolar disorder, including associated depression as well as mania. Lithium has also been reported to be an effective augmentation agent in the treatment of refractory depression (Bauer et al. 2014). The mechanism of lithium in affective disorder remains unclear. It inhibits enzymes that have magnesium as a cofactor such as Glycogen Synthase Kinase 3-beta (GSK3B) that inhibits signaling induced by Brain-Derived Neurotrophic Factor (BDNF), which may have implication in affect regulation (Ge and Jakobsson 2018).

In a meta-analysis 10 augmentation studies ($n = 269$) adding lithium to various antidepressants including selective serotonin reuptake inhibitors, lithium was significantly more effective than placebo (odds ratio = 3.11, 95%CI = 1.80–5.37) (Crossley and Bauer 2007).

In another meta-analysis of nine randomized clinical trials ($n = 237$), lithium augmentation again showed to be more effective than TCA alone or second-generation antidepressant (odds ratio = 2.89 with 95%CI: 1.65, 5.05). Discontinuation due to adverse events was infrequent and did not differ between lithium augmentation group and antidepressant alone group (Nelson et al. 2014).

The STAR*D study showed that up to 16% of nonresponders after two-failed trials at stage 3 could still benefit from lithium augmentation adjunction (Nierenberg et al. 2006).

The additional advantage of lithium augmentation is its well-known effectiveness to reduce suicidal behavior in mood disorders (Lewitzka et al. 2015; Smith and Cipriani 2017).

11.2.5 Augmentation with Triiodothyronine (T3)

Thyroid hormones have been implicated in depression, the study showed that free thyroid hormones concentrations are associated with depression severity and have an impact on final clinical outcome. It can be more efficient to augment and accelerate the treatment of major depressive disorder with triiodothyronine instead of levothyroxine because of individual differences in thyroid hormones metabolism (Berent et al. 2014).

Though its mechanism is not clear, it may desensitize 5-HT 1A inhibitory receptors, directly effect on nuclear receptors affecting gene expression (Nierenberg et al. 2006), and increased brain metabolism. Clinically, supplementation of thyroid hormones is considered to augment and accelerate antidepressant treatment.

A meta-analysis of eight studies (total $N = 292$) supported the efficacy of T_3 augmentation. It found that patients treated with triiodothyronine augmentation were twice as likely to respond as controls (relative response, 2.09; 95% confidence interval [CI], 1.31 to 3.32; $P = 0.002$). This corresponded to a 23.2% absolute improvement in response rates (95%CI: 4.5–41.9%; $P = 0.02$) with a number needed to treat of 4.3 (Aronson et al. 1996).

In stage 3 of STAR-D study, a total of 142 adult outpatients with nonpsychotic major depressive disorder who had not achieved remission or who were intolerant to an initial prospective treatment with citalopram and a second switch or augmentation trial were randomly assigned to augmentation with lithium (up to 900 mg/day; $N = 69$) or with T_3 (up to 50 μg/day; $N = 73$) for up to 14 weeks. After a mean of 9.6 weeks (SD = 5.2) of treatment, remission rates were 15.9% with lithium augmentation and 24.7% with T_3 augmentation, although the difference between treatments was not statistically significant. Lithium was more frequently associated with side effects ($p = 0.045$), and more participants in the lithium group left treatment because of side effects (23.2% vs. 9.6%; $p = 0.027$) (Nierenberg et al. 2006).

The commonly used augmenting agents with strong evidence are summarized below in Table 11.1.

Table 11.1 Augmenting agents with strong evidence used in the treatment of resistant depression

Augmenting agent	Potential mechanism	Comments
Bupropion	Inhibiting the reuptake of both noradrenaline and dopamine	Remission rate of 39% in STAR*D and 27% in VAST-D. Significant weight loss reported in 5% after 12 weeks treatment, and 12% after 36 weeks in VAST-D study (Mohamed et al. 2017; Trivedi et al. 2006a, b)
Lithium	Inhibits enzymes that have magnesium as a cofactor such as Glycogen Synthase Kinase 3-beta (GSK3B) that inhibits signaling induced by Brain-Derived Neurotrophic Factor (BDNF)	More effective than antidepressant (odds ratio = 2.89; 95% CI 1.65–5.05). Remission rate of 16% after two-failed trials in the STAR*D study; Discontinuation due to adverse events was infrequent and did not differ between lithium augmentation group and antidepressant alone group. The additional advantage of lithium augmentation is its well-known effectiveness to reduce suicidal behavior in mood disorders (Nelson et al. 2014; Nierenberg et al. 2006)

(continued)

Table 11.1 (continued)

Augmenting agent	Potential mechanism	Comments
Triiodothyronine	Desensitizes 5-HT 1A inhibitory receptors, directly effects on nuclear receptors affecting gene expression, and increased brain metabolism	Twice as likely to respond as controls (relative response, 2.09; 95%CI: 1.31–3.32, $P = 0.002$) remission rate of 23% after two failed trials in the STAR*D study (Aronson et al. 1996; Nierenberg et al. 2006)
Aripiprazole	Acting as a partial agonist at dopamine D2, D3, and serotonin 5-HT1Areceptors and as an antagonist at 5-HT2Areceptors	Higher remission rate 29.8% than placebo 18.1% ($P < 0.001$) in pooled data; Remission rates were 29% for Augment-ARI, significantly more efficacy than 22% for Switch-BUP ($p = 0.018$), and comparable to 27% for Augment-BUP ($p = 0.46$) in VAST-D study High akathisia in aripiprazole flexible dose 36.6%, fixed dose 14.2%, versus placebo 4.1%; significant weight gain in 4.9% versus 0.7% in pooled data, 9.5% at week 12 and 25.2% at week 36 in VAST-D (Mohamed et al. 2017; Luan et al. 2018)
Quetiapine	Affinity for different serotonergic and noradrenergic receptors, and blocks 5-HT2Areceptors with greater affinity than D2receptors	Higher remission rate 35.8% (231/645) than placebo group 24.5% (78/332) with OR = 1.79; 95%CI = 1.33–2.42; significant sedation 55.5% versus 15.4%; weight gain 4.8% versus 1.5% (Spielmans et al. 2013)
Olanzapine in combination with Fluoxetine	Possible synergistic antidepressant effect via dopamine and norepinephrine	Higher remission rate 31.8% versus control group 24.4% ($P < 0.001$), higher incidence of somnolence, fatigue, peripheral edema, sedation, and hypersomnia (Luan et al. 2017)
Brexpiprazole	Has lower intrinsic activity at the dopamine D2 receptor and tenfold higher affinity for serotonin 5-HT1A and 5-HT2A receptors than aripiprazole	Higher remission 16.5% versus placebo 9.4% and higher response 24.7% versus placebo 14.6%; higher weight gain (7.3% vs. 2%) and akathisia (8.6% vs. 1.0%) in 6 weeks trials; weight gain 25.5% and akathisia 10.0% showed in 52 weeks trials (Thase et al. 2016; Nelson et al. 2016; Hobart et al. 2018a, b)

Developed by author Chen, Peijun in this work

Table 11.2 Some antidepressant augmenting agents used in treatment-resistant depression

Augmenting agent	Proposed mechanism	Comment
Pindolol	5-HT1A receptor antagonist	May speed SSRI onset; ineffective in TRD in larger trials (Pérez et al. 1997)
Buspirone	5-HT1A partial agonist	Used in STAR*D; no consistent placebo-controlled data suggesting efficacy in TRD (Trivedi et al. 2006a, b)
Lamotrigine	Glutamate antagonist	Effective in bipolar depression; uncertain efficacy in TRD (Barbee and Jamhour 2002)
S-Adenosylmethionine	Methyl donor	Effective in the study of SSRI-resistant patients (Papakostas et al. 2010)
Folic acid	Methyl donor	Ineffective in a recent large trial (Bedson et al. 2014)
L-Methylfolate	Methyl donor	Possibly effective in a dose of 15 mg daily (Papakostas et al. 2012)
Metyrapone	Blocks cortisol synthesis	Ineffective in a recent large trial (McAllister-Williams et al. 2016)
Modafinil	Stimulant	Moderately effective in meta-analysis (Goss et al. 2013)
Lisdexamfetamine	Stimulant	Ineffective in phase 3 trials (Papakostas and Ionescu 2015)
Eicosapentaenoic acid (EPA)	Anti-inflammatory	Possibly effective in small randomized trials (Nemets et al. 2002)

From Cowen (2017) Backing into the future: pharmacological approaches to the management of resistant depression. Psychol Med 47(15):2569–2577

11.2.6 Augmentation with Other Agents

Many other agents have been trialed as augmentation agents, were summarized in Table 11.2 (Cowen 2017).

11.3 Comparing Effectiveness of Different Agents and Different Strategies (Switching vs. Augmentation)

In general, optimizing pharmacotherapy consists of switching to different agents, combination with different antidepressants, or augmentation with different class of psychotropic medications. Though direct comparison data are limited, few recent studies shed light into this perspective.

In a randomized trial compared adjunctive quetiapine (300 mg/day) with lithium (0.6–1.2 mmol/L) for 6 weeks in patients ($n = 450$) with major depression who, by history, did not respond to at least one antidepressant, the authors found comparable remission in quetiapine augmentation (32%) and in lithium augmentation (27%). The discontinuation due to adverse events was also comparable with quetiapine (10%) and lithium (8%) (Bauer et al. 2013).

Similar design for an open-label, 6-week, randomized trial ($n = 103$), compared adjunctive aripiprazole (mean dose 3 mg/day) or bupropion (mean dose 199 mg/day), found higher remission rate in aripiprazole augmentation group (55%) than bupropion augmentation (34%). Discontinuation of treatment was comparable for aripiprazole and bupropion (25% vs. 23%) and the incidence of extrapyramidal symptoms and of akathisia was comparable for the two groups (Cheon et al. 2017).

In the next step of randomized clinical trial of STAR-D for patients with nonpsychotic major depressive disorder without remission despite a mean of 12 weeks of citalopram therapy, bupropion-SR augmentation ($n = 565$) was compared with buspirone augmentation ($n = 286$). The bupropion-SR group and the buspirone group have similar rates of HRSD-17 remission (29.7% vs. 30.1%), QIDS-SR-16 remission (39.0% vs. 32.9%). Bupropion SR was associated with a greater reduction (from baseline to the end of this study) in QIDS-SR-16 scores than was buspirone (25.3% vs. 17.1%, $P < 0.04$), a lower QIDS-SR-16 score at the end of this study (8.0 vs. 9.1, $P < 0.02$), and a lower dropout rate due to intolerance (12.5% vs. 20.6%, $P < 0.009$) Trivedi et al. 2006a, b).

In an attempt to compare augment ($N = 269$) and switch ($N = 269$) groups using propensity-score-matched design from STAR-D trial, it found no statistically difference in likelihood of remission (risk ratio, 1.14; 95% confidence level, 0.82–1.58) or response (risk ratio, 1.14; 95% confidence level, 0.82–1.58). Post hoc analysis suggested that augmentation improved outcomes for patients tolerating 12 or more weeks of initial treatment and those with partial initial treatment response (Gaynes et al. 2012).

VAST-D was designed to directly evaluate the comparative effectiveness and side effect profiles of three commonly used "next-step" MDD treatment strategies: switch to another antidepressant, bupropion SR (Switch-BUP); augment current treatment with bupropion SR (Augment-BUP); or augment current treatment with an antipsychotic, aripiprazole (Augment-ARI) (Mohamed et al. 2017).

VAST-D randomized patients with nonpsychotic MDD, who failed at least one antidepressant trial meeting minimal standards for treatment dose and duration into Switch-BUP group ($n = 511$); Augment-BUP group ($n = 506$), or Augment-ARI group ($n = 505$) for 12 weeks of acute treatment and up to 36 weeks for long-term follow-up. At the end of 12 weeks, remission rates were 22% for Switch-BUP, 27% for Augment-BUP and 29% for Augment-ARI. Augment-ARI exceeded Switch-BUP in remission (Odds Ratio [OR]: 1.42; 95%CI: 1.06, 1.89; $p = 0.018$) but comparable between Augment-ARI versus Augment-BUP (OR = 1.11, 95%CI = 0.84–1.48,

p = 0.46), while Augment-BUP appeared to be higher than Switch-BUP but not reached statistically significant difference (OR = 1.31, 95%CI = 0.97–1.75, p = 0.076).

Response rate was greater for Augment-ARI (74%) than for either Switch-BUP (62%; OR: 1.74; 95%CI: 1.33, 2.29; $p < 0.0001$) or Augment-BUP (66%; OR: 1.56; 95%CI: 1.17, 2.05; p = 0.002) while Augment-BUP appeared to be only slightly higher than Switch-BUP without statistically significant difference (OR = 1.16, 95%CI = 0.89–1.50, p = 0.28).

Retention of participants through the 12-week acute phase was greatest for Augment-ARI (80.4%), lower for Augment-BUP (74.7%), and lowest for Switch-BUP (69.1%). Query Withdrawal for lack of response was lower in the two augmentation groups (5.5% Augment-BUP, 2.6% Augment-ARI) than for Switch-BUP (9.6%). Withdrawal for medication side effects was also lower in the two augmentation groups (7.3% and 5.3%, respectively) than for Switch-BUP (10.0%) A total of 207 serious adverse effects occurred for 165 (10.8%) randomized participants with no significant differences for the number of events among treatments and no statistically significant differences for any specific system organ class, or individual SAE.

In studying serious adverse effects (SAE: hospitalization, life-threatening, persistent/significant disability or incapacity, death, or condition that may jeopardize patient and require medical or surgical treatment to prevent one of these outcomes), authors found no statistically significant differences in SAE rates among treatments: 10.2% in Switch-BUP group, 11.3% in Augment-BUP group, and 11.1% in Augment-ARI group. There were significant difference in some specific adverse effects among three treatment groups such as akathisia with rate 4.3%, 5.3%, and 14.9%, respectively; anxiety with rate 24.3%, 22.5%, 16.6%, respectively; somnolence 7.2%, 7.9%, 14.5%, respectively; weight gain of 7% or greater 2.3%, 1.9%, 9.5% at week 12, respectively; 5.2%, 5.2%, 25.2% at week 36, respectively (Chen et al. 2017, unpublished data). One study of OFC for up to 47 weeks reported 55.7% weight gain of 7% or greater (Brunner et al. 2014).

11.4 Future Direction

Options of pharmacotherapy for TRD are more viable over the past 15 years than before due to many advances in clinical trials especially some large trials such as STAR*D and VAST-D. However, many works need to be done to further advance this field.

1. Establish agreement on a standardized, systematic, and feasible definition of TRD.

Inconsistent outcomes of clinical trials are in part due to various definitions of TRD. It is hard to compare different trials without a clearly defined definition. Definition should clearly specify the number of prior treatment attempts, what an adequate dose is, and what an adequate duration is. At the very least, the minimum two-failed adequate treatment is required. Some researchers have proposed staging model to clarify those nonresponding to the first treatment as stage 1 (e.g., insufficient response), and those not responding to two trials as stage 2 (e.g., treatment resistant), and those not responding to three trials as stage 3 (e.g., treatment refractory), and those failed to four or more, or other treatment such as electroconvulsive therapy as stage 4 (e.g., treatment futile). Such clear staging will help design clinical trials and guild treatment planning.

2. Establish safety and tolerability beyond acute treatment phase.

Most clinical trials were conducted in the acute treatment phase. Only a few studies including continuing phase and/or maintenance phase. In real-world, a great number of patients usually continued their treatment for at least 6 months–12 months or longer.

3. Establish individual psychosocial and neurobiological marks.

As biological-genetic variation may account for up to 42% of variance in antidepressant response (Tansey et al. 2013), it is essential to call on more studies to define individual bio-genetic marks to apply precision medicine to clinical practice.

4. Utilize multi-treatment modules.

Like in other medical fields, multi-module treatment has become mainstream and may provide addictive or synergetic effects for MDD, especially TRD such as the combination of psychotherapy and pharmacotherapy in conjunction with brain stimulation therapy such as electroconvulsive therapy, vagus nerve stimulation and transcranial magnetic stimulation; as well as non-traditional therapy such as nutritional supplements, exercise and light therapy.

References

Aronson R, Offman HJ, Joffe RT, Naylor CD (1996) Triiodothyronine augmentation in the treatment of refractory depression: a meta-analysis. Arch Gen Psychiatry 53:842–848

Barbee JG, Jamhour NJ (2002) Lamotrigine as an augmentation agent in treatment-resistant depression. J Clin Psychiatry 63:737–741

Bauer M, Dell'osso L, Kasper S, Pitchot W, Dencker Vansvik E, Köhler J, Jørgensen L, Montgomery SA (2013) J Affect Disord 151(1):209–219. Epub 27 Jun 2013

Bauer M, Adli M, Ricken R, Severus E, Pilhatsch M (2014) Role of lithium augmentation in the management of major depressive disorder. CNS Drugs 28(4):331–342, Apr 2014

Bedson E, Bell D, Carr D, Carter B, Hughes D, Jorgensen A, Lewis H, Lloyd K, McCaddon A, Moat S, Pink J (2014) Folate augmentation of Treatment–Evaluation for Depression (FolATED): randomised trial and economic evaluation. Health Technol Assess 18:1–159

Bennabi D, Aouizerate B, El-Hage W, Doumy O, Moliere F, Courtet P et al (2014) Risk factors for treatment resistance in unipolar depression: a systematic review. J Affect Disord 171C:137–141

Berent D, Zboralski K, Orzechowska A, Gałecki P (2014) Thyroid hormones association with depression severity and clinical outcome in patients with major depressive disorder. Mol Biol Rep 41(4):2419–2425

Blier P, Gobbi G, Turcotte JE, de Montigny C, Boucher N, Hébert C, Debonnel G (2009) Mirtazapine and paroxetine in major depression: a comparison of monotherapy versus their combination from treatment initiation. Eur Neuropsychopharmacol 19:457–465

Blier P, Ward HE, Tremblay P, Laberge L, Hébert C, Bergeron R (2010) Combination of antidepressant medications from treatment initiation for major depressive disorder: a double-blind randomized study. Am J Psychiatry 167:281–288

Brunner E, Tohen M, Osuntokun O, Landry J, Thase ME (2014) Efficacy and safety of olanzapine/fluoxetine combination vs. fluoxetine monotherapy following successful combination therapy of treatment-resistant major depressive disorder. Neuropsychopharmacology 39(11):2549–2559

Bschor T, Kern H, Henssler J, Baethge C (2018) Switching the antidepressant after nonresponse in adults with major depression: a systematic literature search and meta-analysis. J Clin Psychiatry 79(1):16r10749

Carpenter LL, Yasmin S, Price LH (2002) A double-blind, placebo-controlled study of antidepressant augmentation with mirtazapine. Biol Psychiatry 51:183–188

Cheon EJ, Lee KH, Park YW, Lee JH, Koo BH, Lee SJ, Sung HM (2017) Comparison of the efficacy and safety of aripiprazole versus bupropion augmentation in patients with major depressive disorder unresponsive to selective serotonin reuptake inhibitors: a randomized, prospective, open-label study. J Clin Psychopharmacol 37(2):193

Citrome L (2010) Adjunctive aripiprazole, olanzapine, or quetiapine for major depressive disorder: an analysis of number needed to treat, number needed to harm, and likelihood to be helped or harmed. Postgrad Med 122:39–48

Citrome L (2015) Brexpiprazole for schizophrenia and as adjunct for major depressive disorder: a systematic review of the efficacy and safety profile for this newly approved antipsychotic: what is the number needed to treat, number needed to harm, and likelihood to be helped or harmed? Int J Clin Pract 69:978–997

Cowen PJ (2017) Backing into the future: pharmacological approaches to the management of resistant depression. Psychol Med 47(15):2569–2577

Crossley NA, Bauer M (2007) Acceleration and augmentation of antidepressants with lithium for depressive disorders: two meta-analyses of randomized, placebo-controlled trials. J Clin Psychiatry 68:935–940

Demyttenaere K, Van Duppen Z (2019) The impact of (the concept of) treatment-resistant depression: an opinion review. Int J Neuropsychopharmacol 22(2):85–92

Fava M, Rush AJ, Wisniewski SR et al (2006) A comparison of mirtazapine and nortriptyline following two consecutive failed medication treatments for depressed outpatients: a STAR*D report. Am J Psychiatry 163(7):1161–1172

Fornaro M, Martino M, Mattei C, et al (2014) Duloxetine-bupropion combination for treatment-resistant atypical depression: a double-blind, randomized, placebo-controlled trial. Eur Neuropsychopharmacol. 24(8):1269

Gaynes BN, Dusetzina SB, Ellis AR, Hansen RA, Farley JF, Miller WC, Stürmer T (2012) Treating depression after initial treatment failure: directly comparing switch and augmenting strategies in STAR*D. J Clin Psychopharmacol 32(1):114–119

Ge W, Jakobsson E (2018) Systems biology understanding of the effects of lithium on affective and neurodegenerative disorders. Front Neurosci 12:933

Goss AJ, Kaser M, Costafreda SG, Sahakian BJ, Fu CH (2013) Modafinil augmentation therapy in unipolar and bipolar depression: a systematic review and meta-analysis of randomized controlled trials. J Clin Psychiatry 74:1101–1107

Greden JF (2001) The burden of disease for treatment-resistant depression. J Clin Psychiatry 62(Suppl 16):26–31

Gulrez G, Badyal DK, Deswal RS, Sharma A (2012) Bupropion as an augmenting agent in patients of depression with partial response. Basic Clin Pharmacol Toxicol 110(3):227–230

Hobart M, Skuban A, Zhang P et al (2018a) A randomized, placebo-controlled study of the efficacy and safety of fixed-dose brexpiprazole 2 mg/d as adjunctive treatment of adults with major depressive disorder. J Clin Psychiatry 79(4):17m12058

Hobart M, Zhang P, Weiss C, Meehan SR, Eriksson H (2018b) Adjunctive brexpiprazole and functioning in major depressive disorder: a pooled analysis of six randomized studies using the Sheehan disability scale. Int J Neuropsychopharmacol 22(3):173–179

Kautzky A, Dold M, Bartova L, Spies M, Kranz GS, Souery D et al (2019) Clinical factors predicting treatment resistant depression: affirmative results from the European multicenter study. Acta Psychiatr Scand 139:78–88

Kessler D, Burns A, Tallon D, Lewis G, MacNeill S, Round J et al (2018) Combining mirtazapine with SSRIs or SNRIs for treatment-resistant depression: the MIR RCT. Health Technol Assess 22(63)

Lewitzka U, Severus E, Bauer R, Ritter P, Müller-Oerlinghausen B, Bauer M (2015) The suicide prevention effect of lithium: more than 20 years of evidence-a narrative review. Int J Bipolar Disord 3(1):32. https://doi.org/10.1186/s40345-015-0032-2

Luan S, Wan H, Wang S, Li H, Zhang B (2017) Efficacy and safety of olanzapine/fluoxetine combination in the treatment of treatment-resistant depression: a meta-analysis of randomized controlled trials. Neuropsychiatr Dis Treat 13:609–620

Luan S, Wan H, Zhang L, Zhao H (2018) Efficacy, acceptability, and safety of adjunctive aripiprazole in treatment-resistant depression: a meta-analysis of randomized controlled trials. Neuropsychiatr Dis Treat 14:467–477

McAllister-Williams RH, Anderson IM, Finkelmeyer A, Gallagher P, Grunze HC, Haddad PM, Hughes T, Lloyd AJ, Mamasoula C, McColl E, Pearce S (2016) Antidepressant augmentation with metyrapone for treatment-resistant depression (the ADD study): a double-blind, randomised, placebo-controlled trial. Lancet Psychiatry 3:117–127

McIntyre RS, Filteau MJ, Martin L, Patry S, Carvalho A, Cha DS et al (2014) Treatment-resistant depression: definitions, review of the evidence, and algorithmic approach. J Affect Disord 156:1–7

Mohamed S, Johnson GR, Chen P, et al (2017) Effect of antidepressant switching vs. augmentation on remission among patients with major depressive disorder unresponsive to antidepressant treatment: the VAST-D randomized clinical trial. JAMA 318:132–145

Mulder R, Hamilton A, Irwin L, Boyce P, Morris G, Porter RJ, Malhi GS (2018) Treating depression with adjunctive antipsychotics. Bipolar Disord (Suppl 2):17–24. https://doi.org/10.1111/bdi.12701

Nelson JC, Papakostas GI (2009) Atypical antipsychotic augmentation in major depressive disorder: a meta-analysis of placebo-controlled randomized trials. Am J Psychiatry 166(9):980–991

Nelson JC, Baumann P, Delucchi K, Joffe R, Katona C (2014) A systematic review and meta-analysis of lithium augmentation in tricyclic and second generation antidepressants in major depression. J Affect Disord 156:269–275

Nelson JC, Zhang P, Skuban A et al (2016) Overview of short-term and long-term safety of brexpiprazole in patients with major depressive disorder and inadequate response to antidepressant treatment. Curr Psychiatry Rev 12(3):278–290

Nemets B, Stahl Z, Belmaker RH (2002) Addition of omega-3 fatty acid to maintenance medication treatment for recurrent unipolar depressive disorder. Am J Psychiatry 159:477–479

Nierenberg AA, Fava M, Trivedi MH et al (2006) A comparison of lithium and T(3) augmentation following two failed medication treatments for depression: a STAR*D report. Am J Psychiatry 163:1519–1530

Papakostas GI, Shelton RC, Smith J, Fava M (2007) Augmentation of antidepressants with atypical antipsychotic medications for treatment-resistant major depressive disorder: a meta-analysis. J Clin Psychiatry 68:826–831

Papakostas GI, Fava M, Thase ME (2008) Treatment of SSRI-resistant depression: a meta-analysis comparing within- versus across-class switches. Biol Psychiatry 63(7):699–704

Papakostas GI, Mischoulon D, Shyu I, Alpert JE, Fava M (2010) S-adenosyl methionine (SAMe) augmentation of serotonin reuptake inhibitors for antidepressant nonresponders with major depressive disorder: a double-blind, randomized clinical trial. Am J Psychiatry 167:942–948

Papakostas GI, Shelton RC, Zajecka JM, Etemad B, Rickels K, Clain A, Baer L, Dalton ED, Sacco GR, Schoenfeld D, Pencina M (2012) L-methylfolate as adjunctive therapy for SSRI-resistant major depression: results of two randomized, double-blind, parallel-sequential trials. Am J Psychiatry 169:1267–1274

Papakostas GI, Ionescu DF (2015) Towards new mechanisms: an update on therapeutics for treatment-resistant major depressive disorder. Mol Psychiatry 20:1142–1150

Patel K, Allen S, Haque MN, Angelescu I, Baumeister D, Tracy DK (2016) Bupropion: a systematic review and meta-analysis of effectiveness as an antidepressant. Ther Adv Psychopharmacol 6(2):99–144. https://doi.org/10.1177/2045125316629071. Epub 18 Feb 2016

Pérez V, Gilaberte I, Faries D, Alvarez E, Artigas F (1997) Randomised, double-blind, placebo-controlled trial of pindolol in combination with fluoxetine antidepressant treatment. Lancet 349:1594–1597

Pérez V, Salavert A, Espadaler J et al (2017) Efficacy of prospective pharmacogenetic testing in the treatment of major depressive disorder: results of a randomized, double-blind clinical trial. BMC Psychiatry 17(1):250

Rush AJ, Trivedi MH, Wisniewski SR, Stewart JW, Nierenberg AA, Thase ME et al (2006a) STAR*D study team. Bupropion-SR, sertraline, or venlafaxine-XR after failure of SSRIs for depression. N Engl J Med 354(12):1231–1242

Rush AJ, Trivedi MH, Wisniewski SR, Nierenberg AA, Stewart JW, et al (2006b) Acute and longer-term outcomes in depressed outpatients requiring one or several treatment steps: a STAR* D report. Am J Psychiatry 163:1905–1917

Rush AJ, Wisniewski SR, Warden D, Luther JF, Davis LL, Fava M et al (2008) Selecting among second-step antidepressant medication monotherapies: predictive value of clinical, demographic, or first-step treatment features. Arch Gen Psychiatry 65(8):870–880

Rush AJ, Trivedi MH, Stewart JW, Nierenberg AA, Fava M, Kurian BT et al (2011) Combining medications to enhance depression outcomes (CO-MED): acute and long-term outcomes of a single-blind randomized study. Am J Psychiatry 168:689–701

Simon GE, Khandker RK, Ichikawa L, Operskalski BH (2006) Recovery from depression predicts lower health services costs. J Clin Psychiatry 67:1226–1231

Smith KA, Cipriani A. Lithium and suicide in mood disorders: updated meta-review of the scientific literature.Bipolar Disord. 2017 Nov;19(7):575–586

Spielmans GI, Berman MI, Linardatos E, Rosenlicht NZ, Perry A, Tsai AC (2013) Adjunctive atypical antipsychotic treatment for major depressive disorder: a meta-analysis of depression, quality of life, and safety outcomes. PLoS Med 10:e1001403

Tansey KE, Guipponi M, Hu X, Domenici E, Lewis G, Malafosse A et al (2013) Contribution of common genetic variants to antidepressant response. Biol Psychiatry 73(7):679–682. https://doi.org/10.1016/j.biopsych.2012.10.030

Thase ME, Zhang P, Skuban A, Hobart M, Weiss C, Weiller E, Nelson JC (2016) Efficacy of adjunctive brexpiprazole in patients with major depressive disorder: a clinical overview. Curr Psychiatry Rev 12:291–301

Trivedi MH, Fava M, Wisniewski SR, Thase ME, Quitkin F, Warden D, Ritz L, Nierenberg AA, Lebowitz BD, Biggs MM, Luther JF, Shores-Wilson K, Rush AJ (2006a) STAR*D study team. Medication augmentation after the failure of SSRIs for depression. N Engl J Med 354(12):1243–1252

Trivedi MH, Rush AJ, Wisniewski SR et al (2006b) Evaluation of outcomes with citalopram for depression using measurement-based care in STAR*D: implications for clinical practice. Am J Psychiatry 163(1):28–40

Wang SM, Han C, Lee SJ et al (2016) Second generation antipsychotics in the treatment of major depressive disorder: an update. Chonnam Med J. 52(3):159–172

WHO (2018) Depression fact sheet. https://www.who.int/en/news-room/fact-sheets/detail/depression. Accessed 17 Mar 2019

Zisook S, Johnson GR, Tal I, Hicks P, Chen P, Davis L et al (2019) General predictors and moderators of depression remission: a VAST-D report. Am J Psychiatry Published online 4 May 2019

Chapter 12
Individualized Treatment Strategy for Depressive Disorder

Jun Chen and Shaohua Hu

Abstract The goal of treatment for depressive disorders is complete remission of depressive symptoms with full recovery of social function and prevention of recurrence. However, a large proportion of patients do not experience symptomatic remission after the initial treatment, with even lower rates of remission in the longer treatment term. The main objective of individualized treatment applied in psychiatry is to improve precision in disease diagnosis, prognosis, treatment choices, and treatment response. Diverse approaches and techniques, such as genomics, epigenomics, other omics, neural circuit, and artificial intelligence are related to precision psychiatry. Using biology and computational psychiatry tools to find potential biomarkers, and based on precision psychiatry, patients considered to belong to the same endophenotype will be possible to receive biomarkers-based treatment and better prognosis. Especially in the choice of intervention, individualized treatment should be considered. In this review, we present the development of precise treatment in depressive disorders and introduce advances in several domains toward precision medicine and individualized treatment. We pay particular attention to biomarkers and the development of new technologies in depressive disorders, which will help disease complete remission and functional recovery, seek better lives for patients suffered with depressive disorders.

Keywords Depressive disorders · Individualized treatment · Precision medicine · Genomics · Neural circuit · Artificial intelligence

---

J. Chen (✉)
Clinical Research Center and Division of Mood Disorders, Shanghai Mental Health Center, Shanghai Jiao Tong University School of Medicine, Shanghai, China
e-mail: doctorcj2010@gmail.com

S. Hu (✉)
Department of Psychiatry, First Affiliated Hospital, Zhejiang University School of Medicine, Hangzhou, China
e-mail: dorhushaohua@zju.edu.cn

© Springer Nature Singapore Pte Ltd. 2019

Y. Fang (ed.), *Depressive Disorders: Mechanisms, Measurement and Management*, Advances in Experimental Medicine and Biology 1180,
https://doi.org/10.1007/978-981-32-9271-0_12

12.1 Precise Treatment Strategy for Depressive Disorders

12.1.1 Introduction

Several treatment strategies involving pharmacotherapy, psychological therapy, and physical therapy are available for depressive disorders (Malhi and Mann 2018). The main objective of treatment is to improve clinical remission of depression, social function recovery and prevent recurrence (Malhi and Mann 2018). However, many patients with major depressive disorders do not experience symptomatic remission with individualized treatment. Only one-third of patients achieved remission with the first treatment, and even more than one-third patient did not remit after one or more therapies (Rush et al. 2006; Gaynes et al. 2009). Previous studies in depression treatment were concerned with characteristics of individuals associated with response to treatment, but increased research has more focused on moderator and mediator characteristics in recent years (Trivedi 2016). Moderators are variables which predict response to different treatments pre-treatment and mediator variables predict eventual treatment outcomes (Trivedi 2016). To determine the right treatment for the right patient, precision medicine has aroused widespread concern in clinical specialties, which is an emerging approach for targeted pharmacological treatment and prevention. Psychiatry has not yet benefited from the advanced diagnostic and therapeutic technologies, new therapeutic strategies toward precision medicine seem promising for psychiatry (Fernandes et al. 2017). In precision medicine, according to the classifications of disease, patients would receive a pre-established treatment rather than personalized treatment based on individuals' unique features (Dalvie et al. 2016). Precision medicine applied to psychiatry was called "precision psychiatry". The NIMH Research Domain Criteria (RDoC) Project launched by the National Institute of Mental Health (NIMH) has addressed the need for precision medicine for psychiatry (Insel 2014). This project aims to develop a new approach to disease classification based on pathophysiology and provide individual-specific interventions precisely (Insel 2014). Precision psychiatry was propelled by advances in the field of genetics, molecular, and neuroanatomical markers and artificial intelligence, which would be possible to more accurately for diagnosis, prognosis, classifications, treatment choice, and predicting response to treatment (Fernandes et al. 2017).

Herein, we summarize the precise treatment strategies for depressive disorders, and introduce advances in several domains toward precision medicine and propose prospective for the future of precision psychiatry.

12.1.2 Development of Precise Treatment in Depressive Disorders

Different staging models have been proposed in psychiatry, aim to achieve more personalized treatment for patients, a higher level of precision in every stage is desirable (McGorry et al. 2014). From DSM/ICD to precision staging models were used to provide more personalized treatment for patients (Salagre et al. 2018). Treatment based on the DSM/ICD model was according to clinical symptoms and course specifiers. Staging model treatment was based on the clinical course, cognition, functionality, and biomarkers. Furthermore, precision staging model treatment was based on patients' characteristics, including genomics, epigenomics, other omics, and technology (Salagre et al. 2018).

Pharmacogenomics in Depressive Disorders
Pharmacogenetics is a subject that studies genetic background influencing drug response and tolerability. It was shown great variability in response and tolerability among patients with antidepressants treatment (Pare and Mack 1971). Genome-wide association studies were performed for discovering numerous drug-response phenotypes, however, inconsistent findings showed compelling associations between DNA variants and antidepressant drug response (Laje and McMahon 2011). Pharmacometabolomics determines the metabotype or metabolic state of individuals, and the effects of drugs on metabolism, the mechanisms of a drug's action and variation in drug responses were determined by metabolic profiles (Beger et al. 2016). For example, the guidelines recommend avoiding tricyclic antidepressants in patients with poor metabolizer CYP2D6 and CYP2C19 (Relling and Klein 2011). Ultra-rapid CYP2C19 metabolizers should avoid treatment with citalopram, escitalopram, and sertraline, and treatment with paroxetine and fluvoxamine in ultra-rapid CYP2D6 metabolizers patients is also inappropriate (Hamilton 2015). Metabolomics, such as glucose, citrate, and uric acid, have been found as potential serum biomarkers to identify risperidone action in patients with schizophrenia (Xuan et al. 2011).

The monoamine theories of depression have been confirmed for several decades, and the monoaminergic system was the first investigated in antidepressant pharmacogenetics (Slattery et al. 2004). The 5-HT transporter SLC6A4 gene was candidate gene as predictors of treatment efficacy and treatment-related side effects (Fabbri et al. 2014). Serotoninergic genes 5-HT receptors 1A (5-HTR1A) and 2A (5-HTR2A) were also widely investigated. HTR1A blocks could lead to faster antidepressant action and HTR1A blocker, pindolol was proved to increase the response in clinical use (Perez et al. 1997). Furthermore, agonists of HTR2A receptor showed euphoriant effects (Newton et al. 1996). There were other genes coding for proteins involved in the transport and metabolism of antidepressants which were considered having a possible impact on antidepressants pharmacokinetics. The polymorphism of genes coding for CYP enzymes was important in antidepressants pharmacokinetics, particularly CYP2D6 and CYP2C19 genes, which involved in antidepressants metabolism. Although current findings enhance the prospects for genomic approaches, challenges related to heterogeneity and sample availability have plagued pharmacogenomics. Limited power of genetic variants to forecast patient outcomes is a more fundamental limitation of pharmacogenomic studies.

12.1.3 Epigenomics in Depressive Disorders

Pharmacogenetic studies show that the differences in treatment response are attributed to genetic variation (Gonda et al. 2018; Fabbri and Serretti 2015). Genetic and environmental factors occurring across the life span likely underlie the vulnerability for depression (Nestler et al. 2002). According to preliminary statistics, moderate genetic contribution of 37% (95% confidence interval [CI], 31–42%) was shown in a meta-analysis derived from five studies including more than 21,000 depressed subjects (Sullivan et al. 2000), and environmental factors for depression showed a significant contribution of 63% (95% CI, 58–67%) (Sullivan et al. 2000). Childhood trauma was proved associated with poor response to antidepressants (Miller et al. 2015). Epigenetic disruption also provides a potential mechanism to explain the association between gene-environment interactions and neuropsychiatric phenotypes. Epigenetic regulation includes DNA methylation, histone modifications, and non-coding RNAs, and DNA methylation has been one of the most popular and most often studied (Januar et al. 2015). DNA methylation was associated with antidepressant treatment response in major depressive disorders in a previous study (Lisoway et al. 2018). In addition, epigenetic modifications of SLC6A4, BDNF, interleukin-11 (IL11) have been investigated in more than one study as biomarkers for prediction of antidepressant response (Porcelli et al. 2011; Duclot and Kabbaj 2015; Powell et al. 2013). Gene SLC6A4 encoding the serotonin transporter has been associated with antidepressant response (Kraft et al. 2005). Better response following escitalopram treatment was found in depressed patients with higher methylation status of the SLC6A4 gene (Domschke et al. 2014). Increased SLC6A4 expression reduces the availability of serotonin, thus counteracting the serotonergic effects of antidepressants. In addition, hippocampal BDNF signaling plays an important role in the antidepressant effects, whereas its role in the etiology of depression is still not evaluated. Previous studies showed that BDNF expression and signaling could affect antidepressant treatments outcome but BDNF itself does not lead to depressive-like behaviors (Adachi et al. 2008; Castren and Rantamaki 2010). Reduced serum BDNF levels were increased after antidepressant treatment in depressed patients (Sen et al. 2008). Furthermore, clinical evidence indicated that high-serum BDNF levels prior to or early after treatment could predict the better response of antidepressant effect (Dreimuller et al. 2012; Haile et al. 2014). The increase in BDNF levels was observed associated with a reduction of the repressive H3K27 methylation at BDNF promoter 4 in patients with citalopram treatment (Lopez et al. 2013). Inflammatory response genetics also play an important role in the MDD pathogenesis. Increased cytokines levels were found in depressed patients and antidepressant therapy has been associated with decreased inflammatory responses (Miller et al. 2009). DNA methylation in IL11 could predict clinical response to antidepressants and IL11 expression differences were partly mediated by a single-nucleotide polymorphism (rs1126757), which was identified as a predictor of antidepressant response (Powell et al. 2013). In the Genome-Based Therapeutic Drugs for Depression (GENDEP) project, the response of escitalopram treatment could be predicted by interleukin-11 (IL11) gene

(Uher et al. 2010). The field of gene expression and epigenomic biomarkers is still in its infancy stage in precision psychiatry, and targeted therapeutic strategies based on epigenomic biomarkers seem promising in clinical practice.

Neural Circuit Taxonomy for Depressive Disorders
Tremendous advances have been made for understanding the brain circuitry in mental disorders, however, a brain-based taxonomy for depressive disorders is still lacking. Traditional diagnostic categories more focus on symptoms of disease, this taxonomy may be helpful in specify symptoms in terms of potential neural dysfunction, which could ultimately guide clinical practice. Unique biotypes of depression were defined by brain-based taxonomy would differ from traditional diagnostic categories (Williams 2016). The new taxonomy based on brain dysfunctions could help treatment for specific symptoms which these dysfunctions give rise. Six circuits including default mode, salience, negative affect, positive affect (reward), attention, and cognitive control, have been found associated with depression (Williams 2016). Default mode circuit includes the anterior medial prefrontal cortex, posterior cingulate cortex, and angular gyrus (Greicius et al. 2003). Activation of default mode circuit tends to have higher levels of depressive rumination in depressive disorders. Treatment with mindfulness, self-context therapies, Transcranial Magnetic Stimulation (TMS) was suitable for this biotype (Williams 2016). Salience circuit was identified by core nodes in the anterior cingulate cortex, anterior insula, and sublenticular extended amygdala (Seeley et al. 2007). These salience circuit dysfunctions might contribute to so-called anxious avoidance biotypes and treatment with serotonin reuptake inhibitors, Deep Brain Stimulation(DBS) were useful for this type (Williams 2016). Negative affect circuit was closely associated with negative bias and threat dysregulation in depressive disorders and hyperconnectivity between negative affect circuit and the default mode has been found in depression (Sheline et al. 2010). These biotypes as treatment-selection biomarkers may be predictive of better outcomes on cognitive behavior therapy (Williams 2016). The key structures in the positive affect circuit are the anterior cingulate cortex, the orbitofrontal cortex, and the striatum neurons (Haber and Knutson 2010). Anhedonia is a core symptom of major depressive disorders, and previous findings revealed that depressive patients with striatal hypoactivation contribute to the anhedonia biotype (Treadway and Zald 2011). A context-insensitivity biotype of neural dysfunction was found in remitted depression and persisted even during remission (Zhang et al. 2013). Dopamine noradrenaline reuptake inhibitors have better effects on this biotype (Williams 2016). Hypoconnectivity within attention circuit indicates the presence of an inattention biotype in depressive disorders and attention training, stimulants may alternate the attention deficit (Williams 2016). Cognitive dyscontrol biotype was characterized by dysfunction of the cognitive control circuit, cognitive impairment could by treatment with cognitive training and TMS (Williams 2016). Neural circuit taxonomy delivers new insights for precise treatment in depressive patients.

Artificial Intelligence in Depressive Disorders
Recent advances in computational psychiatry present a novel opportunity to analyze human behavior, including data-driven approaches and theory-driven approaches (Huys et al. 2016). To improve treatment outcomes in depressive disorders, replicable, and quantifiable predictors of therapeutic outcomes are needed. However, up to now, no consistent variable has emerged as a predictor of therapeutic effect (Rosenblat et al. 2017). Data-driven approaches can analyze large molecular genomic datasets to improve disease diagnosis, classification, and predict treatment outcomes based on machine-learning methods (Silbersweig and Loscalzo 2017). In the traditional psychiatric classification, the same symptoms always appear in more than one disorders (Borsboom et al. 2011). Machine-learning methods, including a large variety of mathematical/statistical models in the field of artificial intelligence (AI), would automatically classify patients and healthy controls (Orru et al. 2012). There is potential significant value in neuroimaging-based disease diagnoses and prediction of clinical response based on machine-learning analysis. The previous study using MRI data had distinguished schizophrenia from controls and achieved an area under the curve (AUC) of 0.89 between two groups (Huys et al. 2016). Furthermore, several Electroencephalography (EEG) markers have been found in predicting antidepressant response (Iosifescu 2011). In disease diagnosis, AI technologies, such as audio and video analysis, may assist with diagnosis and predict prognosis (Lovejoy et al. 2019). An automated machine-learning speech classifier was performed in predicting psychosis onset and reached an accuracy of 79% (Corcoran et al. 2018). Another study developed systems designed by MIT Media Lab to recognize human emotion, including frustration, confusion, interest, stress, anger, and joy (Picard 2002). In addition, AI may increase the availability and effectiveness of existing treatments, such as Internet-based Cognitive Behavioral Therapy(CBT), CBT chatbots. AI may also enable to rebalance a clinician's workload, personalization of care, and privacy protection of patients (Lovejoy et al. 2019). Essentially, AI applied in psychiatry represents the next stage of precision psychiatry. Advances in AI provide new tools for improving disease diagnosis, monitoring and treatment, and better explaining pathology.

12.2 Individualized Treatment Strategy for Depressive Disorders

12.2.1 Introduction

As we discussed above, precise treatment asks for more progress from the biomarker researches. Unfortunately, up to now, there is no biomarker could be stably used neither in diagnosis nor in treatment of the depressive disorders. Therefore, clinicians still are likely to use some particular features to carry on individualized treatment. Individualized treatment sounds like more clinically relevant, actually, and it is very closely related to the precise treatment. According to the precise treatment, patients' differences can be compensated by adapting the individual dose to genotype in addition to other clinical factors such as gender, weight, age, liver, and kidney function (Kirchheiner et al. 2003). That means patients' feature is very important to influence the selection of antidepressants.

The art of treatment for depressive disorders must be individualized. In order to select the best treatment plan, the physician needs a personalized view to be taken into consideration. Only based on known predicted response or risk of disease, better treatment outcome could be anticipated. In addition, personal histories, family histories, and support networks should be considered when designing an individualized treatment plan.

12.2.2 Specifiers and Clinical Dimensions for Depressive Disorders

DSM-5 Specifiers and Clinical Dimensions

The Diagnostic and Statistical Manual of Mental Disorders, Fifth Edition (DSM-5) is the 2013 update to the Diagnostic and Statistical Manual of Mental Disorders, the taxonomic and diagnostic tool published by the American Psychiatric Association (APA) (US 2013). In the DSM-5, depressive disorders are specified into different categories, which including with melancholic features, with atypical features, with psychotic features, with catatonic features, with anxious distress features, with mixed features, with cognitive dysfunction, with physical symptoms, and with sleep disturbance (Uher et al. 2014; Stein and Reed 2019; Casey et al. 2013). Based on these specifiers and clinical dimensions, DSM-5 want to subdivide depressive disorders into different subtypes, which could be more helpful for clinicians to do the individualized treatment.

ICD-11 Specifiers and Clinical Dimensions

The International Classification of Diseases, the eleventh revision (ICD-11) is the international statistical classification of diseases and related health problems published by the WHO (Selb et al. 2015; First et al. 2015). ICD-11 was accepted by WHO's World Health Assembly (WHA) on 25 May 2019 and will come into effect on 1 January 2022. Just as DSM-5 did before, ICD-11 defines depressive disorders into additional qualifiers which including with prominent anxiety symptoms, with melancholia, current episode perinatal, and with seasonal pattern. Although there is fewer specifiers and clinical dimensions in ICD-11 compared to those in DSM-5, but the individualized treatment is the same core value (Gitlin and Miklowitz 2014).

12.2.3 Clinical Dimensions that Influence the Antidepressant Selection

As discussed above, both DSM-5 and ICD-11 has divided depressive disorders into different specifiers and clinical dimensions, which are important to guide individualized treatment based on clinical features. Unfortunately, up to now, such kind of classification may look not as effective as it designed for. At least, most of the specifiers and clinical dimensions don't have adequate evidence to show the advantage of antidepressant selection. Among them, only 4-clinical dimensions have been proved by meta-analysis with narrow confidence intervals and/or 2 or more RCTs with adequate sample size, preferably placebo-controlled (Milev et al. 2016).

With Psychotic Features
For the psychotic depressive disorders, a meta-analysis (12 studies, $N = 929$) found that combination therapy with an antidepressant plus an antipsychotic is more effective than either treatment alone or placebo. Evidence is limited for treatment with an antidepressant alone or with an antipsychotic alone (Wijkstra et al. 2015).

With Cognitive Dysfunction
For cognitive dysfunction, available evidence suggests that antidepressants have a significant positive effect on psychomotor speed and delayed recall. In a meta-analysis (17 studies, $N = 3653$) found that vortioxetine had the largest effects on processing speed, executive control, and cognitive control (Rosenblat et al. 2015).

With Sleep Disturbances
For sleep disturbances, antidepressants are associated with different insomnia and somnolence rates, mainly depending on their mechanisms of action. Some antidepressants, including agomelatine, mirtazapine, and trazodone, and the atypical antipsychotic, quetiapine, have shown superior effects on subjective or objective sleep measures. However, mirtazapine, quetiapine, and trazodone also have the highest adverse event rates of somnolence and daytime sedation. In a meta-analysis (276 studies) found that agomelatine was the only antidepressant with a lower likelihood of inducing insomnia than placebo (Alberti et al. 2015).

With Somatic Symptoms
For somatic symptoms, there are few comparative studies of antidepressants for somatic symptoms such as pain and fatigue, bupropion was efficacious for fatigue (Thaler et al. 2012). In a meta-analysis (18 studies, $N = 6407$) found that SNRIs, especially duloxetine, was efficacious for painful conditions, including neuropathic pain and fibromyalgia (Lunn et al. 2014).

12.2.4 Factors to Consider in Selecting an Antidepressant

There are many individualized treatment factors that will influence the selection of an antidepressant, which include both clinical features and medication characteristics.

Therefore, selecting an antidepressant definitely involves an individualized needs assessment.

When considering the patient factors, clinical features, and dimensions of depressive disorders should be the first priority. Except for that, comorbid conditions, response and side effects of antidepressants, and even patient preference should be considered. When thinking about medication factors, both efficacy and tolerability should be balanced. Other factors including potential medications interactions, simplicity of use, cost, and availability (Yatham et al. 2006).

12.2.5 Measurement-Based Care

Another important piece of the puzzle for individualized treatment is Measurement-Based Care (MBC). MBC means the systematic use of measurement tools to monitor outcomes and support clinical decision-making (Gaynes et al. 2008; Guo et al. 2015). Using MBC must have a broader idea of treatment outcome, it should include continue evaluation both on symptoms and functional impairment (IsHak et al. 2011). MBC can help physicians to establish a better patient–doctor relationship, to support the long-term individualized treatment for depressive disorders.

12.3 Conclusion

Advances in biomedical "omics" and technologies have now permitted a move from traditional symptom-based categorization toward a biomarker-based precision model. The field of precision medicine applied to psychiatry has struggled to provide a better and more accurate diagnosis, guide treatment choice, predict response to treatment, and minimize adverse effects, thus seek better lives for patients with mental illness. Psychiatry has not beneficiated from the advanced diagnostic and therapeutic technologies till now, but precision psychiatry has considered the complex variability of individuals in clinical phenotypes, genes, and brain function (Trivedi 2016). Genome-wide association studies have revealed that there is no consistent associations between genetic variants and treatment response. Rare gene variants associated with treatment response have been found in few patients, but which have not widespread predictive value as predictors in antidepressant response. The new strategy integrates genetic, clinical, and neurobiological markers to guide clinical practice for depressive disorders is needed. In addition, neural circuit taxonomy has been found associated with the symptoms of depression, the brain-based taxonomy will deliver new insights for clinicians to improve treatment outcomes and close the gap between insights about the neuroanatomical mechanisms of depressive disorders. The advances in pharmacogenomics provide guidance for treatment choice and reduce adverse reactions to treatment.

In the current stage, precise treatment may not be widely used due to the limited basic research findings. But physicians still can use the individualized treatment concept to better understand the relationship between clinical features and potential biomarkers. Specifiers and clinical dimensions for depressive disorders are critical to the selection of treatment. There are many clues that will influence the selection of antidepressants, hence physicians should balance the MBC and their clinical experiences. When measuring the patients' situations, not only the depressive symptoms but also the functions should be measured, in order to get better recovery.

In recent decades, applications of artificial intelligence technology in psychiatry provide an opportunity to integrate multiple forms and sources of data to parse clinical heterogeneity and assist with disease diagnosis, monitoring, and treatment. In psychiatry, new tools used for diagnosis, monitoring, and treatment still face numerous risks and challenges. These will require to develop large cohorts of depressed patients that will help to discover new subtypes of depression and provide precise treatments for each patient. Future studies with well-design and new techniques are warranted to address this issue.

References

Adachi M, Barrot M, Autry AE, Theobald D, Monteggia LM (2008) Selective loss of brain-derived neurotrophic factor in the dentate gyrus attenuates antidepressant efficacy. Biol Psychiatry 63(7):642–649

Alberti S, Chiesa A, Andrisano C, Serretti A (2015) Insomnia and somnolence associated with second-generation antidepressants during the treatment of major depression: a meta-analysis. J Clin Psychopharmacol 35(3): 296–303

Beger RD, Dunn W, Schmidt MA, Gross SS, Kirwan JA, Cascante M, Brennan L, Wishart DS, Oresic M, Hankemeier T, Broadhurst DI, Lane AN, Suhre K, Kastenmuller G, Sumner SJ, Thiele I, Fiehn O, Kaddurah-Daouk R (2016) "Precision medicine pharmacometabolomics task group"-metabolomics society I. Metabolomics enables precision medicine: "A white paper, community perspective". Metabolomics 12(10):149

Borsboom D, Cramer AO, Schmittmann VD, Epskamp S, Waldorp LJ (2011) The small world of psychopathology. PLoS ONE 6(11):e27407

Casey BJ, Craddock N, Cuthbert BN, Hyman SE, Lee FS, Ressler KJ (2013) DSM-5 and RDoC: progress in psychiatry research? Nat Rev Neurosci 14(11):810–814

Castren E, Rantamaki T (2010) The role of BDNF and its receptors in depression and antidepressant drug action: reactivation of developmental plasticity. Dev Neurobiol 70(5):289–297

Corcoran CM, Carrillo F, Fernandez-Slezak D, Bedi G, Klim C, Javitt DC, Bearden CE, Cecchi GA (2018) Prediction of psychosis across protocols and risk cohorts using automated language analysis. World Psychiatry 17(1):67–75

Dalvie S, Koen N, McGregor N, O'Connell K, Warnich L, Ramesar R, Nievergelt CM, Stein DJ (2016) Toward a global roadmap for precision medicine in psychiatry: challenges and opportunities. OMICS 20(10):557–564

Domschke K, Tidow N, Schwarte K, Deckert J, Lesch KP, Arolt V, Zwanzger P, Baune BT (2014) Serotonin transporter gene hypomethylation predicts impaired antidepressant treatment response. Int J Neuropsychopharmacol 17(8):1167–1176

Dreimuller N, Schlicht KF, Wagner S, Peetz D, Borysenko L, Hiemke C, Lieb K, Tadic A (2012) Early reactions of brain-derived neurotrophic factor in plasma (pBDNF) and outcome to acute antidepressant treatment in patients with major depression. Neuropharmacology 62(1):264–269

Duclot F, Kabbaj M (2015) Epigenetic mechanisms underlying the role of brain-derived neurotrophic factor in depression and response to antidepressants. J Exp Biol 218(Pt 1):21–31

Fabbri C, Serretti A (2015) Pharmacogenetics of major depressive disorder: top genes and pathways toward clinical applications. Curr Psychiatry Rep 17(7):50

Fabbri C, Porcelli S, Serretti A (2014) From pharmacogenetics to pharmacogenomics: the way toward the personalization of antidepressant treatment. Can J Psychiatry 59(2):62–75

Fernandes BS, Williams LM, Steiner J, Leboyer M, Carvalho AF, Berk M (2017) The new field of 'precision psychiatry'. BMC Med 15(1):80

First MB, Reed GM, Hyman SE, Saxena S (2015) The development of the ICD-11 clinical descriptions and diagnostic guidelines for mental and behavioural disorders. World Psychiatry 14(1):82–90

Gaynes BN, Warden D, Trivedi MH, Wisniewski SR, Fava M, Rush AJ (2009) What did STAR*D teach us? Results from a large-scale, practical, clinical trial for patients with depression. Psychiatr Serv 60(11):1439–1445

Gaynes BN, Rush AJ, Trivedi MH, Wisniewski SR, Balasubramani GK, McGrath PJ, Thase ME, Klinkman M, Nierenberg AA, Yates WR, Fava M (2008) Primary versus specialty care outcomes for depressed outpatients managed with measurement-based care: results from STAR*D. J Gen Intern Med 23(5):551–560

Gitlin MJ, Miklowitz DJ (2014) Psychiatric diagnosis in ICD-11: lessons learned (or not) from the mood disorders section in DSM-5. Aust NZ J Psychiatry 48(1):89–90

Gonda X, Petschner P, Eszlari N, Baksa D, Edes A, Antal P, Juhasz G, Bagdy G (2019) Genetic variants in major depressive disorder: From pathophysiology to therapy. Pharmacol Ther 194: 22–43

Greicius MD, Krasnow B, Reiss AL, Menon V (2003) Functional connectivity in the resting brain: a network analysis of the default mode hypothesis. Proc Natl Acad Sci USA 100(1):253–258

Guo T, Xiang YT, Xiao L, Hu CQ, Chiu HF, Ungvari GS, Correll CU, Lai KY, Feng L, Geng Y, Feng Y, Wang G (2015) Measurement-based care versus standard care for major depression: a randomized controlled trial with blind raters. Am J Psychiatry 172(10):1004–1013

Haber SN, Knutson B (2010) The reward circuit: linking primate anatomy and human imaging. Neuropsychopharmacology 35(1):4–26

Haile CN, Murrough JW, Iosifescu DV, Chang LC, Al Jurdi RK, Foulkes A, Iqbal S, Mahoney JJ 3rd, De La Garza R 2nd, Charney DS, Newton TF, Mathew SJ (2014) Plasma brain derived neurotrophic factor (BDNF) and response to ketamine in treatment-resistant depression. Int J Neuropsychopharmacol 17(2): 331–336

Hamilton SP (2015) The promise of psychiatric pharmacogenomics. Biol Psychiatry 77(1):29–35

Huys QJ, Maia TV, Frank MJ (2016) Computational psychiatry as a bridge from neuroscience to clinical applications. Nat Neurosci 19(3):404–413

Insel TR (2014) The NIMH research domain criteria (RDoC) project: precision medicine for psychiatry. Am J Psychiatry 171(4):395–397

Iosifescu DV (2011) Electroencephalography-derived biomarkers of antidepressant response. Harv Rev Psychiatry 19(3):144–154

IsHak WW, Greenberg JM, Balayan K, Kapitanski N, Jeffrey J, Fathy H, Fakhry H, Rapaport MH (2011) Quality of life: the ultimate outcome measure of interventions in major depressive disorder. Harv Rev Psychiatry 19(5):229–239

Januar V, Saffery R, Ryan J (2015) Epigenetics and depressive disorders: a review of current progress and future directions. Int J Epidemiol 44(4):1364–1387

Kirchheiner J, Bertilsson L, Bruus H, Wolff A, Roots I, Bauer M (2003) Individualized medicine—implementation of pharmacogenetic diagnostics in antidepressant drug treatment of major depressive disorders. Pharmacopsychiatry 36(Suppl 3):S235–S243

Kraft JB, Slager SL, McGrath PJ, Hamilton SP (2005) Sequence analysis of the serotonin transporter and associations with antidepressant response. Biol Psychiatry 58(5):374–381

Laje G, McMahon FJ (2011) Genome-wide association studies of antidepressant outcome: a brief review. Prog Neuropsychopharmacol Biol Psychiatry 35(7):1553–1557

Lisoway AJ, Zai CC, Tiwari AK, Kennedy JL (2018) DNA methylation and clinical response to antidepressant medication in major depressive disorder: a review and recommendations. Neurosci Lett 669:14–23

Lopez JP, Mamdani F, Labonte B, Beaulieu MM, Yang JP, Berlim MT, Ernst C, Turecki G (2013) Epigenetic regulation of BDNF expression according to antidepressant response. Mol Psychiatry 18(4):398–399

Lovejoy CA, Buch V, Maruthappu M (2019) Technology and mental health: the role of artificial intelligence. Eur Psychiatry 55:1–3

Lunn MP, Hughes RA, Wiffen PJ (2014) Duloxetine for treating painful neuropathy, chronic pain or fibromyalgia. Cochrane Database Syst Rev (1):CD007115

Malhi GS, Mann JJ (2018) Depression. Lancet, 392(10161): 2299–2312

McGorry P, Keshavan M, Goldstone S, Amminger P, Allott K, Berk M, Lavoie S, Pantelis C, Yung A, Wood S, Hickie I (2014) Biomarkers and clinical staging in psychiatry. World Psychiatry 13(3):211–223

Miller AH, Maletic V, Raison CL (2009) Inflammation and its discontents: the role of cytokines in the pathophysiology of major depression. Biol Psychiatry 65(9):732–741

Miller S, McTeague LM, Gyurak A, Patenaude B, Williams LM, Grieve SM, Korgaonkar MS, Etkin A (2015) Cognition-childhood maltreatment interactions in the prediction of antidepressant outcomes in major depressive disorder patients: results from the iSPOT-D trial. Depress Anxiety 32(8):594–604

Milev RV, Giacobbe P, Kennedy SH, Blumberger DM, Daskalakis ZJ, Downar J, Modirrousta M, Patry S, Vila-Rodriguez F, Lam RW, MacQueen GM, Parikh SV, Ravindran AV, Group CDW (2016) Canadian network for mood and anxiety treatments (CANMAT) 2016 clinical guidelines for the management of adults with major depressive disorder: Section 4. Neurostimulation treatments. Can J Psychiatry 61(9):561–575

Nestler EJ, Barrot M, DiLeone RJ, Eisch AJ, Gold SJ, Monteggia LM (2002) Neurobiology of depression. Neuron 34(1):13–25

Newton RA, Phipps SL, Flanigan TP, Newberry NR, Carey JE, Kumar C, McDonald B, Chen C, Elliott JM (1996) Characterisation of human 5-hydroxytryptamine2A and 5-hydroxytryptamine2C receptors expressed in the human neuroblastoma cell line SH-SY5Y: comparative stimulation by hallucinogenic drugs. J Neurochem 67(6):2521–2531

Orru G, Pettersson-Yeo W, Marquand AF, Sartori G, Mechelli A (2012) Using support vector machine to identify imaging biomarkers of neurological and psychiatric disease: a critical review. Neurosci Biobehav Rev 36(4):1140–1152

Pare CM, Mack JW (1971) Differentiation of two genetically specific types of depression by the response to antidepressant drugs. J Med Genet 8(3):306–309

Perez V, Gilaberte I, Faries D, Alvarez E, Artigas F (1997) Randomised, double-blind, placebo-controlled trial of pindolol in combination with fluoxetine antidepressant treatment. Lancet 349(9065):1594–1597

Picard RW (2002) Affective medicine: technology with emotional intelligence. Stud Health Technol Inform 80:69–83

Porcelli S, Drago A, Fabbri C, Gibiino S, Calati R, Serretti A (2011) Pharmacogenetics of antidepressant response. J Psychiatry Neurosci 36(2):87–113

Powell TR, Smith RG, Hackinger S, Schalkwyk LC, Uher R, McGuffin P, Mill J, Tansey KE (2013) DNA methylation in interleukin-11 predicts clinical response to antidepressants in GENDEP. Transl Psychiatry 3:e300

Relling MV, Klein TE (2011) CPIC: clinical pharmacogenetics implementation consortium of the pharmacogenomics research network. Clin Pharmacol Ther 89(3):464–467

Rosenblat JD, Kakar R, McIntyre RS (2015) The cognitive effects of antidepressants in major depressive disorder: a systematic review and meta-analysis of randomized clinical trials. Int J Neuropsychopharmacol 19(2)

Rosenblat JD, Lee Y, McIntyre RS (2017) Does pharmacogenomic testing improve clinical outcomes for major depressive disorder? A systematic review of clinical trials and cost-effectiveness studies. J Clin Psychiatry 78(6):720–729

Rush AJ, Trivedi MH, Wisniewski SR, Nierenberg AA, Stewart JW, Warden D, Niederehe G, Thase ME, Lavori PW, Lebowitz BD, McGrath PJ, Rosenbaum JF, Sackeim HA, Kupfer DJ, Luther J, Fava M (2006) Acute and longer-term outcomes in depressed outpatients requiring one or several treatment steps: a STAR*D report. Am J Psychiatry 163(11):1905–1917

Salagre E, Dodd S, Aedo A, Rosa A, Amoretti S, Pinzon J, Reinares M, Berk M, Kapczinski FP, Vieta E, Grande I (2018) Toward precision psychiatry in bipolar disorder: staging 2.0. Front Psychiatry 9: 641

Seeley WW, Menon V, Schatzberg AF, Keller J, Glover GH, Kenna H, Reiss AL, Greicius MD (2007) Dissociable intrinsic connectivity networks for salience processing and executive control. J Neurosci 27(9):2349–2356

Selb M, Kohler F, Robinson Nicol MM, Riberto M, Stucki G, Kennedy C, Ustun B (2015) ICD-11: a comprehensive picture of health, an update on the ICD-ICF joint use initiative. J Rehabil Med 47(1):2–8

Sen S, Duman R, Sanacora G (2008) Serum brain-derived neurotrophic factor, depression, and antidepressant medications: meta-analyses and implications. Biol Psychiatry 64(6):527–532

Sheline YI, Price JL, Yan Z, Mintun MA (2010) Resting-state functional MRI in depression unmasks increased connectivity between networks via the dorsal nexus. Proc Natl Acad Sci USA 107(24):11020–11025

Silbersweig D, Loscalzo J (2017) Precision psychiatry meets network medicine: network psychiatry. JAMA Psychiatry 74(7):665–666

Slattery DA, Hudson AL, Nutt DJ (2004) Invited review: the evolution of antidepressant mechanisms. Fundam Clin Pharmacol 18(1):1–21

Stein DJ, Reed GM (2019) Global mental health and psychiatric nosology: DSM-5, ICD-11, and RDoC. Braz J Psychiatry 41(1):2

Sullivan PF, Neale MC, Kendler KS (2000) Genetic epidemiology of major depression: review and meta-analysis. Am J Psychiatry 157(10):1552–1562

Thaler KJ, Morgan LC, Van Noord M, Gaynes BN, Hansen RA, Lux LJ, Krebs EE, Lohr KN, Gartlehner G (2012) Comparative effectiveness of second-generation antidepressants for accompanying anxiety, insomnia, and pain in depressed patients: a systematic review. Depress Anxiety 29(6):495–505

Treadway MT, Zald DH (2011) Reconsidering anhedonia in depression: lessons from translational neuroscience. Neurosci Biobehav Rev 35(3):537–555

Trivedi MH (2016) Right patient, right treatment, right time: biosignatures and precision medicine in depression. World Psychiatry 15(3):237–238

Uher R, Payne JL, Pavlova B, Perlis RH (2014) Major depressive disorder in DSM-5: implications for clinical practice and research of changes from DSM-IV. Depress Anxiety 31(6):459–471

Uher R, Perroud N, Ng MY, Hauser J, Henigsberg N, Maier W, Mors O, Placentino A, Rietschel M, Souery D, Zagar T, Czerski PM, Jerman B, Larsen ER, Schulze TG, Zobel A, Cohen-Woods S, Pirlo K, Butler AW, Muglia P, Barnes MR, Lathrop M, Farmer A, Breen G, Aitchison KJ, Craig I, Lewis CM, McGuffin P (2010) Genome-wide pharmacogenetics of antidepressant response in the GENDEP project. Am J Psychiatry 167(5):555–564

US APAD-TFAV (2013) Diagnostic and statistical manual of mental disorders: DSM-5™, vol 25, no 2, 5th edn. Codas, p 191

Wijkstra J, Lijmer J, Burger H, Cipriani A, Geddes J, Nolen WA (2015) Pharmacological treatment for psychotic depression. Cochrane Database Syst Rev (7):CD004044

Williams LM (2016) Precision psychiatry: a neural circuit taxonomy for depression and anxiety. Lancet Psychiatry 3(5):472–480

Xuan J, Pan G, Qiu Y, Yang L, Su M, Liu Y, Chen J, Feng G, Fang Y, Jia W, Xing Q, He L (2011) Metabolomic profiling to identify potential serum biomarkers for schizophrenia and risperidone action. J Proteome Res 10(12):5433–5443

Yatham LN, Kennedy SH, O'Donovan C, Parikh SV, MacQueen G, McIntyre RS, Sharma V, Beaulieu S (2006) Canadian network for mood and anxiety treatments (CANMAT) guidelines for the management of patients with bipolar disorder: update 2007. Bipolar Disord 8(6):721–739
Zhang WN, Chang SH, Guo LY, Zhang KL, Wang J (2013) The neural correlates of reward-related processing in major depressive disorder: a meta-analysis of functional magnetic resonance imaging studies. J Affect Disord 151(2):531–539

Chapter 13
Psychological Treatment for Depressive Disorder

Xiaobai Li and Qi Wang

Abstract Depression is highly prevalent and causes unnecessary human suffering and economic loss. Therefore, its treatment and prevention are of utmost importance. There are several advantages of using psychotherapy either by itself or combined with pharmacological treatment methods in the treatment of depression. First, it is well known that combining biological treatment with psychosocial methods increases the chances of recovery. Second, in some individuals, psychotherapy continues to be the only solution. Third, the use of antidepressants contains some safety risks and side effects, but psychotherapy does not. Fourth, clinically, depressive patients prefer psychotherapy to drug therapy. Use of a depression-focused psychotherapy alone is recommended as an initial treatment choice for patients with mild to moderate depression, with clinical evidence supporting the use of cognitive behavioral therapy (CBT), interpersonal psychotherapy (IPT), psychodynamic psychotherapy (PDP), and problem-solving therapy (PST) in individual and group formats. Important developments took place within the past 20 years in the psychotherapy of depression. In the present chapter, we introduced several key issues, such as, Are all psychotherapies equally effective? Who benefits from psychotherapies? Is telepsychotherapy effective? Finally, we introduce the psychotherapy for special populations.

Keywords Depression · Psychotherapy · Antidepressants

13.1 Introduction

People who suffer from major depression (MDD) experience depressed mood and/or loss interest in activities. MDD is the second major healthcare problem around worldwide induced disability (Kessler et al. 2003). The lifetime prevalence of MDD is 10.8%; 1-month prevalence is 1.3% (Ontario 2017). The 1-year prevalence of MDD

X. Li (✉) · Q. Wang
The Department of Psychiatry, The First Hospital of China Medical University, Shenyang, China
e-mail: 2823725763@qq.com

© Springer Nature Singapore Pte Ltd. 2019

Y. Fang (ed.), *Depressive Disorders: Mechanisms, Measurement and Management*, Advances in Experimental Medicine and Biology 1180,
https://doi.org/10.1007/978-981-32-9271-0_13

is 1.5% (Spijker et al. 2013). MDD affects social function through loss of productivity when unwell during attending work. It also affects the ability of individuals to engage in activities, like parenting and housekeeping. In the general population, more than 20% of people with MDD will develop a chronic course, and among those people, 20–30% fulfill the criteria for chronicity.

Over the past 20 years, the prescriptions of antidepressants have increased which is due to the continuous development of the new type of antidepressant drugs such as commonly used antidepressants in clinics—selective serotonin reuptake inhibitors (SSRI), selective norepinephrine and serotonin reuptake inhibitor (SSNRI) as well as other newer drugs (Nnjjm et al. 2017). However, people with depression tend to be psychotherapists rather than antidepressants (Furukawa 2010). Use of a depression-focused psychotherapy alone is recommended as a choice of initial treatment for mild to moderate depressed patients, with sufficient clinical evidence supporting the use of cognitive behavioral therapy (CBT), interpersonal psychotherapy (IPT), psychodynamic psychotherapy (PDP), and problem-solving therapy (PST) in the form of individuals or groups (Fochtmann 2005). The literature has been shown that these psychological treatments lead to better acute outcomes compared to control (Cuijpers et al. 2008a). Combined psychotherapy and drug therapy are more effective than either one alone (Cuijpers et al. 2013). It is worth mentioning that there is very limited evidence of any treatment intervention for chronic depression, and the combination of psychotherapy and antidepressants may be beneficial for chronic depression (Spijker et al. 2013).

Factors that may suggest the using psychotherapy treatment include significant psychosocial stressors, psychological conflicts, interpersonal difficulties, or most important patient preference (Fochtmann et al. 2005). For perinatal women, the depression-focused psychotherapy alone should be considered as an initial option depending on the severity of symptoms (Müldnernieckowski et al. 2015). When choosing a specific type of psychotherapy, consideration should be given to treatment objectives, patients' preferences, positive response to psychotherapy, and the availability of psychotherapist skilled.

For patients undergoing drug treatment, who are receiving psychotherapy should be monitored systematically to assess their treatment response and adverse reaction. When determining the frequency of treatment, psychiatrists should consider a variety of factors, such as the specific types and objectives of psychotherapy, the severity of symptoms, comorbidity, treatment compliance, the degree of social support, suicidal thoughts, the frequency of visits necessary to establish and maintain therapeutic relationships, etc.

Marriage and the relationship between family members are common in the process of depression, and these problems should be concerned and solved through marital or family therapy.

This chapter mainly focuses on the description of the different evidence-based psychotherapy methods of depression. Then, we introduced several key issues, such as, Are all psychotherapies equally effective? Who benefits from psychotherapies? Is telepsychotherapy effective? Finally, we introduce the psychotherapy for special populations (children and adolescents, elderly people, and perinatal women).

13.2 The Methods of Psychotherapy

13.2.1 Cognitive Behavioral Therapy (CBT)

CBT for depression was developed by Aaron T. Beck in the 1970s (Beck 1993), and is the most widely studied. CBT is an evidence-based strategy in the treatment of depression, which includes highly effective strategy of behavioral activation as well as cognitive intervention and reconstruction. Previous studies indicated that disharmonious marriage, losing job, and recent stressful life events predicted a better response to CBT comparedto drugs (Fournier et al. 2009). Because CBT requires patients to learn new skills, it is more appropriate for patients with cognitive impairments to choose CBT for treatment. Patients who fail to form relationships may also need to modify the "standard" CBT in order to have time to understand that therapists can be confident and agree to participate the process of treatment (Sudak 2012). The UK National Institute for Clinical Excellence (NICE) recommends providing CBT for individuals who relapse or have residual symptoms despite the use of antidepressant therapy (Nel 2014).

CBT is a process which helps individual change erroneous beliefs and behaviors. Cognitive behavioral management of depression includes three stages (Begić 2013):

Behavioral activation. For moderate–to-severe depression patients, behavioral activation is the basic element of CBT. The objective of this stage is teaching the patients to self-monitor and assess current levels of activity. The psychotherapist and patient jointly identify daily activities previously which are being reduced or avoided valued by the patient.

The advantage of behavioral activation is to make patients' withdrawal or avoidance behaviors (such as isolated, timid, withdrawn, unwilling to interact with people, not in contact with an unfamiliar environment, closed themselves in order to obtain a sense of security) which become obvious when patients are engaged in specified activity.

Behavioral activation is also targeted at skill defects (e.g., lack of social skills and self-confidence in communication), which allow patients to encounter appropriate difficulties in social interaction. Such behavior training can ensure future social interaction that will be more rewarding and more satisfying.

Cognitive interventions. The target of this sessions is to identify automatic thinking (the thoughts beneath the surface of consciousness), cognitive errors, and false attributions, and examine to generate more precise ideas, conclusions, and rules. It needs to be emphasized in this part, it is always easy to identify automatic thinking through identifying mood shifts, classically, using logical analysis to see whether the thought is true, partly true, or wrong. Once the patients are able to change the automatic thoughts adroitly, the psychotherapist can shift the target of treatment to uncovering the beliefs which make the individuals negative thinking and vulnerable to stress events.

As for depressed patients, in this session, psychotherapist will help individuals identify: (1) dysfunctional beliefs related to depression; (2) automatic thoughts and cognitive distortions; (3) changes of emotion, behavior, and psychology caused by wrong thoughts; (4) the effect of previous experiences on maintaining the patients' beliefs. Once the patient understands the psychological factors that lead to depressive behavior, technology will be applied in the intermediate stage to help the patient control the symptoms.

Termination. At last, the psychotherapist provides written materials and summarizing the patient's learning in the whole course of treatment. First, looking back what the patient has learned, the patient and psychotherapist together predict and provide solutions for an adverse situation that the patient may be faced in the future. The ultimate goal is to prevent a recurrence.

When CBT is combined with an antidepressant, the treatment effect is more durable than drugs alone (Fava et al. 1998; Lutz et al. 2016; Jakobsen et al. 2016). Compared with patients receiving antidepressants alone, the recurrence rate of combined CBT treatment was reduced by nearly 50% (Hollon et al. 2006). The unipolar depressed patients would have a 61% chance to complete recovery when receiving acute phase CBT treatment alone while patients treated with an antidepressant alone only have a 39% chance to complete recovery (Vittengl et al. 2007).

Different kinds of modifications of CBT were also found to be effective, such as mindfulness-based CBT (MBCT), Internet-based CBT (ICBT), and repetitive negative thinking focused CBT (RNT-CBT). MBCT combined traditional CBT method with mindfulness meditation. It is designed to improve the ability of depressed patients, and to prevent the recurrence of depression in the long term (Tovote et al. 2017). MBCT helps patients become more aware of negative thoughts during the potential relapse period, and help patients to get rid of rumination. Using MBCT, psychotherapists could better participants to process their experiences through mindfulness and meditation, so that they could improve their unwanted feelings (Zhang et al. 2018; Sizoo and Kuiper 2017). ICBT for depression has been widely researched. Several meta-analyses have summarized the evidence, suggesting that ICBT can be a feasible and cost-effective treatment for depression (Andersson and Cuijpers 2009; Richards and Richardson 2012; Cowpertwait and Clarke 2013). In a study concerning primary care, research found ICBT was more effective to treat depression than antidepressants treatment in the long term (Ludman et al. 2007). RNT is a common symptom of depression and is associated with cognitive, behavioral, and emotional functional deficits (Mennin and Fresco 2014). RNT is defined as repetitive, passive/uncontrollable, negative, and relatively abstract thought (Ehring and Watkins 2009). Recently, a few of CBT interventions methods have been developed to specifically target RNT to prevent or reduce psychopathology (Watkins 2015). A recent meta-analysis by Spinhoven showed that RNT-CBT may have a more significant effect on RNT than other psychological interventions, and it may have a specific effect on depressed patients with rumination, anxiety, and content-independent persistent thinking (Spinhoven et al. 2018).

13.2.2 Interpersonal Psychotherapy Therapy (IPT)

IPT was developed in the 1970s, and manualized in 1984 by G. Klerman and his colleagues (Klerman and Weissmann 1987). IPT is a structured psychological intervention method based on interpersonal relationship theory (Jacobson 1955), which is specifically developed for treating depression (Hinrichsen 2001). IPT is based on two clinical practice assumptions (Agras et al. 2000): (1) Depression is a treatable disease, not patient's faults; (2) emotion is related to life events. The recent life events may trigger acute depressive episodes among the vulnerable individuals or trigger emotional episodes that produced interpersonal difficulties and disturbing life events. IPT associates emotional symptoms with current interpersonal difficulties, IPT relieves depressive symptoms and develops social skills to help patients solve current interpersonal crisis (Markowitz 2003).

There are four interpersonal relationships highly associated with depressive episodes (Klerman and Weissman 1986): (1) grief: a stress response to loss, especially to the loss of someone; (2) interpersonal disputes: conflicts with people, such as friends or family member; (3) role transition: significant changes in life, such as loss of job, parenthood or chronic and ineffective diseases; (4) interpersonal deficits: limited social interaction and less interpersonal relationship. The main concern of IPT intervention is the patient's present interpersonal relationships, instead of past relationships or experiences. Traditionally, IPT is scheduled to take place around 12–16 weekly sessions.

Initial phase. A key point of the first phase is to establish a proper framework in view of information obtained from psychological tests and interpersonal relationships surveys. Treatment starts once every week, then transferred once every 2 weeks. The frequency of treatment should be based on the improvement of symptoms and importantly must be agreed by both the psychotherapist and patient.

Middle phase. In this phase of treatment, psychotherapists began to implement treatment strategies suitable for patients' interpersonal problems.

Grief. The depression may be triggered by the loss of a loved one. The psychotherapists help patients to accept losses while positive and/or negative feelings decreased were founded.

Interpersonal disputes. If psychotherapists focus on solving interpersonal disputes, problems may stem from obvious or hidden conflicts with others (such as spouses, colleague, friends, relatives, or offspring). These conflicts are related to situations in which participants have no expectations of the relationship with each other. psychotherapists help patients to examine the relationship and phases of disputes, and possible solutions.

Role transitions. Depression is associated with changes in life status which can include milestones, such as falling in love, marriage, child baby, graduation, promotion, occupation change, or disease diagnosis, etc. Under these circumstances, psychotherapists help patients mourn the loss of their previous roles and assist to accept the challenges brought about by the transition.

Interpersonal deficits. As for depressed patients, especially patients with recurrent depression, they tend to have chronic problems related to the initiation or maintenance of healthy relationships. Social ineptitude always hinders to develop healthy social functions and the maintenance of the lower level of social contact. The indicators of patients with interpersonal deficits may be the issues of vacuity, loneliness, inferiority, or social ineptitude. For individuals who have problems starting or maintaining interpersonal relationships, the objective of IPT is to reduce social avoidance and promote the formation of new social relationships. The experience of stressful events in diary life of individuals could affect negative emotional state as well as the onset and maintenance of depressive behavior. The goal of IPT is assisting patients to recognize the connection between negative emotion and current life events, which is important to relive depression.

Final Phase. The last few weeks is considered the termination during IPT treatment for acute depression. This stage should be discussed throughout the treatment by psychotherapists and patients. At the present stage, the feelings and experience of treatment are fully explored. Main objectives of termination are comprehensively reviewed and the gains from treatment by both patients and psychotherapists are summarized (Rushford 2012). In recent years, IPT has been included in clinical practice guidelines for depression and has been successfully applied to treat MDD (Fochtmann et al. 2005; Park et al. 2014; Qaseem et al. 2016; Damsker et al. 2009; Davidson 2010).

Numerous clinical trials have shown that IPT is effective for treating both acute and recurrent depressed patients such as children and adolescents, perinatal women, elderly people, primary care patients, and depressed patients in developing countries. Generally speaking, because of the need to establish a harmonious psychotherapist—patient relationship, there may be no additional benefits for acutely depressed patients who receive the treatment of IPT. Therefore, patients should be encouraged to take active drug therapy or alternative treatment before IPT (Rushford 2012).

13.2.3 Psychodynamic Psychotherapy (PDP)

PDP's theoretical models come from early conceptualization of psychoanalysis, including object relations theory, self-psychology, and attachment theory. In practice, psychotherapists need to take full consideration of treatment goals and setting changes (Shedler 2010; Leichsenring and Rabung 2009). Gabbard (2010) described the basic principles of PDP as people's spiritual life is unconscious mostly; both childhood experiences and genetic factors mold adults; the empathy of patients to therapists is mainly originated from understanding; the psychotherapists' counter transference provides an important understanding of how the patient induces in others; emotions and behaviors have various functions and are decided by complicated and unconscious forces; psychotherapist help patients to get a sense of reality and uniqueness.

Freud first described the psychodynamic understandings of depression. He investigated the individual's reactions to lover's loss or ideal loss. Freud had tried to interpret why some persons develop mourning facing with loss while others succumb into melancholy. Although mourning involves major disruptions from individual's attitude toward life, it could not be seen as psychopathological. Melancholy comes from the loss of the individual's love, and this process is unconscious, the self-esteem of individuals was found to decrease during the course. Freud pointed out that the melancholy is "an agonizing experience, loss of pleasure in life, loss of ability to love, inhibition of social activities, and reduction of self-perception so as to self-blame and culminates in delusional punishment" (Rangell 2002; Pies 2009).

There are many common features in the treatment of depression using psychotherapy, like providing hope and social support as well as developing a theoretical framework in finding the origin of depression and laying out a scheme to treat depression. However, there are many differences between psychodynamics and other treatments. For instance, relative to CBT, PDP tends to have a stronger emphasis on (1) expression of effect and emotion; (2) focus on defense mechanisms; (3) understanding of repeated patterns regarding behavior, emotions, and relationships; (4) the influence of past experiences on emotions and behaviors on the present; (5) interpersonal context; (6) the relationship between psychotherapist and patients (Luyten and Blatt 2012).

The psychodynamic formulation of depression commonly included five elements those summarized by Busch: (1) narcissistic vulnerability; (2) conflicted anger; (3) severe superego, states of guilt and shame; (4) overidealized and strong devalued expectations of self or other; and (5) characteristic methods of resisting to painful effects (Busch et al. 2012).

13.2.4 Problem-Solving Therapy (PST)

It is no surprise that stressful life events accompanied by ups and downs could negatively affect the individual. These situations, defined in previous studies as negative life events, are leaving stress and discomfort to the individuals who may then develop into psychological problems (Folkman 2013). Marriage failure, academic or business failure, death of relatives, victims of maltreatment, involvement in serious accidents, unemployment, separations, experiencing violence, surviving natural disasters, and the like are defined as adverse life situations (Blackie and Jayawickreme 2015). The people' problem-solving skill levels have the ability to reduce the effects of stressful life situations. Nezu (1987) suggested that the levels of depression could be increased through the interaction of stressful life events with defects in various problem-solving. The development factors of personal depression are the inability to solve problems or manage negative life events. When the skill level of problem-solving increases, the probability of being depressed decreases (Slife and Weaver 1992).

PST is focusing on developing attitudes and skills of problem-solving. The objective of PST is to decrease psychopathology and enhance subjective well-being by assisting people to cope with stressful problems in living. PST was originally put forward by D'Zurilla and Goldfried (1971), and developed by D'Zurilla, Nezu and their colleagues over several years (D'Zurilla and Nezu 2001; Nezu et al. 2013; Maguth Nezu et al. 2008; Nezu et al. 2007). There are mainly three types of PST for depression: (1) social problem-solving therapy (SPST) was outlined in 1980 (Dryden and Kendall 1987), focuses on developing problem-solving skills as well as changing attitudes or beliefs when facing stressful life events. (2) PST for primary care (PST-PC) was formed in the 1990s. PST-PC solves the core elements of the problem and is recommended to be used by trained nurses (Mynorswallis et al. 1995). (3) Self-examination therapy (SET) is aimed at achieving the main goals of life, inputting energy into resolving problems affecting emotions and personal behaviors, and learning to accept and adapt situations that cannot be changed. The key element of SET is the problem-solving skill. SET can be conducted in the formation of self-help, and be applied in group or individual settings (Bowman et al. 1995).

PST usually includes eight sessions (Eskin 2013):

***Sessions I*: *assessment and providing rationale*.** In the first stage, PST should be left to understand and assess the difficulties that individuals are facing at present. Four topics should be considered: (1) How to establish good relationship with clients? (2) How to assess patients' psychological state and behavior characteristics? (3) Informing clients of the basic principles and procedures of PST. (4) Informing patients about the number and duration of the treatment.

***Sessions II*: *problem orientation*.** Individuals' success in problem-solving situations depends on their orientation, which relies on the combination of certain behavioral, cognitive, and emotional characteristics. There are two basic perceptual processes for developing a positive problem orientation which are (1) being aware of the problems; (2) developing a motivational orientation involving the ability to view problems in a positive way.

***Session III*: *defining problems*.** The individual who is aware of his current condition and displaysa clear attitude to it should have the ability to describe and define the problem, in order to facilitate the problem-solving process successfully. Differentiating between complaints (emotional symptoms) and problems (cause) may be helpful in this session.

***Sessions IV*: *goal setting*.** Problem-focused goals and emotion-focused goals.

***Sessions V: generating possible solution alternatives*.** The psychotherapist helps the individuals develop solution alternatives based on problems identified in the previous stages. There are five techniques commonly used in this session, which are brainstorming method, reframing method, advising a friend method, being a hero method, and therapist's advice.

***Sessions VI*: *choosing the appropriate solution*.** The task in this stage is to select a suitable and beneficial solution. The following points should be noted in the quantitative evaluation of the scheme: (1) the solution alternative's ability to solving the problem; (2) the solution's congruence with the goal; (3) the solution's applicability; (4) the solution's acceptability.

***Sessions VII*: *solution implementation*.** Before implementing the solution, the first thing is to discuss and deal with customer fears. Importantly, the motivation of the client is the success of this session. In addition, evaluating and assessing as well as identifying the behavioral characteristics and tendencies of clients were important.

Sessions VIII**: *****assessing the implementation***. Two key points should be emphasized, which are the evaluation of the implementation and evaluation of the treatment effect.

Many previous studies have found that PST is effective in treating depression (Bell and D'Zurilla 2009; pmhdev and PubMedHealth 2008). A recent meta-analysis including 30 randomized trials on PST (with 3530 depression patients) conducted by Cuijpers et al. (2018a) found that PST is an effective psychotherapy for adult depression. Eskin et al. (2008) observed the efficacy of PST used to treat adolescents and adults with depression. Six sessions of individual PST were given to patients. The results showed that PST was an effective psychotherapy for the treatment of adolescent and young adults depression than usual care. However, in the UK NICE guidelines, PST is not be suggested conduct separately to treat depression. The psychotherapy method is recommended to be included in family CBT together with emotive behavior therapy (Cuijpers et al. 2018b).

13.3 Mindfulness-Based Intervention (MBI)

Researches on MBI for depression have significantly increased in the past 10 years. A premise of this psychotherapy is that individual will decrease the reaction to unhappy things but more introspection by practicing mindfulness, which in turn leads to positive emotions and behavior outcomes (Morgan 2003). Although it is difficult to achieve mindfulness, once it is achieved, it is ultimately beneficial. This skill results in increased cognitive flexibility and insight as well as strong self-regulation ability.

Because of the exact therapeutic effects of mindfulness exercises on regulating emotions, mindfulness therapy attracts many medical staff, clients, neuropsychologist, and so on. In the West, mindfulness spreads rapidly and is popular among the public, which is attributed to the success of standardized and effective MBI. The most widely used in healthcare institutions and for the public are mindfulness-based stress reduction (MBSR) and mindfulness-based cognitive therapy (MBCT). In some developed Western countries, mindfulness practices have been incorporated into cognitive behavioral practices.

MBSR: MBSR is the most famous MBI, which benefits in improving psychological symptoms. This psychotherapy was developed in the early 1980s by Kabat-Zinn (1982). MBSR could decrease stress levels of individuals by meditation practices. MBSR is an 8-week treatment program which consists of (1) group meditation class (2–2.5 h) per week. In the class, there is a well-trained teacher who teaches individuals to master mindfulness skill, (2) daily home practice guided by audio (45–60 min/day), (3) mindfulness retreat. The course content focuses on learning how to concentrate on body feeling, using various physical and mental meditation exercises. Common meditation training methods include sitting meditation, body scanning, yoga, Tai Chi, gentle stretching exercises, and so on (Nyklicek et al. 2016).

MBCT: MBCT was originally outlined and developed by Teasdale and colleagues to prevent recurrence of depression (Teasdale et al. 2000). This psychotherapy is applied to depressed patients, to let them recognize negative emotion but not judge or responding immediately to these. Mindfulness enhanced the individuals' internal awareness; when combined with principles of CBT, this psychotherapy could help people get rid of maladaptive patterns of repetitive negative thinking which facilitate in the occurrence of depression. Because MBCT was initially developed, there are a few of the randomized trials regarding the efficacy of MBCT that have shown that this psychotherapy effectively decreased the rates of recurrence of depression (Nyklicek et al. 2016; Schuver and Lewis 2016; Strauss et al. 2014).

Others: ***Retreats and residential programs***. In practice, this program is a very cost-effective and convenient way to deliver intensive or moderate mindfulness exercises, which have shown promising effects on improving depression and anxiety (Cohen et al. 2017).

Brief mindfulness interventions. The time of this psychological intervention is relatively short. Even so, previous study showed that there are several significant changes of immediate effects on psychological and neuroendocrine after brief mindfulness interventions when individuals encounter social stress (Creswell et al. 2014). Considering that depression and anxiety may be an acute stress response, brief mindfulness interventions may be effective to improving the levels of depression.

Internet and smartphone mindfulness-based interventions. A recent meta-analysis included 15 randomized trials (2360 participants) by Spijkerman et al. (2016) found that, compared to control, online MBI had a potential beneficial impact on depression ($g = 0.29$), anxiety ($g = 0.22$), stress ($g = 0.51$), and subjective well-being ($g = 0.23$).

MBI not only can be used for depression, anxiety, stress to a life event, and suicidal ideation but also treat drug-dependent depression and addiction (Forkmann et al. 2014; Hosseinzadeh and Barahmand 2014). Previous study showed that the curative effect of MBCT may be better than psychoeducation in treating unremitting or recurrent depression (Chiesa et al. 2012). Moreover, MBCTcan be used concomitantly with first-line drugs to enhance the curative effect.

MBCT is a simple and easy way of psychotherapy, which has been found to be as effective as antidepressants in term of preventing depression from recurrence, which can be used for individuals who do not want or cannot tolerate antidepressant treatments. Importantly, as psychotherapy, MBCT will increase compliance with antidepressant treatment (Kuyken et al. 2015). In addition, previous literature has shown that MBCT is as cost-effective as antidepressants when comparing the levels of total health, the quality of social services, and the losses of productivity (Kuyken et al. 2015). MBI has a positive impact on depressed patients and their comorbid conditions and is a useful tool to address clinical psychological problems in primary care nursing practice regardless of whether or not medication is used (Maxwell and Duff 2016).

13.3.1 Others

Nondirective supportive therapy (NDST): The effects of NDST for depression have been examined in many literature. A meta-analysis of 31 randomized controlled trials conducted by Cuijpers et al. (2012a) found that, NDST has a considerable effect ($g = 0.58$, 95% CI: 0.45–0.72) on depression, the extra-therapeutic factors (the effect of usual care or waiting list) were responsible for 33.3% of the overall improvement of symptoms, specific factors [the effects of NDST vs. other therapies (CBT, supportive therapy, acceptance and commitment therapy, bibliotherapy, and behavioral activation)] for 17.1%, and nonspecific factors (the effects of NDST vs. usual care or waiting list control) for 49.6%.

Music therapy: Simply, the psychotherapist uses music therapy to help patients improve, restore or maintain health (Bruscia 1995). Previous literature indicated that different forms of music therapy can lead to changes in biological and behavioral aspects. A review by Aalbers (2017) showed that music therapy can benefit patients in the short term. Applying music therapy to patients' usual care can help improve depression. Music therapy is conducive to the rehabilitation of personal and social functions. For example, it has the function of improving cognitive behavior, improving emotion, language function, and interpersonal relationship of depressed patients (Maratos et al. 2008).

Reminiscence therapy: Reminiscence often refers to memory recall which usually includes daydreaming, telling stories, or memory cherishing, which can help adults, especially the elderly, improve depression and increase self-esteem and subjective well-being (Woods et al. 2018). For individuals who have rich life experiences, the opportunity to systematically and structurally review and rethink one's life, self-examine or redefine the self can be achieved by the goal of facing the distress of life and resisting unhealthy emotions such as depression and anxiety. This therapy can be used in targeting psychological problems in elderly adults with and without cognitive impairments (Huang et al. 2015). A meta-analysis conducted by Bohlmeijer et al. (2003) showed that reminiscence therapy exerted significant effects (overall effect size of 0.84) on the improvement of depression in general elderly population. The effect will increase in old subjects with higher depressive symptoms ($d = 1.23$), as compared to no high depressive symptoms ($d = 0.37$).

Interestingly, reminiscence therapy has also been shown to reduce depression symptoms in young (around 25 years) and middle-aged (around 50 years) adults. Therefore, reminiscence therapy provides a useful tool for young and middle-aged depressed patients to assemble insights from the past experiences and practice to use them in the present, so as to improve mood and behavior better and prepare for the future. Compared to older adults (>45 years), however, younger adults (18–45 years) show a greater tendency for bitterness when remembering negative feelings. Young people usually recall recent events or experiences; they could benefit from contextualized experiences, further balancing negative with positive memories; this process may help them improve depressive mood (Hallford 2014).

Emotion-focused therapy (EFT): EFT has been studied for treating depression and emotional trauma, and may also help improve anxiety, interpersonal problems as well as eating disorders (Greenberg 2014). The emotion management schemes are automatic and usually unconscious, yet hold tremendous impact on the individual's emotional processing. In depression, the common emotional schemes involve worthlessness and secondary hopelessness. EFT does not try to extinguish or regulate emotional problems clearly, on the contrary, problematic emotions may be the root cause for habitual maladjustment (Robinson et al. 2012; Greenberg and Watson 2006). A review study showed that EFT can help individuals regulate their emotions and reduce their depression level, with the extension of treatment time (Robinson et al. 2012).

13.4 Several Important Issues in Psychotherapy

13.4.1 Are All Psychotherapies Equally Effective?

In recent years, psychological treatment has attracted more and more attention to the treatment of mental disorders. Many different types of psychotherapy are designed for targeting group, and there are growing high-quality evidences showing their curative effect. An important problem that needs to be considered is that all these psychotherapies' effects seem to be equally or about equally effective, judging from the evidence available; there is no one type of psychotherapy significantly more effective than others (Barth et al. 2013). There are over 100 clinical comparative studies since 20 years to directively compare two or more psychotherapies for treating depression. It is a pity that these trials pose a big issue that is, low statistical power bringing about the conclusion.

Cuijpers (2016) selected three meta-analyses of psychotherapies (CBT, IPT, and nondirective counseling) of depression for analysis. The largest trial comparing CBT with another psychotherapy consists of 178 depressed patients and had enough power to detect a differential effect size ($d = 0.42$). There are no trials included in the three meta-analyses and they had enough power to detect effect sizes ($d < 0.34$), some trails very close to the threshold ($d = 0.24$) for detecting a clinical effect size. It is worth mentioning that meta-analysis can be found in some low-power individual trials, which may have a large risk of bias. If we only analyze these trials with low risk of bias, it may lead to a lack of enough studies to detect clinically relevant effects. In Cuijpers' s study, he found many individual trials extremely underpowered and had no enough power to detect clinically relevant effect sizes (Cuijpers 2016). From this study we are still not sure however clinically relevant differences between these therapies, the reasons may be related to the psychological treatment schemes, treatment time and frequency as well as treatment quality. High-quality studies are also needed in the future to clarify the psychotherapies' effect.

13.4.2 Who Benefits from Psychotherapies?

In the past, many studies have examined the effect of psychotherapy on different target groups of people, including adolescent, adults, elderly, and patients with comorbid somatic diseases etc. Specific psychosis subgroups have also been studied, like threshold depression, major depression, bipolar depression, chronic depression, perinatal depression, and so on. There are a few studies that compared the efficacy of psychotherapy for different subgroups. For example, previous study found the comparable efficacy of psychotherapies between elderly group versus younger adults group (Gould et al. 2012), student populations group versus adults group (Davies et al. 2014), sub-threshold depression group versus major depression group (Cuijpers et al. 2014a), primary care group versus specialized mental healthcare group (Cuijpers et al. 2009), chronic depression group versus other populations with depression group (Jobst et al. 2016).

A recent meta-analysis including 256 randomized trials conducted by Cuijpers et al. (2018a) compared the efficacy of psychotherapy among different target groups of depressed patients; the research results show that there is no necessary connection between the specific characteristics (older adults, women, patients with comorbid somatic diseases, and individuals with the self-rating Depression Scale score being higher than the cut-off value) and disease outcome.

13.4.3 The Predictors of Response to Psychotherapy

A study indicated that up to 50% of depressed patients would not respond to psychotherapy (Cuijpers et al. 2014b). Fischer et al. found depressed patients with elevated hypothalamic–pituitary–adrenal (HPA) axis (major stress-responsive systems and the change of HPA axis activity is the most significant feature of depression) activity may have lower responses to psychological therapies (Fischer and Cleare 2017). In clinical terms, depression and anxiety often occur in comorbidity. Feske et al. (1998) found that comorbidity anxiety may reduce the effect of psychotherapy on depression. Ogrodniczuk et al. (2004) found alexithymia (difficulty identifying feelings) may reduce depressed patients' response to psychotherapy, causing residual symptoms to persist. A systematic review by Van et al. showing that depressed patients who were married, may benefit more from CBT. However, the researchers did not find that sex and age could affect the efficacy of CBT (Van et al. 2008). There is an interesting study that found that depressed and anxious patients with higher intelligence could derive more benefits from long-term psychological intervention than short-term psychological intervention (Knekt et al. 2014). It is well understood that these patients with higher social support are more effective in psychotherapy, and that long-term psychological intervention is more effective than short-term psychological intervention (Lindfors et al. 2014).

The structural and functional neuroimaging also showed a potential for developing the prognostic makers of psychotherapy response in the treatment of depression. A functional MRI study by Carl et al. found sustained activation in the anterior cingulate cortex of MDD patients who were more responsive to behavioral activation therapy (Koch et al. 2009). A result of a meta-analysis indicated that the distributed fronto striatal–limbic network of functional and structural regions could predict the response of drugs and CBT to depression (Carl et al. 2016).

13.4.4 Group Versus Individual: Which is More Effective?

Group counseling and personal counseling are two common ways of psychological intervention, which is more effective is the topic most concerned by clients and psychotherapists. Assuming that the two ways are equally effective, clients are more likely to choose group psychotherapy. However, due to the different mechanisms of the two methods, the curative effect of the two may be comparable. For example, in individual psychotherapy, the psychotherapist–client relationship is more intimate, but group psychotherapy focuses more on cohesion and social support (McDermut et al. 2010).

Previous studies on individual psychotherapy are more numerous. The results of comparisons between group psychological intervention and individual psychological intervention are not yet consistent, some of which have found that individual psychotherapy is more effective (Wierzbicki and Bartlett 1987), while some studies found that the two were equally effective (O'Shea et al. 2015). It is worth mentioning that although the sample size is large, the statistical difference between group psychological intervention and individual psychological intervention is relatively small. Another meta-analysis conducted by Cuijpers et al. (2008b) showed that in the short term, the effect of individual psychological intervention for depression may be better. During the 1–6 month follow-up, no indications were found that group psychological intervention and individual psychological intervention have a difference in effectiveness in treating depression. The difference between the two methods was small (d = 0.20).

13.4.5 Is Telepsychotherapy Effective?

Telepsychotherapy can be delivered by telephone, web, videotelephony, and computer videoconferencing (Kaplan 1997), among these, telephone and web are the most common forms of delivery. In recent years, more and more clinical practice and clinical research have begun to pay attention to telepsychotherapy. Internet and mobile-based interventions (IMI) is a cost-effective and maneuverable psychotherapy, which can overcome the barriers of psychological treatment, such as stigma, poor economic level, low perceived need, etc. In clinical practice, evidence-based psychotherapy like CBT and IPT, could be delivered by web or telephone, depending on patient's needs and condition (Shahnavaz and Hedman-Lagerlof 2018; Holmqvist et al. 2014). Telepsychotherapy can provide some conveniences for patients, such as access to psychological services widely and timely, sharing sensitive information more privately, access to social support widely (Zhao et al. 2017). A recent meta-analysis by Josephine (2017) found the levels of depression were significantly decreased by IMI intervention when compared to waitlist conditions (Josephine et al. 2017).

The use of mobile phones has been popularized around the world. Mobile apps are also accessible every day. In recent years, more and more apps have been developed for medical and healthcare purposes, especially mental health. Specially designated apps can be used for mental health education, behavioral training, increasing the compliance of taking medicine, monitoring disease symptoms, etc. (Capon et al. 2016). A prospective 18-month follow-up study indicated that the prognosis of patients with depression improved through CBT intervention delivered by the (Watts et al. 2013). Anyway, there is still insufficient evidence-based medical evidence to prove the efficacy of the apps in treating depression although the apps show potential for depression and anxiety treatment and strong operability and accessibility. Long-term longitudinal data are needed in the future to prove telepsychotherapies' validity before it can be applied widely in psychiatric practice (Donker 2013).

13.4.6 Is a Combination of Antidepressants and Psychotherapy Superior to Each Alone?

There are many literature showing the short-term effects of antidepressants and psychotherapies on depression in acute and maintenance phases. Cuijpers et al. (2013) conducted a meta-analysis which directly compared the efficacy of psychotherapies and antidepressants for depressive and anxiety disorder. The overall effect size showed that there was no significant difference between psychotherapy and antidepressants ($g = 0.02$, 95%CI: −0.07–0.10). However, antidepressant treatment was more effective than psychotherapies for dysthymia ($g = 0.30$), and psychotherapies were more effective than antidepressant treatment for obsessive compulsive disorder ($g = 0.64$). Moreover, psychotherapy was more effective than tricyclic antidepressants treatment ($g = 0.21$). Accumulating evidences indicate that a combination of antidepressants and psychotherapy is superior for depression than each alone. Another meta-analysis also conducted by Cuijpers et al. showed that the efficacy of combined psychological and drug therapy is better than each therapy alone for depressed outpatients and elderly depression patients (Fu et al. 2013).

To explore the long-term efficacy of combined therapy, Karyotaki et al. conducted a meta-analysis, which showed that the efficacy of combined treatment is better than monotherapy at 6 months or longer post-randomization in MDD patients in the acute phase (OR = 2.93, 95%CI: 2.15–3.99). Moreover, as for the maintenance treatment, the efficacy of the combined treatment is better than antidepressants alone at 6 months or longer post-randomization (OR = 1.61, 95%CI: 1.14–2.27). The results indicated that combined treatment results in a superior enduring effect than antidepressants treatment or psychotherapies (Cuijpers et al. 2012b). Another meta-analysis found that psychotherapy was more efficacious than antidepressants for quality of life (g = 0.21), and the combined treatment performed significantly better than each treatment alone for both functioning and quality of life in MDD patients, yielding small effect size (g = 0.32–0.39) (Karyotaki et al. 2016).

13.4.7 Neurobiological Foundations of Psychotherapy

13.4.7.1 Brain Functional Imaging

In the past, functional imaging studies of depression have found that metabolism of frontal and temporal lobes, insula, and basal ganglia are decreased. The decrease in hippocampal metabolism is related to the severity of depression (Gelenberg 2010; Zaremba et al. 2018). An early study reported that the metabolic activity of the dorsal, medial, and ventral prefrontal cortex were decreased after CBT intervention, but researcher found the activity of the hippocampus and dorsal cingulate cortex were significantly increased (Goldapple et al. 2004). Using positron emission tomography (PET), Karlsson et al. (2010) found 5-HT1A binding in dorsolateral and ventrolateral prefrontal cortex, ventral anterior cingulate cortex, inferior temporal, and angular gyrus as well as insular cortex were increased significantly after treatment for 18 weeks in those depressed patients who underwent intervention by short-term psychodynamic psychotherapy; however, these change were not found in those depressed patients who receiving fluoxetine treatment.

In studied functional magnetic resonance imaging (fMRI) of depressed patients, Fu et al. found after 16 weeks of CBT intervention that the activity of the anterior and posterior cingulate gyri, superior frontal gyrus, inferior parietal cortex, and precuneus of medication-free depressed patients were significantly decreased to normal compared to healthy volunteers. The results from this study also showed that the increased activity of the dorsal anterior activity could be a predictor of treatment response to CBT (Fu et al. 2008). Buchheim et al. study the effects of long-term PDT on 16 unmedicated outpatients. They performed MRI scanning before and 15 months after treatment. The results from this study indicated that the activity of the left anterior hippocampus/amygdala and medial prefrontal cortex in depressed patients were higher than healthy individuals before PDT; those changes can be reversed by 15 months of psychotherapy. Importantly, the decrease in brain activity is positively correlated with the improvement of depression (Buchheim et al. 2012).

A system review by Barsaglini et al. concluded that psychotherapy mainly affects (top to down) frontal lobe and improve the dysfunctional thinking process;

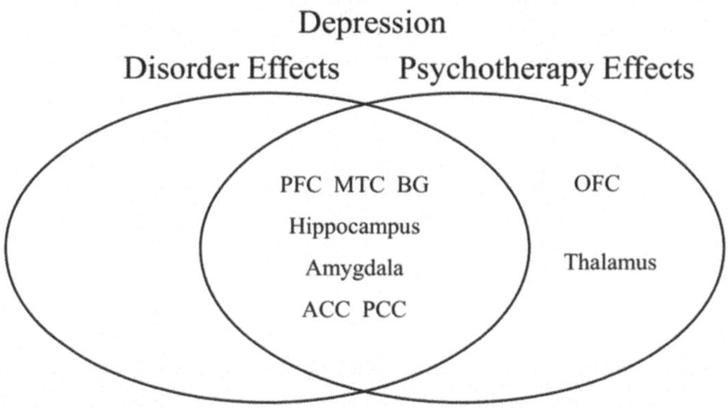

Fig. 13.1 The overlap between regions that are functionally altered at baseline and regions that have been reported to show psychotherapy-related changes (Barsaglini et al. 2014). ACC: anterior cingulate corte; BG: basal ganglia; OFC: orbitofrontal cortex; PFC: prefrontal cortex; PCC: posterior cingulate cortex; MTC: medial temporal cortex

antidepressant mainly affects (bottom to up) ventral and limbic regions and regulates emotions. The author summarized the psychotherapy-related changes of the brain area shown in Fig. 13.1 (Barsaglini et al. 2014).

13.4.7.2 Systemic Inflammation and Oxidative Stress

More and more studies have proved that long-term system inflammation and related changes in inflammatory cytokines is an important mechanism of depression. A recent system review and meta-analysis conducted by O'Toole et al. (2018) analyzed the effects of psychotherapy on inflammatory biomarker. In these trial (19 randomized controlled trials) that were included in meta-analysis, most psychotherapies (CBT, MBI, imagery) can improve inflammatory factors to a certain extent, including C-reactive protein (CRP), interleukin (IL)-6, tumor necrosis factor (TNF)-α, IL-1β, IL-1α, IL-2, and interferon (IFN)-γ. After meta-analysis, the author found that psychotherapy could reduce the levels of pro-inflammatory factors ($g = 0.15$) from pre- to post-intervention. However, this biological effect did not last until the follow-up period ($g < -0.01$). Among those inflammatory factors, just the level of CRP was reduced significantly following psychotherapy. Unfortunately, although this study found that psychotherapy can reduce inflammatory factors, it did not set the subjects strictly (Jobst et al. 2018).

The relationship between oxidative stress and depression has been confirmed. Kaufmann et al. had observed the oxidative anti-oxidative systems in 49 MDD patients (24.06 ± 3.52 of age), and in response to psychotherapy (CBT or cognitive narrative psychotherapy), the researchers found that the levels of thiobarbituric acid reactive species (TBARS) and nitric oxide (NO) in MDD patients were higher than health control, while the levels of total thiol content (SH) was higher than control. Those abnormal changes among MDD patients were improved after psychological therapy (seven weekly structured sessions) and during the follow-up (6 months). Those results indicating that oxidative systems associated with MDD, and the improvement of these oxidative stress indexes may be an important mechanism of cognitive psychotherapy (O'Toole et al. 2018).

13.4.7.3 Cellular Biological Markers

The interaction between cyclic adenosine phosphate response element-binding protein (CREB) and brain-derived neurotrophic factor (BDNF) is an important signal transduction pathway in neuronal plasticity and cellular resilience. They play an important role in the pathogenesis of depression. Koch et al. found that the expression of CREB was increased in T lymphocytes of depressed patients (free dugs) who respond (reduction of ≥50% of baseline) to psychotherapy (6 weeks of IPT). However, there are no significant changes in the levels of BDNF in the plasma between responders and nonresponders. The results indicated that CREB associates with psychotherapies' response independent from drug treatment or BDNF levels (Kaufmann et al. 2015).

13.5 Psychotherapy with Special Populations

13.5.1 Children and Adolescents

It is reported that the incidence of minor depression in children and adolescents is 2.2%, the incidence of moderate to severe depression in adolescents is 9%, and the cumulative incidence rate of any depression between 9 and 16 years old is 9.5% (Costello et al. 2003; Rushton et al. 2002). Because the symptoms of depression in children and adolescents are not very typical, depression is often neglected in children and adolescents, even in medical institutions.

Irritability is the most common symptom of depression in children and adolescents, and somatic complaints, social withdrawal, and despair are also common clinical manifestations (Ryan et al. 1987). Compared with children, adolescents are more likely to suffer from fatigue, somnolence, weight loss, and hopelessness. Moreover, substance abuse is prevalent among adolescents. The separation anxiety and attention-deficit/hyperactivity are common in children. Among depressed adolescents, females are risk factors for suicides (Yorbik et al. 2004).

Clinically, children and adolescent depression are mainly treated with SSRIs [fluoxetine (Jane Garland et al. 2016)], and psychotherapy. A health technology assessment report by Canadian Agency for Drugs and Technology in Health suggested that concluded that antidepressant should not be prescribed outside of a comprehensive treatment approach insist of psychological and social support, mental health education, problem-focused interventions, suicide prevention (Jane Garland et al. 2016). Although medications are effective in treating depression, physicians need to be alert to drugs that may increase the risk of suicide (Christiansen et al. 2016).

Psychotherapy is a safe and effective way to treat children and adolescent depression. A recent review conducted by Weersing et al. (2017) observed the effect of psychosocial intervention depression in children and adolescents. The results of this study showed that the effect of psychological intervention on depression in children is worse than that in treating adolescents. The evidence of psychotherapy for child is relatively weak. However, CBT seems to be effective in treating children's depression. CBT and IPT are effective in treating adolescent depression and had been confirmed by many clinical trials.

Several studies that supported the value of family therapy were combined with individual therapy for depressed adolescents and children. A result from previous study showed that individual CBT combined with participating family meeting monthly could significantly improve the levels of depression in 4–7th-grade children (Brunstein Klomek and Stanley 2007). Fristad et al. (2003) found that compared with usual care, multifamily psychoeducation (such as increased parents' recognition of children's emotions; improved parent–child relationship; and increased utilization of services by families) for families of 8–11 years old children, improved emotional regulation of children and home environment.

It is worth mentioning that the suicide rate of young people has been increasing, be it developed or developing countries (Rajewska-Rager et al. 2015). The prevalence of suicide attempts among adolescents in the United States was 1.3–3.8% in males and 1.5–10.1% in females (Lewinsohn et al. 2010). A study from China showed that the 17.5% of adolescents have a suicidal idea, and 7.3% tried to commit suicide among 1529 adolescents (the average age is 14.74) (Li et al. 2016). An early study including 107 adolescent depression patients showed that the suicidal ideation significantly reduced after three treatments (CBT, family therapy, supportive therapy) but there are no differences between different treatments (Li et al. 2016). Multisystemic therapy is developed for suicidal adolescents. Huey et al. found that multisystemic therapy was more effective than emergency hospitalization in reducing suicidal ideation of psychiatrically disturbed youngsters (Huey et al. 2004).

13.5.2 Elderly People

Geriatric depression (GD) is the most common mental disorder in late adulthood (Tavares and Barbosa 2018), affecting 1–10% of older adults (≥60 years) who may experience chronic disease, pain, social isolation, etc. (Raue et al. 2017). Antidepressants and psychotherapy, alone or in combination, are commonly used in the treatment GD. Studies have shown that the efficacy of antidepressants is moderate. Old people usually have underlying diseases which could increase vulnerability to side effects. In addition, some elderly patients prefer psychotherapy to drug therapy (Raue et al. 2010).

Psychotherapy has been proven effective in the treatment of GD. A meta-analysis conducted by Huang et al. (2015) indicated that psychotherapy is effective for GD, but the magnitude of the effect varies widely (SMD: 0.05–1.36, $P < 0.0001$) with the type of control group (placebo, supportive therapy, usual care, waitlist). PST and CBT had the strongest evidence for GB treatment, followed by ITP (Alexopoulos et al. 2008; Hobbs et al. 2018; Mufson et al. 2018). Before psychotherapy for GD, psychotherapists should systematically assess patients' symptoms, medical history, psychosocial factors, stressors, cognitive function, suicide risk, etc. (McGovern et al. 2014).

13.5.3 Psychotherapies for Chronic Medical Conditions

The patients with chronic illnesses like hypertension, diabetes, rheumatoid arthritis, coronary heart disease, Parkinson's disease et al., are more likely to suffer from depression. CBT has been widely investigated in patients with heart failure and Parkinson's disease, and the results show that treatment is effective. CBT focused on enhancing self-care and functioning under chronic medical conditions (Jeyanantham et al. 2017; Dobkin et al. 2011). A meta-analysis including 19 randomized trials conducted by Baumeister showed psychotherapy positive effects on short- and medium-term depression severity (SMD = −0.41; − 0.42, respectively) and depression remission (OR = 2.88; 2.49, respectively) in patients with diabetes mellitus (Baumeister et al. 2014).

13.5.4 Psychotherapies for Acute Medical Illness

Clinically, post-stroke depression (PSD) is common among stroke survivors. The incidence of PSD was 25.4, 17.6, and 12.4%, during 2 weeks, 3 months, and 1 year after stroke (Wang et al. 2017). Visser et al. (2016) found that PST seems to improve task-oriented coping and quality of life after stroke during a follow-up period of more than 6 months. In a randomized controlled study, researchers found that CBT can improve fatigue and sleep quality as well as decrease the levels of depression (Nguyen et al. 2017). However, due to the uneven quality of the studies, there also have been meta-analysis that did not produce positive results for CBT. According to a recent meta-analysis conducted by Wang et al. (2018), CBT alone (SMD = −0.76) or combined with antidepressants (SMD = −0.95) can effectively treat PSD. CBT also had higher remission (RR = 1.76) and response rate (RR = 1.41), with the improvement of anxiety symptoms, neurological function, and activities of daily living. Hummel et al. (2017) also assessed the efficacy of CBT in the treatment of GD with acute conditions, such as fractures, falls, and neurological conditions as well as cardiovascular events. The levels of depression were significantly improved after patients intervention by 15 sessions of CBT, with the improvement of physical functioning.

13.5.5 Psychotherapies for Cognitive Impairment

Cognitive impairment commonly occurs in the elderly and includes a range of mild memory impairment and lack of concentration, and executive dysfunction (such as organizational and planning capabilities decline, and initiation difficulties) as well as more prominent dementing disorders. A system review by Simon et al. (2015) found that GD for patients with cognitive deficits could benefit from CBT. The results from a previous study suggested that PST could decrease the levels of depression and improve the executive dysfunction in GD patients (Arean et al. 2010).

Dementia is one of the most common neurodegenerative diseases in the elderly. Previous literature showing that reminiscence therapy could be effective in the improvement of emotion and cognitive functions in elderly people with mild to moderate dementia (Huang et al. 2015; Bohlken et al. 2017).

13.5.6 Psychotherapies for Suicide Risk

Suicide is the important public problem for older adults. Elderly people have higher rates of completed suicide than any other age group. MDD is the most common suicide-related disease. Chronic medical disease, functional impairment, psychological, and social stress as well as drug abuse are also high risk of suicide (Conwell and Thompson 2008). Gustavson et al. (2016) found after PST intervention for 12 and 36 weeks, suicidal ideation was significantly reduced in the elderly. Heisel et al. (2009) found that IPT was effective in reducing suicide ideation, death ideation, and depression. A study by Kiosses et al. (2015) found both home-delivered problem adaptation therapy and supportive therapy could reduce the suicidal idea in adults with cognitive impairment. A review by Coon et al. showed that CBT is suitable for the elderly with a mood disorder, suicidal ideation or suicidal behavior (Coon and Gallagher-Thompson 2004).

13.5.7 Psychotherapies for Women

Perinatal women. Perinatal depression (PD) is the important health concern for women and families. Between 10 and 16% of women were diagnosed with MDD during pregnancy or the first year after childbirth while the incidence of MDD in normal adult women was 5% (Claridge 2014). Women with PD often have difficulty in emotional interaction with their infants. This may slow down the development of children's cognition, behavior, social mood, and sports skills (Kingston et al. 2012). PD affects not only children but also the mental health of their spouses.

PD can be treated by medication or individual psychotherapy. Some medications (such as SSRI fluoxetine) are effective in treating PD; however, concerning the adverse effect on fetus and infant while breastfeeding, the vast majority of women prefer psychotherapy to drug therapy (Kim et al. 2010).

A system review including 40 randomized controlled trials by Sockol (2015) found that CBT could reduce the incidence of postpartum depression. The intervention was conducted during the postpartum period which seems more effective than the antenatal period. The authors also found that individual psychotherapy was more effective than group psychotherapy to reduce depression. The conclusion from this meta-analysis was that CBT interventions could be effective to treat and prevent PD. IPT has been demonstrated efficacy to reduce the levels of depressive symptoms and the incidence of depressive episodes in perinatal women. IPT also reduced the levels of anxiety symptoms, increased social support, increased environmental adaptability, improved interpersonal relationship (Sockol 2018). PD is associated with low income and unemployment; Shannon et al. had found that brief IPT intervention is acceptable to low-income PD women and is beneficial for the improvement of depressive symptoms and increased social support. It is a pity that the attendance of pregnant women is relatively low, although brief IPT is simple and operable (Lenze and Potts 2017). A meta-analysis conducted by Claridge (2014) aims to observe the intervention effects of systemically oriented psychotherapy; this study included 24 studies involving 1540 pregnant women; researcher found a large and positive effect size ($d = 1.14$) for systemically oriented psychotherapies in the treatment of PD. The severity of depression, the type of treatments, and the methods of Psychological evaluation could be effective.

Psychotherapy during pregnancy. The results from a randomized controlled trial conducted by Le et al. (2011) showed that CBT can be an effective psychotherapy for treating postpartum depression in high-risk Latinas women. The research divided 217 pregnant women into CBT group ($n = 112$) and usual care group ($n = 105$); intent-to-treat analyses found 8 weeks of CBT treatment (consists of three individual booster sessions) could prevent PD; the incidence of MDD in the CBT group was 7.8%, slightly lower than the usual care group (9.6%).

Austin et al. (2008) observed the effectiveness of antenatal group CBT in the primary care settings for mild-to-moderate depressed pregnant women and pregnant women who are at risk of PD or perinatal anxiety. A total of 80 pregnant women received the CBT intervention (6 weekly 2 h sessions) successfully. The researchers found that there was no significant difference in effects between CBT group and control group. The reasons for negative results may involve low symptom level at baseline. Grote et al. (2009) found that enhanced brief IPT ameliorated depression during pregnancy, prevented the recurrence of depression and improved social function 6 months after delivery. In order to investigate the effects of IPT for adolescent pregnant women, two trials were carried out in New York City public school. The objective of the first trial ($n = 14$) is improving depression by delivery CBT during health class, and objective of the second trial ($n = 11$) is improving depression by delivery CBT after school. After 12 weeks of intervention, the researchers found that depressive symptoms in adolescent pregnant women decreased by 50%, 13/14 girls' symptom level decreased in trial 1, and depressive symptoms in adolescent pregnant women decreased by 40%, and 10/11 girls' symptom level decreased in trial 2. The results indicating IPT seems to be effective for adolescent pregnant women (Miller et al. 2008).

Psychotherapy during the postpartum. An early study assessed the long-term effect of three psychotherapies on mothers' emotions in routine primary care. The subjects were randomly assigned to receive usual care, CBT, or PDP treatment. Researchers found that all of these three methods could improve the maternal mood at 4.5 months after intervention, but just the PDP effect was higher than the control group. It is a pity that women would not benefit from psychotherapy after 9 months of intervention, and psychotherapy did not reduce subsequent episodes of postpartum depression (Cooper et al. 2003). Lau and Htun (2017) found that Internet-based CBT significantly improved the levels of depression and anxiety as well as relieved stress reaction among postpartum women. The results from a meta-analysis show that IPT alone or combined with antidepressants could shorten recovery time in postpartum depressed women, and prolong relapse time (Miniati et al. 2014).

A recent meta-analysis that includes nine randomized controlled trial conducted by Lin et al. (2018) aims to assess the effect of self-help psychotherapy for treating and preventing postpartum depression; the researchers found that self-help psychotherapy showed a large overall effect on patients in posttreatment ($g = 0.51$) and followed up (3 weeks to 6 months, $g = 0.32$). Self-help psychotherapy is more effective than control in promoting recovery from postpartum depression. However, there is no significant difference between self-help psychotherapy and control, in terms of preventing postpartum depression.

13.5.8 Patients with Chronic Depression

Chronic depression is a common disability disorder which is often not adequately treated. The severity of chronic depression varies from dysthymic disorder to severe chronic MDD. The comorbidity rate of chronic depression is high and often leads to functional impairment. Chronic depression patients usually have poor interpersonal relationships, cognitive biases, and coping style disorders, and are in a state of chronic stress. Because of their symptoms, accompanied by despair and passive emotional concentration, these patients even pose a major challenge to experienced clinicians (Klein and Santiago 2003). There is very limited evidence of any therapeutic intervention for chronic depression, and psychotherapy combined with antidepressants may be beneficial for this disorder (Spijker et al. 2013).

A system review and meta-analysis that included six randomized controlled trials, comprising 1510 chronic depression patients conducted by Negt et al. (2016) found that cognitive behavioral analysis system of psychotherapy (CBASP) could improve the chronic depression better than usual care. Compared with usual care and IPT,

CBT showed moderate-to-high effect sizes ($g = 0.64$–0.75). Anyway, there are no significant differences between CBASP and antidepressant treatment ($g = -0.29$ to 0.02). CBASP can increase the effect of antidepressant treatment when combined ($g = 0.49$–0.59). The results showed that CBASP was effective for chronic depression. In an early study, ten chronic depressed patients received a 12-week CBT; the results showed that 60% of the patients could respond to psychotherapy. Chronic depression increased the levels of arousal to negative emotion. The results of this study also showed that the left amygdala reactivity increased during the implicit processing of emotional expression after psychotherapy (Klein et al. 2014).

Oxytocin is a neuropeptide associated with social behavior. The level of oxytocin is significantly decreased due to the interpersonal dysfunction of patients with chronic depression (Jobst et al. 2015). In a recent pilot study, the researchers found the depressive symptoms significantly decreased (the response rate was 50%) after 10 weeks of psychological intervention by Cognitive Behavioral Analysis System of Psychotherapy (CBASP), with the reduction of the plasma levels of oxytocin. Moreover, the results from the study indicated that oxytocin levels could predict the outcome of psychotherapy in chronic depression (Jobst et al. 2018).

13.6 Summary and Future Prospects

For patients with mild to moderate depression, depression-focused psychotherapy alone is recommended as the initial treatment option, with clinical evidence supporting the use of CBT, IPT, PDP, and PST in individual and in group formats. These therapies have been shown to produce better acute outcomes than the control group (routine care, waiting lists, and placebos). Psychotherapy combined with medication is more effective than either one alone.

It is worth mentioning that most psychotherapies have been developed in a developed country like North America, Europe, and Australia. Therefore, it is not known if these therapies are also effective in developing countries and underdeveloped countries. Many children, adults, and the elderly in low-income countries are at greater risk of depression and are in need of high-quality care. In fact, access to mental health care in low-income communities is often limited, attitudinal, and there are systemic barriers which may lead to inadequate medical utilization. Anyway, when individuals in low-income countries receive evidence-based nursing, their disease outcomes are usually positive. A meta-analysis study by Cuijpers et al. (2018b) indicated that the effectiveness of psychotherapies developed in the Western countries may be better than or not superior to that of non-Western countries, but may be equally effective and can therefore also be used in these latter countries. The field must make progress in spreading effective psychotherapies to the community. In addition, clinicians and psychotherapist, as well as service providers, should provide evidence-based psychological treatments to low-income individuals and families. The Psychotherapy should be flexible, sensitive, and convenient. Intensive outreach is required to gain initial trust and participation in community services.

Another question needs to be paid attention to. Although most meta-analyses have so far found that psychotherapy has a moderate to large impact on depression, several meta-analyses point out that the impact of psychotherapy on adult depression may be overestimated (Cuijpers et al. 2010a, b, 2011). Recently, a meta-analysis conducted by Cuijpers et al. (2018c) showed that if all studies are included and there are no further analyses of risk of bias, publication bias, and type of control group, the effects of psychotherapies for depression may be considerably overestimated. In that meta-analysis, researchers adjust these methodological issues, then, they found that the effectiveness of psychotherapy decreased significantly from moderate to small. For several types of psychotherapies (PST, ITP, and behavioral activation), there is insufficient evidence to determine whether they lead to clinically relevant effects because too few low-risk studies are available. As for PDP, the effect of the psychotherapy did not exceed the threshold of clinical significance and was not significant after adjusting for bias risk. High-quality researches are needed in the future to clarify the psychotherapies' effect.

References

Aalbers S, Fusar-Poli L, Freeman RE et al (2017) Music therapy for depression. Cochrane Database Syst Rev 11:CD004517

Agras WS, Walsh BT, Fairburn CG et al (2000) A multicenter comparison of cognitive-behavioral therapy and interpersonal psychotherapy for bulimia nervosa. Arch Gen Psychiatry 57(5):459–466

Alexopoulos GS, Raue PJ, Kanellopoulos D et al (2008) Problem solving therapy for the depression-executive dysfunction syndrome of late life. Int J Geriatr Psychiatry 23(8):782–788

Andersson G, Cuijpers P (2009) Internet-based and other computerized psychological treatments for adult depression: a meta-analysis. Cogn Behav Ther 38(4):196–205

Arean PA, Raue P, Mackin RS et al (2010) Problem-solving therapy and supportive therapy in older adults with major depression and executive dysfunction. Am J Psychiatry 167(11):1391–1398

Austin MP, Frilingos M, Lumley J et al (2008) Brief antenatal cognitive behaviour therapy group intervention for the prevention of postnatal depression and anxiety: a randomised controlled trial. J Affect Disord 105(1–3):35–44

Barsaglini A, Sartori G, Benetti S et al (2014) The effects of psychotherapy on brain function: a systematic and critical review. Prog Neurobiol 114:1–14

Barth J, Munder T, Gerger H et al (2013) Comparative efficacy of seven psychotherapeutic interventions for patients with depression: a network meta-analysis. PLoS Med 10(5):e1001454

Baumeister H, Hutter N, Bengel J (2014) Psychological and pharmacological interventions for depression in patients with diabetes mellitus: an abridged Cochrane review. Diabet Med 31(7):773–786

Beck AT (1993) Cognitive therapy: nature and relation to behavior therapy. J Psychother Pract Res 2(4):342–356

Begić D (2013) Cognitive behavioral therapy for depression. Psychiatr Clin North Am 35(1):99–110

Bell AC, D'Zurilla TJ (2009) Problem-solving therapy for depression: a meta-analysis. Clin Psychol Rev 29(4):348–353

Blackie LER, Jayawickreme E (2015) Adverse life experiences as unique situations. Eur J Pers 29(3):385–386

Bohlken J, Weber SA, Siebert A et al (2017) Reminiscence therapy for depression in dementia: an observational study with matched pairs. Geropsych J Gerontopsychology Geriatr Psychiatry 30(4):145–151
Bohlmeijer E, Smit F, Cuijpers P (2003) Effects of reminiscence and life review on late-life depression: a meta-analysis. Int J Geriatr Psychiatry 18(12):1088–1094
Bowman D, Scogin F, Lyrene B (1995) The efficacy of self-examination therapy and cognitive bibliotherapy in the treatment of mild to moderate depression. Psychother Res 5(2):131–140
Brunstein Klomek A, Stanley B (2007) Psychosocial treatment of depression and suicidality in adolescents. CNS Spectr 12(2):135–144
Bruscia KE (1995) Case studies in music therapy. Notes 52(1):129
Buchheim A, Viviani R, Kessler H et al (2012) Changes in prefrontal-limbic function in major depression after 15 months of long-term psychotherapy. PLoS ONE 7(3):e33745
Busch F, Rudden M, Shapiro T (2012) Psychodynamic treatment of depression. Psychiatr Clin North Am 35(1):111–129
Capon H, Hall W, Fry C et al (2016) Realising the technological promise of smartphones in addiction research and treatment: an ethical review. Int J Drug Policy 36:47–57
Carl H, Walsh E, Eisenlohrmoul T et al (2016) Sustained anterior cingulate cortex activation during reward processing predicts response to psychotherapy in major depressive disorder. J Affect Disord 203:204–212
Chiesa A, Mandelli L, Serretti A (2012) Mindfulness-based cognitive therapy versus psycho-education for patients with major depression who did not achieve remission following antidepressant treatment: a preliminary analysis. J Altern Complement Med 18(8):756–760
Christiansen E, Agerbo E, Bilenberg N et al (2016) SSRIs and risk of suicide attempts in young people—a Danish observational register-based historical cohort study, using propensity score. Nord J Psychiatry 70(3):167–175
Claridge AM (2014) Efficacy of systemically oriented psychotherapies in the treatment of perinatal depression: a meta-analysis. Arch Womens Ment Health 17(1):3–15
Cohen JN, Jensen D, Stange JP et al (2017) The immediate and long-term effects of an intensive meditation retreat. Mindfulness 8(4):1–14
Conwell Y, Thompson C (2008) Suicidal behavior in elders. Psychiatr Clin North Am 31(2):333–356
Coon DW, Gallagher-Thompson D (2004) Cognitive behavioral therapy with suicidal older adults. Behav Cogn Psychother 32(4):481–493
Cooper PJ, Murray L, Wilson A, et al (2003) Controlled trial of the short- and long-term effect of psychological treatment of post-partum depression. I. Impact on maternal mood. Br J Psychiatry 182:412–419
Costello EJ, Mustillo S, Erkanli A et al (2003) Prevalence and development of psychiatric disorders in childhood and adolescence. Arch Gen Psychiatry 60(8):837–844
Cowpertwait L, Clarke D (2013) Effectiveness of web-based psychological interventions for depression: a meta-analysis. Int J Ment Health Addict 11(2):247–268
Creswell JD, Pacilio LE, Lindsay EK et al (2014) Brief mindfulness meditation training alters psychological and neuroendocrine responses to social evaluative stress. Psychoneuroendocrinology 44(6):1–12
Cuijpers P (2016) Are all psychotherapies equally effective in the treatment of adult depression? The lack of statistical power of comparative outcome studies. Evid Based Ment Health 19(2):39–42
Cuijpers P, Straten AV, Warmerdam L et al (2008a) Psychological treatment of depression: a meta-analytic database of randomized studies. BMC Psychiatry 8(1):36
Cuijpers P, Straten A, Warmerdam L (2008b) Are individual and group treatments equally effective in the treatment of depression in adults? Eur J Psychiatry 22(1):38–51
Cuijpers P, van Straten A, van Schaik A et al (2009) Psychological treatment of depression in primary care: a meta-analysis. Br J Gen Pract 59(559):e51–60
Cuijpers P, van Straten A, Bohlmeijer E et al (2010a) The effects of psychotherapy for adult depression are overestimated: a meta-analysis of study quality and effect size. Psychol Med 40(2):211–223

Cuijpers P, Smit F, Bohlmeijer E et al (2010b) Efficacy of cognitive-behavioural therapy and other psychological treatments for adult depression: meta-analytic study of publication bias. Br J Psychiatry 196(3):173–178
Cuijpers P, Andersson G, Donker T et al (2011) Psychological treatment of depression: results of a series of meta-analyses. Nord J Psychiatry 65(6):354
Cuijpers P, Driessen E, Hollon SD et al (2012a) The efficacy of non-directive supportive therapy for adult depression: a meta-analysis. Clin Psychol Rev 32(4):280–291
Cuijpers P, Reynolds CF, Donker T et al (2012b) Personalized treatment of adult depression: medication, psychotherapy, or both? A systematic review. Depress Anxiety 29(10):855–864
Cuijpers P, Sijbrandij M, Koole SL et al (2013) The efficacy of psychotherapy and pharmacotherapy in treating depressive and anxiety disorders: a meta-analysis of direct comparisons. World Psychiatry 12(2):137–148
Cuijpers P, Koole SL, van Dijke A et al (2014a) Psychotherapy for subclinical depression: meta-analysis. Br J Psychiatry 205(4):268–274
Cuijpers P, Karyotaki E, Weitz E et al (2014b) The effects of psychotherapies for major depression in adults on remission, recovery and improvement: a meta-analysis. J Affect Disord 159:118–126
Cuijpers P, de Wit L, Kleiboer A et al (2018a) Problem-solving therapy for adult depression: an updated meta-analysis. Eur Psychiatry 48:27–37
Cuijpers P, Karyotaki E, Reijnders M et al (2018b) Psychotherapies for depression in low- and middle-income countries: a meta-analysis. World Psychiatry 17(1):90–101
Cuijpers P, Karyotaki E, Reijnders M et al (2018c) Was eysenck right after all? A reassessment of the effects of psychotherapy for adult depression. Epidemiol Psychiatr Sci 1–10
Damsker JI, Macklis R, Brady LW et al (2009) Canadian network for mood and anxiety treatments (CANMAT) clinical guidelines for the management of major depressive disorder in adults. III. Pharmacotherapy. J Affect Disord 117(1):S54–S64
Davidson JR (2010) Major depressive disorder treatment guidelines in America and Europe. J Clin Psychiatry 71(Suppl E1):e04
Davies EB, Morriss R, Glazebrook C (2014) Computer-delivered and web-based interventions to improve depression, anxiety, and psychological well-being of university students: a systematic review and meta-analysis. J Med Internet Res 16(5):e130
Dobkin RD, Menza M, Allen LA et al (2011) Cognitive-behavioral therapy for depression in Parkinson's disease: a randomized, controlled trial. Am J Psychiatry 168(10):1066–1074
Donker T (2013) Smartphones for smarter delivery of mental health programs: a systematic review. J Med Internet Res 15(11):e247–e247
Dryden W, Kendall PC (1987) Advances in cognitive-behavioral research and therapy. J Cogn Psychother 1(4):263–264
D'Zurilla TJ, Goldfried MR (1971) Problem solving and behavior modification. J Abnorm Psychol 78(1):107–126
D'Zurilla TJ, Nezu AM (2001) Problem-solving therapies. Guilford Press, pp 211–245
Ehring T, Watkins ER (2009) Repetitive negative thinking as a transdiagnostic process. Int J Cogn Ther 1(3):192–205
Eskin M (2013) Processes and sessions of problem-solving therapy. Probl Solving Ther Clin Pract 147–180
Eskin M, Ertekin K, Demir H (2008) Efficacy of a problem-solving therapy for depression and suicide potential in adolescents and young adults. Cogn Ther Res 32(2):227–245
Fava GA, Rafanelli C, Grandi S et al (1998) Prevention of recurrent depression with cognitive behavioral therapy: preliminary findings. Arch Gen Psychiatry 55(9):816–820
Feske U, Frank E, Kupfer DJ et al (1998) Anxiety as a predictor of response to interpersonal psychotherapy for recurrent major depression: an exploratory investigation. Depress Anxiety 8(4):135–141
Fischer S, Cleare AJ (2017) Cortisol as a predictor of psychological therapy response in anxiety disorders-Systematic review and meta-analysis. J Anxiety Disord 47:60–68

Fochtmann LJ, Alan MD, Gelenberg J (2005) Guideline watch: practice guideline for the treatment of patients with major depressive disorder. Focus 3(1):1–15

Folkman S (2013) Stress: appraisal and coping. Springer, New York, pp 1–460

Forkmann T, Wichers M, Geschwind N et al (2014) Effects of mindfulness-based cognitive therapy on self-reported suicidal ideation: results from a randomised controlled trial in patients with residual depressive symptoms. Compr Psychiatry 55(8):1883–1890

Fournier JC, Derubeis RJ, Shelton RC et al (2009) Prediction of response to medication and cognitive therapy in the treatment of moderate to severe depression. J Consult Clin Psychol 77(4):775–787

Fristad MA, Goldberg-Arnold JS, Gavazzi SM (2003) Multi-family psychoeducation groups in the treatment of children with mood disorders. J Marital Fam Ther 29(4):491–504

Fu CH, Williams SC, Cleare AJ et al (2008) Neural responses to sad facial expressions in major depression following cognitive behavioral therapy. Biol Psychiatry 64(6):505–512

Fu CH, Steiner H, Costafreda SG (2013) Predictive neural biomarkers of clinical response in depression: a meta-analysis of functional and structural neuroimaging studies of pharmacological and psychological therapies. Neurobiol Dis 52(4):75–83

Furukawa TA (2010) 'Third wave' cognitive and behavioural therapies versus treatment as usual for depression. Cochrane Database Syst Rev 10(9):CD008705

Gabbard GO (2010) Long-term psychodynamic psychotherapy: a basic text. Am Psychiatr Pub 639–640

Gelenberg AJ (2010) Depression symptomatology and neurobiology. J Clin Psychiatry 71(1):e02

Goldapple K, Segal Z, Garson C et al (2004) Modulation of cortical-limbic pathways in major depression: treatment-specific effects of cognitive behavior therapy. Arch Gen Psychiatry 61(1):34–41

Gould RL, Coulson MC, Howard RJ (2012) Cognitive behavioral therapy for depression in older people: a meta-analysis and meta-regression of randomized controlled trials. J Am Geriatr Soc 60(10):1817–1830

Greenberg L (2014) The therapeutic relationship in emotion-focused therapy. Psychotherapy (Chic) 51(3):350–357

Greenberg LS, Watson JC (2006) Emotion-focused therapy for depression. Am Psychol Assoc 15–25

Grote NK, Swartz HA, Geibel SL et al (2009) A randomized controlled trial of culturally relevant, brief interpersonal psychotherapy for perinatal depression. Psychiatr Serv 60(3):313–321

Gustavson KA, Alexopoulos GS, Niu GC et al (2016) Problem-solving therapy reduces suicidal ideation in depressed older adults with executive dysfunction. Am J Geriatr Psychiatry 24(1):11–17

Hallford D (2014) Cognitive-reminiscence therapy for the treatment of depression in young adults. Deakin University. http://hdl.handle.net/10536/DRO/DU:30067334

Heisel MJ, Duberstein PR, Talbot NL et al (2009) Adapting interpersonal psychotherapy for older adults at risk for suicide: preliminary findings. Prof Psychol Res Pr 40(2):156–164

Hinrichsen GA (2001) Comprehensive guide to interpersonal psychotherapy. Br J Psychiatry 180(3):286

Hobbs MJ, Joubert AE, Mahoney AEJ et al (2018) Treating late-life depression: comparing the effects of internet-delivered cognitive behavior therapy across the adult lifespan. J Affect Disord 226:58–65

Hollon SD, Stewart MO, Strunk D (2006) Enduring effects for cognitive behavior therapy in the treatment of depression and anxiety. Annu Rev Psychol 57(1):285–315

Holmqvist M, Vincent N, Walsh K (2014) Web- vs. telehealth-based delivery of cognitive behavioral therapy for insomnia: a randomized controlled trial. Sleep Med 15(2):87–195

Hosseinzadeh AN, Barahmand U (2014) Effectiveness of mindfulness-based cognitive therapy for co-morbid depression in drug-dependent males. Arch Psychiatr Nurs 28(5):314–318

Huang HC, Chen YT, Chen PY et al (2015a) Reminiscence therapy improves cognitive functions and reduces depressive symptoms in elderly people with dementia: a meta-analysis of randomized controlled trials. J Am Med Dir Assoc 16(12):1087–1094

Huang AX, Delucchi K, Dunn LB et al (2015b) A systematic review and meta-analysis of psychotherapy for late-life depression. Am J Geriatr Psychiatry 23(3):261–273

Huey SJ Jr, Henggeler SW, Rowland MD et al (2004) Multisystemic therapy effects on attempted suicide by youths presenting psychiatric emergencies. J Am Acad Child Adolesc Psychiatry 43(2):183–190

Hummel J, Weisbrod C, Boesch L et al (2017) AIDE-acute illness and depression in elderly patients. Cognitive behavioral group psychotherapy in geriatric patients with comorbid depression: a randomized, controlled trial. J Am Med Dir Assoc 18(4):341–349

Jacobson E (1955) The interpersonal theory of psychiatry 3(1):149–156

Jakobsen H, Andersson G, Havik OE et al (2016) Guided internet-based cognitive behavioral therapy for mild and moderate depression: a benchmarking study. Internet Interventions 15(7):1–8

Jane Garland E, Kutcher S, Virani A et al (2016) Update on the use of ssris and snris with children and adolescents in clinical practice. J Can Acad Child Adolesc Psychiatry 25(1):4–10

Jeyanantham K, Kotecha D, Thanki D et al (2017) Effects of cognitive behavioural therapy for depression in heart failure patients: a systematic review and meta-analysis 22(6):731–741

Jobst A, Sabass L, Palagyi A et al (2015) Effects of social exclusion on emotions and oxytocin and cortisol levels in patients with chronic depression. J Psychiatr Res 60:170–177

Jobst A, Brakemeier EL, Buchheim A et al (2016) European Psychiatric Association Guidance on psychotherapy in chronic depression across Europe. Eur Psychiatry 33:18–36

Jobst A, SabaãŸ L, Hall D et al (2018) Oxytocin plasma levels predict the outcome of psychotherapy: a pilot study in chronic depression. J Affect Disord 227:206–213

Josephine K, Josefine L, Philipp D et al (2017) Internet- and mobile-based depression interventions for people with diagnosed depression: a systematic review and meta-analysis. J Affect Disord 223:28–40

Kabat-Zinn J (1982) An outpatient program in behavioral medicine for chronic pain patients based on the practice of mindfulness meditation: theoretical considerations and preliminary results. Gen Hosp Psychiatry 4(1):33–47

Kaplan EH (1997) Telepsychotherapy. Psychotherapy by telephone, videotelephone, and computer videoconferencing. J Psychother Pract Res 6(3):227–237

Karlsson H, Hirvonen J, Kajander J et al (2010) Research letter: psychotherapy increases brain serotonin 5-HT1A receptors in patients with major depressive disorder. Psychol Med 40(3):523–528

Karyotaki E, Smit Y, Holdt HK et al (2016) Combining pharmacotherapy and psychotherapy or monotherapy for major depression? A meta-analysis on the long-term effects. J Affect Disord 194:144–152

Kaufmann FN, Gazal M, Mondin TC et al (2015) Cognitive psychotherapy treatment decreases peripheral oxidative stress parameters associated with major depression disorder. Biol Psychol 110:175–181

Kessler RC, Berglund P, Demler O et al (2003) The epidemiology of major depressive disorder: results from the national comorbidity survey replication (NCS-R). JAMA 289(23):3095–3105

Kim DR, O'Reardon JP, Epperson CN (2010) Guidelines for the management of depression during pregnancy. Curr Psychiatry Rep 12(4):279–281

Kingston D, Tough S, Whitfield H (2012) Prenatal and postpartum maternal psychological distress and infant development: a systematic review. Child Psychiatry Hum Dev 43(5):683–714

Kiosses DN, Rosenberg PB, McGovern A et al (2015) Depression and suicidal ideation during two psychosocial treatments in older adults with major depression and dementia. J Alzheimers Dis 48(2):453–462

Klein DN, Santiago NJ (2003) Dysthymia and chronic depression: introduction, classification, risk factors, and course. J Clin Psychol 59(8):807–816

Klein JP, Becker B, Hurlemann R et al (2014) Effect of specific psychotherapy for chronic depression on neural responses to emotional faces. J Affect Disord 166:93–97

Klerman GL, Weissman MM (1986) The interpersonal approach to understanding depression. Guilford Press, pp 429–456

Klerman GL, Weissmann MM (1987) Interpersonal psychotherapy (IPT) and drugs in the treatment of depression. Pharmacopsychiatry 20(1):3–7

Knekt P, Saari T, Lindfors O (2014) Intelligence as a predictor of outcome in short- and long-term psychotherapy. Psychiatry Res 220(3):1019–1027

Koch JM, Hinze-Selch D, Stingele K et al (2009) Changes in CREB phosphorylation and BDNF plasma levels during psychotherapy of depression. Psychother Psychosom 78(3):187–192

Kuyken W, Hayes R, Barrett B et al (2015) The effectiveness and cost-effectiveness of mindfulness-based cognitive therapy compared with maintenance antidepressant treatment in the prevention of depressive relapse/recurrence: results of a randomised controlled trial (the PREVENT study). Health Technol Assess 386(9988):63–73

Lau Y, Htun TP (2017) Therapist-supported internet-based cognitive behavior therapy for stress, anxiety, and depressive symptoms among postpartum women: a systematic review and meta-analysis 19(4):e138

Le HN, Perry DF, Stuart EA (2011) Randomized controlled trial of a preventive intervention for perinatal depression in high-risk Latinas. J Consult Clin Psychol 79(2):135–141

Leichsenring F, Rabung S (2009) Effectiveness of long-term psychodynamic psychotherapy: a meta-analysis. JAMA 301(9):931–932

Lenze SN, Potts MA (2017) Brief Interpersonal psychotherapy for depression during pregnancy in a low-income population: a randomized controlled trial. J Affect Disord 210:151–157

Lewinsohn PM, Rohde P, Seeley JR (2010) Adolescent suicidal ideation and attempts: prevalence, risk factors, and clinical implications. Clin Psychol Sci Pract 3(1):25–46

Li D, Bao Z, Li X et al (2016) Perceived school climate and chinese adolescents' suicidal ideation and suicide attempts: the mediating role of sleep quality. J Sch Health 86(2):75–83

Lin PZ, Xue JM, Yang B et al (2018) Effectiveness of self-help psychological interventions for treating and preventing postpartum depression: a meta-analysis. Arch Womens Ment Health 21(5):491–503

Lindfors O, Ojanen S, Jaaskelainen T et al (2014) Social support as a predictor of the outcome of depressive and anxiety disorder in short-term and long-term psychotherapy. Psychiatry Res 216(1):44–51

Ludman EJ, Simon GE, Tutty S et al (2007) A randomized trial of telephone psychotherapy and pharmacotherapy for depression: continuation and durability of effects. J Consult Clin Psychol 75(2):257–266

Lutz W, Schiefele AK, Wucherpfennig F et al (2016) Clinical effectiveness of cognitive behavioral therapy for depression in routine care: a propensity score based comparison between randomized controlled trials and clinical practice. J Affect Disord 189(5):150–158

Luyten P, Blatt SJ (2012) Psychodynamic treatment of depression. Psychiatr Clin North Am 35(1):111–129

Maguth Nezu C, D'Zurilla TJ, Nezu AM (2008) Problem-solving therapy: theory, practice, and application to sex offenders. Wiley, pp 103–123

Maratos A, Gold C, Wang X, et al (2008) Music therapy for depression. Cochrane Database Syst Rev 108(1):CD004517

Markowitz JC (2003) Interpersonal psychotherapy for chronic depression. J Clin Psychol 59(8):847–858

Maxwell L, Duff E (2016) Mindfulness: an effective prescription for depression and anxiety. J Nurse Pract 12(6):403–409

McDermut W, Miller IW, Brown RA (2010) The efficacy of group psychotherapy for depression: a meta-analysis and review of the empirical research. Clin Psychol Sci Pract 8(1):98–116

McGovern AR, Kiosses DN, Raue PJ et al (2014) Psychotherapies for late-life depression. Psychiatr Ann 44(3):147–152

Mennin DS, Fresco DM (2014) Emotion regulation therapy. Guilford Publishers, pp 169–487

Miller L, Gur M, Shanok A et al (2008) Interpersonal psychotherapy with pregnant adolescents: two pilot studies. J Child Psychol Psychiatry 49(7):733–742

Miniati M, Callari A, Calugi S et al (2014) Interpersonal psychotherapy for postpartum depression: a systematic review. Arch Womens Ment Health 17(4):257–268

Morgan D (2003) Mindfulness-based cognitive therapy for depression: a new approach to preventing relapse. Psychother Res 13(1):123–125

Mufson L, Rynn M, Yanes-Lukin P et al (2018) Stepped care interpersonal psychotherapy treatment for depressed adolescents: a pilot study in pediatric clinics. Adm Policy Ment Health 45(3):417–431

Müldnernieckowski Ł, Cyranka K, Smiatekmazgaj B et al (2015) Psychotherapy for pregnant women with psychiatric disorders. Psychiatr Pol 49(1):49–56

Mynorswallis LM, Gath DH, Lloydthomas AR et al (1995) Randomised controlled trial comparing problem solving treatment with amitriptyline and placebo for major depression in primary care. BMJ 310(6977):441–445

Negt P, Brakemeier EL, Michalak J et al (2016) The treatment of chronic depression with cognitive behavioral analysis system of psychotherapy: a systematic review and meta-analysis of randomized-controlled clinical trials. Brain Behav 6(8):e00486

Nel PW (2014) The NICE guideline on the treatment of child and adolescent depression: a meta-review of the evidence for individual CBT. Eur J Psychother Couns 16(3):267–287

Nezu AM (1987) A problem-solving formulation of depression: a literature review and proposal of a pluralistic model. Clin Psychol Rev 7(2):121–144

Nezu AM, Nezu CM, D'Zurilla TJ (2007) Problem-solving skills training. Encycl Stress 37(8):227–231

Nezu AM, Nezu CM, D'Zurilla TJ (2013) Problem-solving therapy: a treatment manual. Springer Pub 235–252

Nguyen S, Wong D, McKay A et al (2017) Cognitive behavioural therapy for post-stroke fatigue and sleep disturbance: a pilot randomised controlled trial with blind assessment. Neuropsychol Rehabil 1–16

Nnjjm VDS, Wingen M, Vermeeren A et al (2017) Driving performance of depressed patients who are untreated or receive long-term antidepressant (SSRI/SNRI) treatment. Pharmacopsychiatry 50(5):182–188

Nyklicek I, van Son J, Pop VJ et al (2016) Does Mindfulness-based cognitive therapy benefit all people with diabetes and comorbid emotional complaints equally? Moderators in the DiaMind trial. J Psychosom Res 91:40–47

O'Toole MS, Bovbjerg DH, Renna ME et al (2018) Effects of psychological interventions on systemic levels of inflammatory biomarkers in humans: a systematic review and meta-analysis. Brain Behav Immun. https://doi.org/10.1016/j.bbi.2018.04.005 (in press)

Ogrodniczuk JS, Piper WE, Joyce AS (2004) Alexithymia as a predictor of residual symptoms in depressed patients who respond to short-term psychotherapy. Am J Psychother 58(2):150–161

Ontario HQ (2017) Psychotherapy for major depressive disorder and generalized anxiety disorder: a health technology assessment. Ont Health Technol Assess Ser 17(15):1–167

O'Shea G, Spence SH, Donovan CL (2015) Group versus individual interpersonal psychotherapy for depressed adolescents. Behav Cogn Psychother 43(1):1–19

Park SC, Oh HS, Oh DH et al (2014) Evidence-based, non-pharmacological treatment guideline for depression in Korea. J Korean Med Sci 29(1):12–22

Pies RW (2009) Depression and the pitfalls of causality: implications for DSM-V. J Affect Disord 116(1–2):1–3

pmhdev, PubMedHealth (2008) Problem-solving therapy for depression in adults: a systematic review. Res Soc Work Pract 18(2):117–131

Qaseem A, Barry MJ, Kansagara D (2016) Nonpharmacologic versus pharmacologic treatment of adult patients with major depressive disorder: a clinical practice guideline from the American college of physicians. Ann Intern Med 164(5):350–359

Rajewska-Rager A, Sibilski P, Lepczynska N (2015) Risk factors for suicide among children and youths with bipolar spectrum and early bipolar disorder. Psychiatr Pol 49(3):477–488

Rangell L (2002) The theory of psychoanalysis: vicissitudes of its evolution. J Am Psychoanal Assoc 50(4):1109–1137
Raue PJ, Schulberg HC, Lewis-Fernandez R et al (2010) Shared decision-making in the primary care treatment of late-life major depression: a needed new intervention? Int J Geriatr Psychiatry 25(11):1101–1111
Raue PJ, McGovern AR, Kiosses DN et al (2017) Advances in psychotherapy for depressed older adults. Curr Psychiatry Rep 19(9):57
Richards D, Richardson T (2012) Computer-based psychological treatments for depression: a systematic review and meta-analysis. Clin Psychol Rev 32(4):329–342
Robinson AL, Mccague B, Whissell C (2012) Single-case investigation of an emotion-focused therapy group for anxiety and depression. Int J Integr Psychother 3(1):59–81
Rushford N (2012) Interpersonal psychotherapy. Encycl Hum Behav 34(3):472–478
Rushton JL, Forcier M, Schectman RM (2002) Epidemiology of depressive symptoms in the national longitudinal study of adolescent health. J Am Acad Child Adolesc Psychiatry 41(2):199–205
Ryan ND, Puig-Antich J, Ambrosini P et al (1987) The clinical picture of major depression in children and adolescents. Arch Gen Psychiatry 44(10):854–861
Schuver KJ, Lewis BA (2016) Mindfulness-based yoga intervention for women with depression. Complement Ther Med 26:85–91
Shahnavaz S, Hedman-Lagerlof E (2018) Internet-based cognitive behavioral therapy for children and adolescents with dental anxiety: open trial. J Med Internet Res 20(1):e12
Shedler J (2010) The efficacy of psychodynamic psychotherapy. Am Psychol 65(2):98–109
Simon SS, Cordas TA, Bottino CM (2015) Cognitive behavioral therapies in older adults with depression and cognitive deficits: a systematic review. Int J Geriatr Psychiatry 30(3):223–233
Sizoo BB, Kuiper E (2017) Cognitive behavioural therapy and mindfulness based stress reduction may be equally effective in reducing anxiety and depression in adults with autism spectrum disorders. Res Dev Disabil 64:47–55
Slife BD, Weaver CA (1992) Depression, cognitive skill, and metacognitive skill in problem solving. Cogn Emot 6(1):1–22
Sockol LE (2015) A systematic review of the efficacy of cognitive behavioral therapy for treating and preventing perinatal depression. J Affect Disord 177:7–21
Sockol LE (2018) A systematic review and meta-analysis of interpersonal psychotherapy for perinatal women. J Affect Disord 232:316–328
Spijker J, Van SA, Bockting CL et al (2013) Psychotherapy, antidepressants, and their combination for chronic major depressive disorder: a systematic review. Can J Psychiatry 58(7):386–392
Spijkerman MPJ, Pots WTM, Bohlmeijer ET (2016) Effectiveness of online mindfulness-based interventions in improving mental health: a review and meta-analysis of randomised controlled trials. Clin Psychol Rev 45:102–114
Spinhoven P, Klein N, Kennis M et al (2018) The effects of cognitive-behavior therapy for depression on repetitive negative thinking: a meta-analysis. Behav Res Ther 106:71–85
Strauss C, Cavanagh K, Oliver A et al (2014) Mindfulness-based interventions for people diagnosed with a current episode of an anxiety or depressive disorder: a meta-analysis of randomised controlled trials. PLoS ONE 9(4):e96110
Sudak DM (2012) Cognitive behavioral therapy for depression. Psychiatr Clin North Am 35(1):99–110
Tavares LR, Barbosa MR (2018) Efficacy of group psychotherapy for geriatric depression: a systematic review. Arch Gerontol Geriatr 78:71–80
Teasdale JD, Segal ZV, Williams JM et al (2000) Prevention of relapse/recurrence in major depression by mindfulness-based cognitive therapy. J Consult Clin Psychol 68(4):615–623
Tovote KA, Schroevers MJ, Snippe E et al (2017) What works best for whom? Cognitive behavior therapy and mindfulness-based cognitive therapy for depressive symptoms in patients with diabetes. PLoS ONE 12(6):e0179941
Van HL, Schoevers RA, Dekker J (2008) Predicting the outcome of antidepressants and psychotherapy for depression: a qualitative, systematic review. Harv Rev Psychiatry 16(4):225–234

Visser MM, Heijenbrok-Kal MH, Van't Spijker A et al (2016) Problem-solving therapy during outpatient stroke rehabilitation improves coping and health-related quality of life: randomized controlled Trial. Stroke 47(1):135–142

Vittengl JR, Clark LA, Dunn TW et al (2007) Reducing relapse and recurrence in unipolar depression: a comparative meta-analysis of cognitive-behavioral therapy's effects. J Consult Clin Psychol 75(3):475–488

Wang Z, Zhu M, Su Z et al (2017) Post-stroke depression: different characteristics based on follow-up stage and gender-a cohort perspective study from Mainland China. Neurol Res 39(11):996–1005

Wang SB, Wang YY, Zhang QE et al (2018) Cognitive behavioral therapy for post-stroke depression: a meta-analysis. J Affect Disord 235:589–596

Watkins E (2015) Psychological treatment of depressive rumination. Curr Opin Psychol 4:32–36

Watts S, Mackenzie A, Thomas C et al (2013) CBT for depression: a pilot RCT comparing mobile phone vs. computer. BMC Psychiatry 13:49

Weersing VR, Jeffreys M, Do MT et al (2017) Evidence base update of psychosocial treatments for child and adolescent depression. J Clin Child Adolesc Psychol 46(1):11–43

Wierzbicki M, Bartlett TS (1987) The efficacy of group and individual cognitive therapy for mild depression. Cogn Ther Res 11(3):337–342

Woods B, O'Philbin L, Farrell EM et al (2018) Reminiscence therapy for dementia. Cochrane Database Syst Rev 3:CD001120

Yorbik O, Birmaher B, Axelson D et al (2004) Clinical characteristics of depressive symptoms in children and adolescents with major depressive disorder. J Clin Psychiatry 65(12):1654–1659; quiz 1760–1651

Zaremba D, Enneking V, Meinert S et al (2018) Effects of cumulative illness severity on hippocampal gray matter volume in major depression: a voxel-based morphometry study. Psychol Med 48(14):1–8

Zhang Z, Zhang L, Zhang G et al (2018) The effect of CBT and its modifications for relapse prevention in major depressive disorder: a systematic review and meta-analysis. BMC Psychiatry 18(1):50

Zhao D, Lustria MLA, Hendrickse J (2017) Systematic review of the information and communication technology features of web- and mobile-based psychoeducational interventions for depression. Patient Educ Couns 100(6):1049–1072

Chapter 14
Internet-Based Management for Depressive Disorder

Zuowei Wang, Zhiang Niu, Lu Yang and Lvchun Cui

Abstract The advances in the Internet and related technologies may lead to changes in professional roles of psychiatrists and psychotherapists. The application of artificial intelligence (AI) and electronic measurement-based care (eMBC) in the treatment of depressive disorder has addressed more interest. AI could play a role in population health management and patient administration as well as assist physicians to make a decision in the real-world clinical practice. The eMBC strengthens MBC through web/mobile devices and telephone consulting services, to monitor disease progression, and customizes the MBC interface in electronic medical record systems (EMRs).

Keywords Depressive disorder · Therapy · Artificial intelligence (AI) · Measurement-Based Care (MBC)

14.1 Introduction

The China Internet Network Information Center (CNNIC) reported that the number of Internet users in China has reached 829 million. Seeking health and medical information has become one of the most popular purposes of surfing the Internet. Then in the last decade or so, electronic mental health (eMH) was also growing rapidly. The term eMH refers to mental health information and services provided through the Internet and related technologies. The current Internet-based services of mental health mainly include health education, symptom assessment and monitoring of recurrence, and management and intervention of chronic disease. The advances in the Internet and related technologies (e.g., mobile application and WeChat) may lead to changes in professional roles of psychiatrists and psychotherapists. For example,

Z. Wang (✉)
Hongkou District Mental Health Center, Shanghai 200083, China
e-mail: wzwhk@163.com

Z. Niu · L. Yang · L. Cui
Division of Mood Disorders, Shanghai Mental Health Center, Shanghai Jiao Tong University School of Medicine, Shanghai, China

© Springer Nature Singapore Pte Ltd. 2019

Y. Fang (ed.), *Depressive Disorders: Mechanisms, Measurement and Management*, Advances in Experimental Medicine and Biology 1180,
https://doi.org/10.1007/978-981-32-9271-0_14

the Internet-delivered Interpersonal Psychotherapy and Cognitive Behavior Therapy have been supplied for patients with depressive disorder. This chapter would introduce the application of artificial intelligence (AI) and electronic measurement-based care (eMBC) for the clinical therapy of depressive disorder.

14.2 Application of Artificial Intelligence (AI) in Depressive Disorder

Artificial intelligence (AI) is a branch of computer science that seeks to understand the nature of intelligence and produce new ways of responding to problems similar to human intelligence. The application of AI technologies in the medical field has made rapid progress, but the application in clinical practice has not yet been generalized. The overall goal of these technologies is to use computer algorithms to find relevant information from the medical data and then support clinical decisions. Specifically, AI could play a role in population health management and patient administration at the macro level as well as assisting physicians making diagnosis, selecting therapy, and predicting outcome in the real-world clinical practice (Watson et al. 2019). For depressive disorder, current research on AI mainly focuses on early diagnosis, personalized treatment, and prognosis evaluation.

14.3 AI in the Diagnosis of Depressive Disorder

14.3.1 AI in the Early Diagnosis of Depressive Disorder

The lifestyle and environmental factors relevant to the depressive disorder are multidimensional and heterogeneous. The advantage of AI is that it can reduce multidimensional variables to lower dimension and provide prediction models of the onset of disease. The US National Health and Nutrition Examination Survey (NHANES) database formed two cluster solutions for the early diagnosis of depressive disorder through two machine learning (ML) methods (self-organizing maps and supervised boosted regression) (Dipnall et al. 2017).

The progress of AI technology makes it possible to obtain dynamic data of high-risk individuals and patients with depressive disorder, which could reduce the cost of long-term follow-up of these individuals. The current development trend mainly lies in human–computer interaction (wearable smart devices) and app software on mobile phones. In 2019, the American Psychological Association (APA) has reported a framework capable of detecting depressive disorder named artificial intelligence mental evaluation (AiME). This framework is used with an artificial intelligence algorithm to help a psychiatrist identify depressive symptoms and make a faster preventative intervention (Victor et al. 2019). The AI-enabled app with empathetic and evidence-driven conversational features has also been developed to analyze in-app user feedback and evaluate the performance of an ML classifier. These emerging AI technologies can allow for a comprehensive assessment of the patient's lifestyle and environmental status, leading to better early identification of depressive disorder (Ferrari et al. 2018).

14.3.2 AI in the Differential Diagnosis of Depressive Disorder

The development of AI technology could help processing high-dimensional MRI imaging data to identify between major depression (MDD) and bipolar depression (BPD). A study has used pattern recognition and functional magnetic resonance imaging (fMRI) to find that patterns of whole-brain neural activity to happy faces in patients with BPD were significantly less distinct from those for neutral faces either in patients with MDD and healthy control (Mourao-Miranda et al. 2012). The ML techniques such as support vector machines (SVM) and relevance vector machines (RVM) have been applied to predictive classification of individual MRI data with variable success. The doctors can make accurate predictions by using brain scan data to identify depressive disorder in healthy people (Mwangi et al. 2012).

14.3.3 AI in the Exploration of Specific Type in Depressive Disorder

AI technology also provides a new way to stratify subtypes of depressive disorder (Diniz et al. 2016). The STAR*D (Sequenced Treatment Alternatives to Relieve Depression) study analyzed with three ML methods (naïve Bayes classifier, SVM, and random forest) successfully built a risk prediction model incorporating baseline sociodemographic and clinical features for treatment-resistant depression (TRD) (Perlis 2013). For other special subtypes of depressive disorder, such as postpartum depression, adolescent depression and late life depression, most studies are still small samples, and the use of AI technique to integrate big and heterogeneous data to obtain more solid conclusions will be the trend in the future study of depressive disorder.

14.4 AI in the Treatment of Depressive Disorder

14.4.1 AI in the Personalized Treatment for Depressive Disorder

The STAR*D study found that first-episode depressive patients only had a 33% chance of remission after standardized treatment with antidepressants, and more than half of the patients still need further intensive treatment (Trivedi et al. 2006).

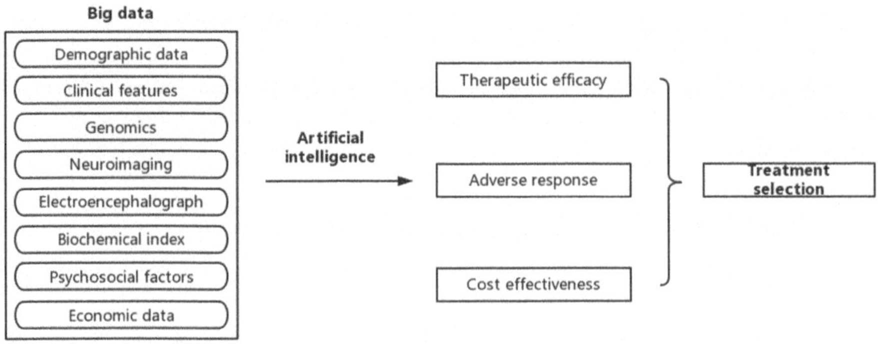

Fig. 14.1 Artificial intelligence translates big data and novel predictive variables into real-world treatment recommendations

The physicians do not have a proven mechanism to assess whether patients with depressive disorder respond to a specific treatment. The goal of precision medicine is to provide evidence-based answers to "*What works for whom*?" (Cohen and Derubeis 2018). AI, with its strong function of handling huge data, can detect complex high-dimensional interactions that might inform predictions, translating big data and novel predictive variables into real-world treatment recommendations to improve outcomes in depressive disorder (Fig. 14.1) (Horvitz and Mulligan 2015).

14.4.2 AI in the Prediction of Treatment Response

The AI technologies can be applied to investigate and identify biomarkers that help to predict outcomes (both therapeutic efficacy and adverse response) in patients with MDD (Lam et al. 2016).

1. Genomics: The big data approaches are applied to identify potentially significant variants related to treatment response and extract the most promising predictors (Abbasi 2017). The Genome-wide association study (GWAS) and ML approach (LASSO regression, classification-regression trees, and SVM) could be integrated to produce a variety of gene models that predict the response to certain treatments (Maciukiewicz et al. 2018).
2. Neuroimaging: Different forms of AI such as random forest, improved support vector machine models, or sparse linear or nonlinear regression models could produce advanced feature selection strategies, improving the individualized prediction accuracy to reach a level that is sufficient to assist pretreatment clinical decisions (Redlich et al. 2016; Abbott et al. 2016).
3. Electroencephalography (EEG): A list of EEG features was found to discriminate various treatment responses of depressive disorder. ML methods such as LASSO regression and SVM have been previously used in several EEG

applications to establish automated EEG classification, differentiating responders from nonresponders (Khodayarirostamabad et al. 2013).

4. Predictive model including multiple clinical and biological markers: Depressive disorder is a typical disease that conforms to the "bio-psychology-society" medical model. Using single measurement may have limited predictive ability and cannot describe the relationship between these three factors while combining various kinds and a large amount of information into a statistical model through AI may reflect the disease situation in the real world and instruct the treatment regiment (Chekroud et al. 2016).

14.4.3 AI in the Prediction of Cost-Effectiveness

Clinicians require real-world data to produce evidence-based instructions and further improve their treatment decisions. Simultaneously, policy-makers and stakeholders increasingly demand cost-effectiveness evidence in order to support their conclusions and decisions. Economic data and information collected from the diagnosis and treatment of the depressive disorder are of great practical value as treatment recommendation is based on both outcomes and costs. Using various AI technologies to process data collected at baseline can produce a decision support system in the context of health interventions (Bremer et al. 2018). Patients with depressive disorder can be individually allocated to the most appropriate treatment type according to their cost effective ratio. This will lead to improving better outcomes for individuals and reduce healthcare costs.

14.5 AI in the Prognosis of Depressive Disorder

14.5.1 AI in the Prediction of the Recurrence

At present, the prediction of recurrence of depressive disorder in clinical practice mainly adopts the "one-size-fits-all" method, and the duration of depression maintenance treatment recommended is determined according to the number of depressive episodes. A prediction model based on AI technology will improve the predicting performance, which will greatly enhance the utilization of electronic health data of patients with depressive disorder, identifying risk factors with the ability to predict automatically (Aziz 2016).

Dynamic evaluation and monitoring is the key component in the prediction. Smartphone app is an information-gathering tool of great potential as it is disposable and widely used. Through self-recording daily mood scores, detecting light exposure (using the installed sensor) and daily worn activity trackers, digital log data of activity, sleep and heart rate were collected . Passive digital phenotypes can be processed into features and a mood prediction algorithm can be developed based on AI technology (Cho et al. 2019). This portable prediction tool is like an early warning device and is a benefit for euthymic patients to detect the recurrence signs automatically and then take effective measures.

14.5.2 AI in the Prediction of Functional Outcomes

AI technology SVM can be used to integrate the complex scale with numerous items that reflect social functions and basic demographic information and build a prediction model that could augment effective therapeutic strategies. The results of neuroimaging measures can be also included to produce a more precise model aiming at the prevention of social functioning impairments in patients with recent onset depressive disorder, but the performance of the model still remains to be improved (Koutsouleris et al. 2018). The application of AI modeling in predicting the prognosis of depressive symptoms is still in its infancy, and more studies based on AI analysis at big dataset ought to be implemented.

14.6 Application of Electronic Measurement-Based Care (EMBC) in Depressive Disorder

Most modern medical care uses objective measures to guide and evaluate treatment. For example, blood pressure measurements are often used to screen for hypertension to determine if treatment is needed and to guide treatment. In contrast, current mental health services are mainly based on subjective assessments of clinicians. The integration of evidence-based practices (EBPs) into medical settings is a successful strategy to improve clinical outcomes. EBPs have developed considerable empirical support and demonstrated a growing appeal to clinicians. Measurement-based care (MBC) is considered a core component of numerous EBPs and can be added to the treatment of depression. The Sequenced Treatment Alternatives to Relieve Depression (STAR*D) study found that MBC can be successfully integrated into clinical practice and improve patient outcomes (Trivedi et al. 2007).

MBC is the systematic use of measurement tools to monitor clinical outcomes and support clinical decision-making, which has been shown to improve outcomes, e.g., symptom remission and treatment adherence. Although MBC has clinical benefits, studies have shown that less than 20% of clinicians and psychiatrists routinely assess patients' symptoms before each face-to-face consultation (Lewis et al. 2015). The main reasons include time restrictions (Oslin et al. 2019). Except for the time limitation, funding constraints and limited technical resources are also issues of concern to many clinicians and institutions (Waldrop and Mcguinness 2017). With the development of information technology, Enhanced Measurement-Based Care (eMBC) based on network/mobile devices and telephone guidance services was proposed, which could help overcome the challenges of MBC implementation.

1. Definition of eMBC

The affective disorders team of the University of British Columbia (UBC) first proposed the concept of eMBC. The eMBC strengthens MBC through web/mobile devices and telephone consulting services, employs assessment tools based on web and mobile applications (APPs) to monitor disease progression, and customizes the MBC interface in electronic medical record systems (EMRs). Through the use of self-management strategies, eMBC could create collaboration between clinicians and patients. The main difference between eMBC and MBC is that technical tools are applied in eMBC to improve patient participation, both in the assessment and in the outcomes tracking; eMBC is also easier to operate and more acceptable for clinicians and patients.

2. Benefits of Implementing eMBC

The eMBC helps overcome the challenges of MBC implementation through web/mobile devices and telephone guidance counseling services. With mobile devices, patients can monitor their symptoms routinely and more frequently, and have a better understanding of their depression experiences (Bauer et al. 2018). Completing self-reports on mobile devices is inexpensive and less time-consuming, which enables clinicians to obtain clinical data between two clinical visits (Bauer et al. 2018). In addition, reminders (such as smartphone reminders, emails, text messages, etc.) can be set up on the mobile apps hosting the assessment tool to help remind patients to complete their scales routinely. If the depressive symptoms do not improve, the clinician will receive feedback to help with the intervention. The shortage of mental health resources could lead to the failure to seek "face-to-face" mental health services. The eMBC helps to fill the gap in mental health services reaching patients through the Internet and telephone, which could be an effective way to distribute clinical resources and to improve mental health services in remote areas (CMHA-BC 2016).

3. Assessment Tools Applied for eMBC

The eMBC has been applied to monitor symptom change; assessment across particular domains would provide clinicians with additional information to guide treatment process and identify intervention targets. These additional domains include functional impairment, quality of life, side effects, and adherence, which can be incorporated into eMBC, e.g., Quick Inventory for Depressive Symptomatology, Self-Rated (QIDS-SR), Patient Health Questionnaire-9 (PHQ-9), Sheehan Disability Scale (SDS), Intensity and Burden of Side Effects Rating (FIBSER), and Patient Assessment Questionnaire (PAQ) for treatment adherence. In the implementation of eMBC, the patient self-rating scales are often used to monitor the disease process. Research has shown that patient self-rating scale is comparable to the clinician's rating scale in monitoring the effectiveness of treatment and could avoid bias due to the rater.

4. Application of eMBC in Depressive Disorder

A community mental health center evaluated the feasibility and acceptability of using mobile devices for symptom assessment based on the clinical data, the results showed that mobile devices were more private and convenient than filling out paper forms (Goldstein et al. 2011). Bauer et al. analyzed 547 patients who completed PHQ-2 and PHQ-9 routinely with smart phones to monitor their depressive symptoms and found that eMBC, as a new approach to symptom assessment, could provide clinicians with reliable results (Bauer et al. 2018). The preliminary evidence had also supported that Internet-based cognitive behavioral therapy (CBT) reduced the depressive symptoms in patients with depression (Topolovec-Vranic et al. 2010). Recently, an Evidence-Based Practices (EBP) project, based on CBT, was conducted by the affective disorders team of UBC. Patients were asked to complete PHQ-9 and other self-rating questionnaires; the results were sent to their doctors by fax and mail. The project also provided health education and telephone guidance services to patients on DVD or printed matter. The study further supported that the Internet-based CBT and telephone guidance services significantly improved the clinical outcome (CMHA-BC 2016). A systematic review of remote MBC including 36 studies suggested that eMBC could be an effective intervention to improve mental health services (Goldberg et al. 2018).

A Canada–China Implementation Project: Enhanced Measurement-Based care Effectiveness in Depression (EMBED) adapts and develops information technologies (i.e., smart phone apps, online training programs, electronic medical records, and Internet-delivered self-management interventions with telephone and WeChat (SMS) coaching) to implement eMBC for depression in community mental health centers in Shanghai of China. The aims of EMBED includes identifying barriers and facilitators to MBC in China, adapting to an eMBC program for the Chinese healthcare setting, comparing the effectiveness of eMBC versus standard MBC for physician implementation and clinical outcomes, and building knowledge and capacity for scale-up in China and beyond.

References

Abbasi J (2017) 23andMe, big data, and the genetics of depression. JAMA 317(1):14–16

Abbott CC, Loo D, Sui J (2016) Determining electroconvulsive therapy response with machine learning. JAMA Psychiatry 73(6):545–546

Aziz HA (2016) Handling big data in modern healthcare. Lab Med 47(4):e38–e41

Bauer AM, Baldwin SA, Anguera JA et al (2018) Comparing approaches to mobile depression assessment for measurement-based care: prospective study. J Med Internet Res 20(6):e10001

Bremer V, Becker D, Kolovos S et al (2018) Predicting therapy success and costs for personalized treatment recommendations using baseline characteristics: data-driven analysis. J Med Internet Res 20(8):e10275

Chekroud AM, Zotti RJ, Shehzad Z et al (2016) Cross-trial prediction of treatment outcome in depression: a machine learning approach. Lancet Psychiatry 3(3):243–250

Cho CH, Lee T, Kim MG et al (2019) Mood prediction of patients with mood disorders by machine learning using passive digital phenotypes based on the circadian rhythm: prospective observational cohort study. J Med Internet Res 21(4):e11029

CMHA-BC (2016) Bounce back: reclaim your health. Annual Report, 2015–2016. Canadian Mental Health Association, Vancouver, BC

Cohen ZD, Derubeis RJ (2018) Treatment selection in depression. Annu Rev Clin Psychol 14:209–236

Diniz BS, Lin CW, Sibille E et al (2016) Circulating biosignatures of late-life depression (LLD): towards a comprehensive, data-driven approach to understanding LLD pathophysiology. J Psychiatr Res 82:1–7

Dipnall JF, Pasco JA, Berk M et al (2017) Why so GLUMM? Detecting depression clusters through graphing lifestyle-environs using machine-learning methods (GLUMM). Eur Psychiatry 39:40–50

Ferrari P, Parisi MM, Colombo R et al (2018) Depression and mania induce pro-inflammatory activation of macrophages following application of serum from individuals with bipolar disorder. Clin Psychopharmacol Neurosci 16(1):103–108

Goldberg SB, Buck B, Raphaely S et al (2018) Measuring psychiatric symptoms remotely: a systematic review of remote measurement-based care. Curr Psychiatry Rep 20(10):81

Goldstein LA, Connolly Gibbons MB, Thompson SM et al (2011) Outcome assessment via handheld computer in community mental health: consumer satisfaction and reliability. J Behav Health Serv Res 38(3):414–423

Horvitz E, Mulligan D (2015) Policy forum. Data, privacy, and the greater good. Science 349(6245):253–255

Khodayarirostamabad A, Reilly JP, Hasey GM et al (2013) A machine learning approach using EEG data to predict response to SSRI treatment for major depressive disorder. Clin Neurophysiol 124(10):1975–1985

Koutsouleris N, Kambeitz-Ilankovic L, Ruhrmann S et al (2018) Prediction models of functional outcomes for individuals in the clinical high-risk state for psychosis or with recent-onset depression: a multimodal, multisite machine learning analysis. JAMA Psychiatry 75(11):1156–1172

Lam RW, Milev R, Rotzinger S et al (2016) Discovering biomarkers for antidepressant response: protocol from the Canadian biomarker integration network in depression (CAN-BIND) and clinical characteristics of the first patient cohort. BMC Psychiatry 16:105

Lewis CC, Scott K, Marti CN et al (2015) Implementing measurement-based care (iMBC) for depression in community mental health: a dynamic cluster randomized trial study protocol. Implement Sci 10:127

Maciukiewicz M, Marshe VS, Hauschild AC et al (2018) GWAS-based machine learning approach to predict duloxetine response in major depressive disorder. J Psychiatr Res 99:62–68

Mourao-Miranda J, Almeida JR, Hassel S et al (2012) Pattern recognition analyses of brain activation elicited by happy and neutral faces in unipolar and bipolar depression. Bipolar Disord 14(4):451–460

Mwangi B, Ebmeier KP, Matthews K et al (2012) Multi-centre diagnostic classification of individual structural neuroimaging scans from patients with major depressive disorder. Brain 135(Pt 5):1508–1521

Oslin DW, Hoff R, Mignogna J et al (2019) Provider attitudes and experience with measurement-based mental health care in the VA implementation project. Psychiatr Serv 70(2):135–138

Perlis RH (2013) A clinical risk stratification tool for predicting treatment resistance in major depressive disorder. Biol Psychiatry 74(1):7–14

Redlich R, Opel N, Grotegerd D et al (2016) Prediction of individual response to electroconvulsive therapy via machine learning on structural magnetic resonance imaging data. JAMA Psychiatry 73(6):557–564

Topolovec-Vranic J, Cullen N, Michalak A et al (2010) Evaluation of an online cognitive behavioural therapy program by patients with traumatic brain injury and depression. Brain Inj 24(5):762–772

Trivedi MH, Rush AJ, Wisniewski SR et al (2006) Evaluation of outcomes with citalopram for depression using measurement-based care in STAR*D: implications for clinical practice. Am J Psychiatry 163(1):28–40

Trivedi MH, Rush AJ, Gaynes BN et al (2007) Maximizing the adequacy of medication treatment in controlled trials and clinical practice: STAR*D measurement-based care. Neuropsychopharmacology 32(12):2479–2489

Victor E, Aghajan ZM, Sewart AR et al (2019) Detecting depression using a framework combining deep multimodal neural networks with a purpose-built automated evaluation. Psychol Assess

Waldrop J, Mcguinness TM (2017) Measurement-based care in psychiatry. J Psychosoc Nurs Ment Health Serv 55(11):30–35

Watson DS, Krutzinna J, Bruce IN et al (2019) Clinical applications of machine learning algorithms: beyond the black box. BMJ 364:l886

Chapter 15
Potential Anti-Depressive Treatment Maneuvers from Bench to Bedside

Min Cai, Huaning Wang and Xia Zhang

Abstract Numerous antidepressants are available for the treatment of the major depressive disorder (MDD). Unfortunately, the disadvantages of these antidepressive medications, including inadequate treatment response, the therapeutic lag between drug administration and the onset of symptoms alleviation, and the safety consideration limit their clinical use and accelerate the exploration of advanced antidepressants with novel action mechanisms/newer targets, with fewer side effects. In this chapter, a series of compounds showing clinical potent in the treatment of MDD has been reviewed based on their reported results from different phase clinical trials. Although the majority of these strategies currently only lead to a systematic approach in the aspects of treatment resistant depression, some of them would be a routine clinical practice which is usable in the treatment of MDD, such as ketamine. Additionally, beyond the mechanism of action for novel therapeutic molecules involving glutamatergic, opiate, cholinergic receptors, and neuroplasticity, some supplemental procedures such as polyunsaturated fatty acids were also included in this chapter due to their solid property against MDD.

Keywords Novel pharmacological antidepressants · Ketamine · Neurotransmitters · Anti-inflammation

---

M. Cai · H. Wang (✉)
Department of Psychiatry, Xijing Hospital, Fourth Military Medical University, Xi'an, Shaanxi, China
e-mail: xskzhu@fmmu.edu.cn

X. Zhang (✉)
Departments of Psychiatry and Cellular and Molecular Medicine, University of Ottawa, Institute of Mental Health Research at the Royal, Ottawa, ON, Canada
e-mail: Xia.Zhang@theroyal.ca

© Springer Nature Singapore Pte Ltd. 2019

Y. Fang (ed.), *Depressive Disorders: Mechanisms, Measurement and Management*, Advances in Experimental Medicine and Biology 1180,
https://doi.org/10.1007/978-981-32-9271-0_15

15.1 Introduction

Traditional treatments including monoamine oxidase inhibitors, tricyclic antidepressants and the subsequently developed SSRI/SNRIs for depression typically take weeks to achieve measurable improvement and months to get full remission of symptoms in depressed patients (Rush et al. 2006). Besides such lag time for treatment effectiveness, up to 30% MDD individuals are refractory to treatments despite receiving multiple distinct medications (Mrazek et al. 2014). This unbenefited practice for the current treatments emphasizes the requirement of better, faster-responded medications to treat this life-threatening mental illness.

15.2 *Ketamine* and Other Compounds Targeting Glutamatergic Signaling

15.2.1 Ketamine

Ketamine is a noncompetitive *N*-methyl-D-aspartate (NMDA) glutamate receptor antagonist. It has been used for procedural sedation and anesthesia introduction in some emergency medical settings for more than 50 years. This NMDAR-blocking agent was observed in 2000 that it elicited a rapidly clinical improvement against core depressive symptoms even within hours after treatment following a single administration of subanesthetic dose to medication-resistant depression patients (Berman et al. 2000). Even though the plasma redistribution half-life of ketamine is about 4 min, these antidepressant effects tend to emerge 1–2 h after the ketamine has bated and persist for two weeks or longer (Mion and Villevieille 2013). Subsequently, some placebo-controlled trials have demonstrated the rapid, robust, and relatively sustained antidepressant effect of ketamine in unipolar and bipolar depression patients (Zarate et al. 2006, 2012; Ibrahim et al. 2011). These studies introduced an intravenously infuse ketamine with a dose of 0.5 mg/kg over 40 min and demonstrated that the antidepressant effect was detectable within 4 h following the treatment and this effect sustained for 3–7 days. Moreover, the response rate at 24 h after treatment is 40–60% (Diazgranados et al. 2010).

Particularly, the supplement of ketamine might induce transient perceptual disturbances, dissociation, dysphoria, euphoria, or anxiety during infusion (Aan Het Rot et al. 2012). Also, some physical side effects such as nausea, dizziness, and moderate blood pressure and heart rate increase have been reported. These adverse effects generally occurred within an hour after stopping ketamine infusion and disappeared completely within two hours. Amounts of open-label and series case reports demonstrated the comparable safety and tolerability of a single supplementation of ketamine (Blier et al. 2012).

However, some limitations refuse this antidepressant maneuver realistically from bench to bedside. The first is all early placebo-controlled ketamine researches actually based on the unblinding of treatment status due to the acute adverse effect of ketamine. A recent trial partially mitigated this limitation by induced midazolam (a benzodiazepine anesthetic) as an active comparator; this indicates an optimized blinding method for trials demonstrating the robust rapid antidepressant effect of ketamine in the future (Szymkowicz et al. 2013).

Another thing is that the long-term safety and efficacy of repeated ketamine dosing, which looks to be more beneficial than single ketamine infusion, still need to be established. Some open-label case reports show that the repeated ketamine supplementation safely extended the antidepressant effect of ketamine to several months, even some unpublished trials suggest that this benefit may be extended for more than a year (Blier et al. 2012; Szymkowicz et al. 2013). In consistence with this view, a limited pilot data indicates that up to six administrations with low-dose ketamine infused once, twice, or three times per week develop a maintaining and prolonging treatment response (Murrough et al. 2013; Rasmussen et al. 2013; Shiroma et al. 2014; Segmiller et al. 2013). Moreover, some other studies reported the various routes and dosage of ketamine administration, such as 0.2 mg/kg intravenous bolus and 50 mg intranasal and 0.25 mg/kg intramuscular injection, elicited a similar rapid antidepressant effect (Larkin and Beautrais 2011; Lapidus et al. 2014; Chilukuri et al. 2014). To date, the durability and efficacy of repeated ketamine infusion remain to need more well designed trials to determine.

The third limitation is dated that there remains a large gap in knowledge regarding the potential long-term safety issues. In a systemic review, Brooke Short et al. (2018) aggregated and analyzed the safety of ketamine in depression and comments on experience by systematically reviewed 228 published papers. Their finding is that acute side effects induced by single-dose use in depression are common and just generally transient and resolve spontaneously; the administration with a high dose of ketamine is relevant with potentially serious toxic effects, including urological, hepatic, craving or dependence, and cognitive changes. These subjects use ketamine for the treatment of chronic pain or use recreationally but not in depression patients. The authors indicate that almost all randomized controlled trials evaluated the safety of one single ketamine session but followed up very short term. As an increasingly used practice, the safety of long-term, repeated ketamine dosing is still uncertain, which is essential before ketamine could be used as a clinical maneuver for the treatment of depression.

Furthermore, as a dissociative anesthetic, ketamine is capable of producing dramatic psychotomimetic effects, and these effects have been related to its antidepressant efficacy. More than the glutamate system, ketamine interacts with some extra neurotransmitter systems, including mu, delta, and kappa opioid receptors. Currently, ketamine is used as an acute or chronic analgesic medication based on its property that inhibits opioid receptors. This effect was reversed by *kappa* opioid receptor antagonists, indicating a *mu* or *delta* opioid mechanism in ketamine's antinociceptive effects. Here too, even though the specific receptor system or systems responsible for mediating dissociative effects are still unknown, a low dose (25 mg) of the opioid receptor antagonist naltrexone can augment the psychoactive effects of lower subanesthetic doses (~0.4 mg/kg/h) of ketamine, but not higher subanesthetic doses (~0.6 mg/kg/h) in healthy subjects (Krystal et al. 2006). Recently, Nolan Williams et al. (2018) further demonstrated that opioid receptors are determinable for ketamine's acute antidepressant effect.

Despite further studies that should be established to expand the understanding of the opioid effects of subanesthetic dose ketamine, these findings present an additional challenge to the current mechanism of action of ketamine that underlies its potent antidepressant properties and to the potent of side effects like addiction and psychoactivation.

15.2.2 Other Medications Targeting Glutamate Signaling

The glutamate N-methyl-D-aspartate receptor (NMDAR) antagonist ketamine was recently repurposed as a rapidly acting antidepressant, catalyzing the vigorous investigation of glutamate-signaling modulators as novel therapeutic agents for depressive disorders. As plenty of preclinical and translation studies have investigated the mechanisms of action of glutamate-targeting antidepressants, some candidate glutamate-modulating compounds are currently in human testing for major depressive disorders to validate these results in humans. Some of them present important milestones in drug development in this area but still need more detail.

15.2.2.1 *D*-Cycloserine

Beyond an antituberculotic drug, *D*-Cycloserine (DCS) was reported to have antidepressive-like effect as early as the 1950s. DCS acts as a partial agonist at the glycine site of NMDAR in low dose, whereas it seems to be a functional NMDAR antagonist at high doses (Heresco-Levy et al. 2013). As an adjunction therapy to standard antidepressant medication in patients with TRD, a single-site-6-week trial of gradually titrated high-dose DCS introduced a superior benefit compared with placebo (Kantrowitz et al. 2015). Moreover, a small open-label study recently showed good tolerability of 8-week-constant administration of high-dose DCS (titrated to 1000 mg/day) in patients with treatment-reflector bipolar depression after ketamine treatment (Zhang et al. 2008).

15.2.2.2 GLYX-13 and NRX-1074

Acting as a partial agonist at the glycine-binding site of the NMDAR, GLYX-13 is a tetrapeptide (Thr-Pro-Pro-Thr) although the precise manner of which compound regulated NMDAR functioning is not fully understood. In a recent company-invested

RCT, a single i.v. infusion of GLYX-13 elicits a promising antidepressant effect compared with placebo in 120 patients with MDD, even though this effect is mixed (Preskorn et al. 2015). Another four-dose trial that investigated the two middle, but not the highest or the lowest doses, had a significantly greater antidepressant effect than placebo. However, no difference was detected in the ratio of patients who experienced the improvement of symptoms among the placebo group and any of the dose treatment groups. Encouragingly, this compound was well tolerated and did not show any evidence of risk of the pyschomimetic symptoms. In 2014, GLYX-13 has been granted Fast Track status by the US Food and Drug Administration (FDA) for the MDD treatment. NRX-1074 is a purportedly higher potency and an orally active analog of GLYX-13. Recently, the oral formulation of this compound is anticipated to advance to a phase II trial.

15.2.2.3 CERC-301 (MK-0657)

MK-0657 is the oral NR2B-selective antagonist, which is known as CERC-301. A small proof-of-concept study was conducted in patients with TRD and investigated that the oral administration of 4–8 mg MK-0657 for 12 days presented favorable tolerability but a mixed efficacy profile (Ibrahim et al. 2012). A randomized, double-blind placebo-controlled trial designed by FDA in 2013 for the treatment of MDD was initiated to assess the adjunctive antidepressant effects of two doses of CERC-301 (NCT02459236). The designation of this study included two intermittent dose supplementations 7 days apart to clarify the intermittent administration of NMDAR antagonist that is more effective than on a daily basis. The outcomes of this trial will be anticipated in the near future.

15.2.2.4 AV-101 (4-Chlorokynurenine (4-Cl-KYN))

Behaving as a potent and selective antagonist targeting at the obligatory glycine-binding site on NMDAR NR1 subunits, 7-Chlorokynurenic acid (7-Cl-KYNA) has been used to probe NMDAR function but its brain penetrance is poor. Another prodrug 4-chlorokynurenine (4-Cl-KYN; now known as AV-101) readily crosses the blood–brain barrier and could be converted to 7-Cl-KYNA in astrocytes. Currently, no data for 4-Cl-KYN in humans have been reported.

15.2.2.5 Dextromethorphan-Containing Compounds

Dextromethorphan is a nonselective NMDAR antagonist, also is an active ingredient in several over-the-counter cough suppressants. Some combination product containing dextromethorphan plus quinidine was sold as an approved medication for the treatment of pseudobulbar affect. AVP-786 was considered as a next-generation combination, as it contains a deuterium-modified form of dextromethorphan to enhance its pharmacokinetic profile (Mullard 2016). In 2014, Avanir Pharmaceuticals announced a phase II program using AVP-786 as a treatment in MDD. Following this trial, a separate pilot study of dextromethorphan plus quinidine in patients with TRD is currently underway and a phase III trials of AXS-05 (manufactured by Axsome Therapeutics), an oral combination of dextromethorphan plus bupropion began in 2016.

15.2.2.6 Sarcosine (N-Methylglycine)

Sarcosine is a naturally occurring GlyT1 inhibitor, it inhibits glycine reuptake from the synaptic cleft, which subsequently enhances synaptic glycine levels and increases NMDAR activity. A small RCT participated 40 MDD patients to 6 weeks of treatment with sarcosine or citalopram and the result showed that sarcosine performed a superior reduction in depressive symptoms compared with citalopram (Huang et al. 2013).

15.2.2.7 AMPAR Modulators

AMPAR plays a crucial role in synaptic plasticity; some previous studies were a major focus on the drug development of positive allosteric modulators of the AMPAR signaling for CNS disorders, including depression. Two early phase human studies about AMPAR potentiator in depression have been published (Ward et al. 2010). These reports showed promising initial results of the evaluation of the safety and preliminary efficacy for Org 26,576 in patients with MDD.

15.2.2.8 Metabotropic Glutamate Receptor Modulators (MGluRs)

The development of antidepressant drug focusing on mGluRs is in relatively early stages, but these studies have represented a particularly promising avenue. Classes of compounds behaving as mGluR2/3 and mGluR5 antagonists have so far received the most attention. Beyond the rapid, ketamine-like antidepressant effects in preclinical models, a phase I study of the mGluR2/3 antagonist BCI-632 has been completed but the data has not yet been published. A recent phase IIb clinical trial of the selective mGluR5 negative allosteric clinical trial of the selective mGluR5 negative allosteric modulator basimglurant (Hoffman-La Roche) has been completed; the study tested this manufacture as an adjunctive therapy to current antidepressant medications in MDD patients but the result failed to separate the compound from placebo on its primary end point (Jain et al. 2013).

Finally, even though some studies have aimed to demonstrate the solid antidepressant effect of glutamate-signaling compounds, there are still some fundamental questions regarding the future prospects of the development of this line of drugs, including the concern about efficacy, safety and tolerability.

15.3 Antidepressive Candidates Worked as GABA Receptor Modulators

Although the neurobiology of depression has not yet been fully calculated, increasing experimental and clinical evidence points to a relevance between MDD and GABAergic deficits (Luscher et al. 2011). Patients subjected to MDD show reduced central nervous system (CNS) GABA concentrations (Hasler et al. 2007; Sanacora et al. 2004), which play a vital role in the neural regulation of stress. In terms of the exploration of antidepressants targeting at GABA, mainly $GABA_A$ receptors, there is already some clinical evidence showed that regulation of $GABA_A$ receptors elicits the promising antidepressant effect. Alprazolam and adinazolam treatment induce antidepressant responses similar to widely prescribed antidepressants in MDD patients, but only within the first 2 weeks of treatment (Amsterdam et al. 1986; Petty et al. 1995). Eszopiclone, a positive allosteric modulator (PAM) of $GABA^A$ receptors, is used as a co-therapy in patients comorbid with MDD and insomnia and shows not only the sleep improvement but also a faster onset of antidepressant response and a greater magnitude of the antidepressant effect (Fava et al. 2006). Further studies—for instance, studies using nonsedating $\alpha 2/\alpha 3$-selective PAMs are expected to further validate the GABA hypothesis of depression and these compounds could be fast-acting antidepressants in potential.

15.4 Cholinergic System Modulation

15.4.1 Scopolamine

Cholinergic system modulation has also been demonstrated to elicit antidepressant properties, which was evident by initial studies. This was evident from initial studies with physostigmine showing that it exacerbated depressive symptoms or induced depressive symptoms in manic patients. A pilot study assessing the effect of the cholinergic system in the cognitive deficiency in relation with MDD investigated that the antimuscarinic drug scopolamine had antidepressant benefits. Another randomized placebo-controlled clinical trial determined the efficacy of the *I.V.* administration of 4 μg/kg scopolamine compared with placebo (Furey and Drevets 2006). As following, two subsequent RCTs that measured the efficacy of multiple infusions separated by 3–5 days reported similar results (Drevets and Furey 2010).

15.5 Triple Reuptake Inhibitors

Amitifadine

Amitifadine (EB1010, formerly DOV 21947) is a triple reuptake inhibitor of serotonin, norepinephrine, and dopamine. In a small clinical trial accessing a small sample of MDD patients, this agent showed a significant antidepressant response. A 6-week, multicenter, randomized, double-blind, parallel, placebo-controlled trial assessed the efficacy and tolerability of amitifadine (25 or 50 mg) in 63 MDD patients. At the end point, this medication elicited significant improvement in HDRS and CGI-S scores compared with placebo (Marks et al. 2012). However, the methodology of this study was questioned by a later publication. Only half of the patients on amitifadine supplement group confirmed the trial although no serious side effects were reported (Tran et al. 2012).

15.6 Psychedelic Agents as Potential Antidepressants

In some small sample studies, naturally occurring hallucinogens like psilocybin , N,N-dimethyltryptamine (DMT), and D-lysergic acid diethylamide (LSD) have been reported promising effect for treatment of MDD and anxiety disorders (Dos Santos et al. 2016b). One open-label trial that measured the efficacy of two doses of psilocybin with psychological support, shows promising reduction of patient-reported depressive symptoms (Carhart-Harris et al. 2016). One RCT evaluated the effect of two different doses (1–3 and 22–30 mg/kg) of psilocybin in patients with cancer, finding significant alleviation in depressive symptoms with high dose supplementation of psilocybin (Griffiths et al. 2016). An open-label trial accessing ayahuasca also investigated a significant reduction of depressive symptoms and a good tolerance and minimal side effects (Osorio Fde et al. 2015). These studies suggest a potent of using hallucinogenic compounds to treat depressive symptoms in patients with terminal illnesses like cancer (Dos Santos et al. 2016a). However, it should be pointed out that more RCTs are needed to determine the treatment potential of these compounds balancing therapeutic potential with recreational abuse potential (Carhart-Harris and Goodwin 2017).

15.7 Other Miscellaneous Agents

Unlike the serotonergic and noradrenergic systems, the miscellaneous agents like histamine H3 antagonists, captodiamine, and other compounds may also elicit promising antidepressant efficacy due to their pharmacological profile. However, this antidepressant effect is just proved in animal studies (Tiligada et al. 2011).

15.8 Medications or Agents Modulating Central Inflammation

Accumulating preclinical studies have highlighted the potential therapeutic opportunity of targeting central inflammatory processes against depression. Beyond this, some of these therapies have exerted efficaciousness in human patients, and others are also promising targets for future clinical investigation (Miyaoka et al. 2012).

An open-label study in patients with unipolar psychotic depression indicated that using minocycline as an adjunctive therapy to SSRIs is a safe and working maneuver for the treatment of depressive and psychotic symptoms. Another randomized, double-blind, placebo-controlled trial of minocycline enrolled HIV patients with mild-to-moderate comorbid depression showed a significant alleviation in both depression symptoms and time to symptom improvement (Emadi-Kouchak et al. 2016). However, a follow-up study failed to demonstrate the efficacy of minocycline on cognitive symptoms in this patient population despite the reduction of depressive symptoms. More than these, additional placebo-controlled studies for assessing the efficacy and safety of minocycline as an adjunctive treatment in medication-resistant MDD (Husain et al. 2015) and bipolar depression (Husain et al. 2017) are planned or ongoing.

Beyond agents mentioned above, some anti-inflammatories like polyunsaturated fatty acids (PUFAs), N-acetyl cysteine (NAC), infliximab, aspirin, and COX-2 inhibitors such as celecoxib have been used in the field of MDD treatment as an adjunctive procedure or monotherapy (Fond et al. 2014). A meta-analysis analyzing anti-inflammatory agents in unipolar depression found that PFUAs were effective avenues as treatments for MDD if the eicosapentaenoic acid concentration was >60%. Furthermore, this review also demonstrated the significant effects of celecoxib and aspirin for the treatment of MDD. Another trial assessing the supplementation of infliximab (5 mg/kg) at weeks 0, 2, and 6 demonstrated that patients with elevated C-reactive protein levels were benefited greatly from this treatment procedure (Tran et al. 2012).

15.8.1 Riluzole

Some case reports and open-label trials have determined the efficacy and safety of riluzole as a mono or adjunctive therapy for the treatment of TRD and generalized anxiety disorder (Zarate et al. 2004). Moreover, a recent parallel, randomized, double-blind, placebo-controlled trial demonstrate that riluzole exerts a greater alleviation of depression symptoms and decreased time to symptom reduction as a combination treatment with citalopram when compared with citalopram alone (Salardini et al. 2016).

15.8.2 Some Probiotics

Also, some studies have investigated the antidepressant properties of probiotics in MDD patients. A randomized, double-blind, placebo-controlled trial that administrated MDD patients with either a placebo or capsules containing a mix of three probiotic strains (Lactobacillus acidophilus, Lactobacillus casei, and Bifidobacterium bifidum) daily for a period of 8 weeks demonstrated a reduction in self-reported symptoms of depression in probiotic-treated patients at the end of the treatment period as quantified by the Beck Depression Inventory, compared to the placebo-treated group (Aizawa et al. 2016). Interestingly, this antidepressive effect of probiotic treatment also seemed to be in correlation with the anti-inflammatory effect, as probiotic-treated patients exhibited decreased serum levels of CRP. Additionally, some other clinical trials also report positive results of probiotics on mood even though not focusing on MDD patients (Steenbergen et al. 2015).

15.9 Neurostimulation Treatment Procedures for MDD

15.9.1 Acupuncture

Acupuncture has a long history as a treatment used for many chronic diseases in China, Japan, and Korea. Even in modern China, contemporary acupuncture practice is commonly undertaken as part of the medical care system in hospitals. Nevertheless, acupuncture is used as a part of traditional medicine and a mix of hospital and private practice in the United Kingdom, respectively. A Cochrane database of systematic reviews published by Smith CA et al. in 2018 assessed the effectiveness of acupuncture for the treatment of individuals with depression. As an update on the previous versions, this review includes 7104 participants from 64 studies, confirms the assessment of acupuncture versus no treatment/usual treatment (Viereck et al. 2017). Evidence supplied in this review suggest that acupuncture (manual and electro-) may moderately reduce the severity of depression by end of treatment although it is unclear whether the data show differences between groups in the risks of adverse events. In studies comparing acupuncture versus medication, the results show that acupuncture may confer small benefit in reducing the severity of depression by end of treatment. Furthermore, studies focusing on substantial variation resulting from the use of different classes of medications and different procedures of acupuncture stimulation demonstrated lower risks of adverse effects following acupuncture compared with medication alone, as evaluated by the Montgomery–Asberg Depression Rating Scale (MADRS).

Also, some studies compare acupuncture plus medication versus medication monotherapy and suggest that this combination treatment maneuver is highly beneficial in alleviating the severity of depression.

15.9.2 Transcranial Direct Current Stimulation (TDCS)

As a noninvasive brain stimulation technique that is less costly than transcranial magnetic stimulation (TMS), transcranial directcurrent stimulation has not been demonstrated in correlation with seizures (Bikson et al. 2016). tDCS is a maneuver applied a weak, direct current through electrodes that are placed on the scalp to induce alterations in cortical activity and excitability. The current induced by tDCS is triggered at the dorsolateral prefrontal cortex, which is considered as a treatment target for mood regulation (Nitsche and Paulus 2000). A trial involving patients with MDD has determined the superiority of tDCS plus sertraline over the monotherapy of the medication or tDCS only (Meron et al. 2015). Based on this, a study published in New England Journal of Medicine established a non-inferiority trial to compare the safety and efficacy of escitalopram versus tDCS for treating depression. In this 3-group trial, tDCS did not show non-inferiority to escitalopram in the intention-to-treat and per-protocol analyzes in reducing depression symptoms (Brunoni et al. 2017). However, there is no available standardized tDCS treatment, for instance, the amperage and duration of application of direct current; some optimized procedure of tDCS may elicit antidepressant effect which is not weaker or even safer than current selective serotonin reuptake inhibitors but needs further studies (Donde et al. 2017).

15.9.3 Advanced Treatment Maneuvers-Related TMS

Based on the consideration that TMS has made steady progress in coming into being a robust treatment modality, some studies support the optimized protocols like high-frequency TMS (HF-TMS, $\geq$5 Hz) over the left dorsolateral prefrontal cortex (DLPFC), 3000 pulses per session and low-frequency TMS (LF-TMS) ($\leq$1 Hz) over the right DLPFC, and more than 1200 pulses per session also exert solid antidepressant efficacy (Kellner et al. 2016a, b). But no deference was detected between these two HF-TMS and LF-TMS involving 1 and 10 Hz protocols (Dell'Osso et al. 2015). Moreover, a recent review also detailed some new developments for depression treatment, such as synchronized TMS (sTMS), deep TMS (dTMS), and low-frequency magnetic stimulation (LFMS). A multisite trial is underway currently to explore the efficacy of LFMS [NCT01654796, Trial of Low Field Magnetic Stimulation Augmentation of Antidepressant Therapy in Treatment-Resistant Depression (RAPID)] (Ibrahim et al. 2011). Recently, a randomized non-inferiority trial confirmed theta burst versus high-frequency repetitive transcranial magnetic stimulation in MDD patients (THREE-D); the results show that this newer intermittent theta burst stimulation (iTBS) treatment procedure is non-inferior to 10 Hz rTMS for the treatment of TRD. And the number of dropouts, side effect, safety, and tolerability profiles are similar between these two TMS procedures. More importantly, iTBS can be delivered in 3 min versus 37.5 min for a standard 10 Hz treatment session. Thus, this maneuver could coherently be a potential treatment selection for MDD in the future (Blumberger et al. 2018).

15.9.4 Vagus Nerve Stimulation (VNS)

Vagus nerve stimulation (VNS) has been shown to be efficacious for patients with treatment-resistant depression in the long-term management and is approved by FDA as an adjunctive procedure for TRD (Rush et al. 2000). In APA accommodation, it is the treatment option for patients who have no response to at least four adequate trials for depression treatments including ECT (Practice guideline for the treatment of patients with major depressive disorder (revision) 2000). In a post-market surveillance study required by the FDA, the adjunction of VNS resulted in superior outcomes in both effectiveness and mortality reduction after a 5-year period, in comparison with usual antidepressants alone for patients with a chronic, sever course of TRD. However, whether this treatment maneuver could become a regular therapeutic option for MDD patients still needs more further exploration (Aaronson et al. 2017).

15.10 Other Potential Treatment Procedures

15.10.1 Whole-Body Hyperthermia or Hypothermia

Beyond the neurotransmitter theory, warm-sensitive thermosensory signaling pathways projecting from the skin (and other epithelial linings) to determinable subcortical and cortical brain regions may affect neural activity and behavior in ways correlating with the treatment of MDD (Raison et al. 2014). Importantly, these brain areas are most implicated in registering and reacting to pleasant thermal signals that show attenuated activity in patients with MDD (Rolls et al. 2008). Furthermore, MDD patients display disabilities in thermoregulation characterized by upregulated core body temperature and reduced capacity to sweat (Hanusch et al. 2013). Based on these considerations and some animal studies demonstrating that whole-body heating activated subdivisions of the dorsal raphe nucleus implicated in antidepressant responses, an unpublished clinical trial found that the whole-body heating elicited an acute antidepressive-like response. To conduct a preliminary examination of the correlation of these findings, a small open trial of whole-body hyperthermia in 16 medically healthy adults who uffered from MDD investigated that a single session of WBH exposure significantly alleviated the depressive symptoms when evaluated at 5 days after treatment, even persisted across 6 weeks of follow-up. These open-label trials suggest WBH as a safe, rapid-acting antidepressant maneuver with a prolonged therapeutic benefit, but further studies are still required to identify both the optimal temperature and number and timing of treatments before this procedure could promise largest- and longest-lasting clinical response for most patients (Janssen et al. 2016).

15.10.2 Potential Omega-3 Polyunsaturated Fatty Acids

Some epidemiological studies have found that supplementation of fish is associated with lower risk of depression (Nemets et al. 2006). Some studies investigated that EPA act as an antidepressive agent (Su et al. 2014). In 2009, Martins et al. published a meta-analysis study base on 28 clinical trials which supply evidence that EPA may be more efficacious in the treatment of depression than DHA (Martins 2009). Furthermore, the food intake of or supplementation with fish oil/ω-3 PUFA has been determined to protect youngsters (15–25 years) from MDD (Nemets et al. 2006; Rice et al. 2016). However, a recent meta-analysis using 35 RCTs from 1980 to 2014 argued that further RCTs are needed to study populations with diagnosed or clinically significant depression of adequate duration using EPA-predominant ω-3 PUFA formulations (Hallahan et al. 2016).

15.10.3 The Potential Antidepressive Effect of Micronutrient (Zinc, Magnesium, and Selenium)

Some empirical evidence indicate the association between several micronutrients—zinc, magnesium, and selenium—and depression. Collectively, some clinical studies support an inverse association between zinc supplementation and alleviate depressive symptoms. Less evidence is available regarding the relevance between magnesium, selenium deficiency, and depression.

15.10.3.1 Zinc

As a crucial trace element, zinc is important for many biochemical and physiological processes in relation to brain function as well as cellular metabolism. The association of zinc and depression was first investigated in the 1980s, and then a number of human researches extensively studied this relevance. A 17-study observed meta-analysis indicated that the serum zinc concentration was about 0.12 g/mL lower in depressed subjects than that in control subjects. Some intervention studies with dietary or supplemental zinc have shown antidepressant-like and mood-enhancing activities of zinc (Lai et al. 2012). Some randomized controlled trials that demonstrated the supplement of zinc to antidepressant drugs elicited more improvement in depressive symptoms than using antidepressants along (Siwek et al. 2009). Another study suggested that women who received zinc and multivitamin supplements experienced greater decrease in depression-dejection scores of the Profile of Moods State (POMS) assessment than those who only received multivitamin (Sawada and Yokoi 2010). Zinc deficiency is associated with the increased vulnerability to psychological stress, however, further prospective studies are needed to determine the dietary zinc or other supplementation method that is beneficial to depression.

15.10.3.2 Magnesium

Magnesium is one of the most crucial micronutrients for the proper activity of many biochemical and physiological processes, including DNA replication, transcription, and translation. It is a bivalent intracellular cation that acts as a coenzyme or an activator for over 300 enzymatic systems, many of which are essential for proper brain function. Magnesium levels are important for brain function. The shortage of magnesium plays an important role in the pathogenesis of Alzheimer's disease, diabetes, stroke, hypertension, migraines, and attention deficit hyperactivity deficit (Grober et al. 2015). Plenty of studies have demonstrated the relevance of magnesium and various brain regions in the limbic system, thus indicating a possible role for magnesium in the etiology and progression of depression. Although the result of studies tried to demonstrate the correlation of depressive symptoms and magnesium levels and magnesium intake are complex; some intervention studies have suggested a beneficial role of magnesium supplementation in the treatment of depression. A recent randomized clinical trial in adults diagnosed with mild-to-moderate depression showed that the supplementation of 248 mg of magnesium per day for 6 weeks resulted in a clinically significant alleviation of depressive symptoms compared with placebo treatment, as measured by a reduction of 6 points of Patient Health Questionnaire-9 (PHQ-9) scale (Tarleton et al. 2017). Consistently, another randomized controlled trial enrolled 60 MDD individuals with hypomagnesemia found that days of supplementation of 500 mg magnesium oxide led to a significant improvement in Beck Depression Inventory scores as compared to matched individuals but received placebo. However, intervention studies among postpartum women (Fard et al. 2017) and an elderly MDD population with hypomagnesemia found a controversial result of magnesium supplements of 328 mg/day and 50 mg/day, respectively (Barragan-Rodriguez et al. 2008). Moreover, there is still no observed correlation between the effect of magnesium supplementation and pretreatment with magnesium or depressive status. Additional intervention studies might be set to clarify whether magnesium consumption can offer any benefits in the treatment of depression.

15.10.3.3 Selenium

Selenium is a vital trace element that is critical for the proper functioning of some seleno proteins involved in antioxidant defenses in the brain. It has been estimated that one in seven people have low dietary selenium intake, and selenium deficiency has been implicated in a variety of conditions, such as renal disease (Iglesias et al.

2013) and obesity (Alasfar et al. 2011). Moreover, given its neuromodulatory role in brain function, recent studies have found the correlation between selenium levels and depression. Intervention human studies of selenium supplementation have reported inconsistent results. A randomized control study that enrolled 166 Iranian women demonstrated that 8-week-selenium-supplementation during pregnancy was associated with increased selenium serum levels in correlation with lower scores on the Edinburgh Postnatal Depression Scale (EPDS) compared to those who received placebo (Mokhber et al. 2011). However, another study reported the results of a randomized control trial to determine the effect of selenium consumption on mood using the Profile of Moods States-Bipolar Form (POMS-BI) questionnaire and found that the administration of selenium significantly increased plasma selenium levels but no influence of mood scores was detected after 6 months of supplementation (Rayman et al. 2006).

15.11 Conclusion

In this chapter, we listed some of the potential maneuvers that have been approved with promising efficacy against MDD in some clinical trials. Like the story of clozapine, the promising treatment efficacy of ketamine on MDD reverse engineering was used to try and decipher new treatment procedures, such as agents triggering at the glutamatergic system. Moreover, with developed neurostimulation techniques comes the opportunity to try them in treatment targeting at a more rapid, lower side effect rate. However, there is an unmet need for more trials to test the efficacy of these agents alone or in combination with currently FDA approved serotonergic antidepressants.

References

Aan Het Rot M, Zarate CA Jr, Charney DS, Mathew SJ (2012) Ketamine for depression: where do we go from here? Biol Psychiatry 72(7):537–547

Aaronson ST, Sears P, Ruvuna F et al (2017) A 5-year observational study of patients with treatment-resistant depression treated with vagus nerve stimulation or treatment as usual: comparison of response, remission, and suicidality. Am J Psychiatry 174(7):640–648

Aizawa E, Tsuji H, Asahara T et al (2016) Possible association of Bifidobacterium and Lactobacillus in the gut microbiota of patients with major depressive disorder. J Affect Disord 202:254–257

Alasfar F, Ben-Nakhi M, Khoursheed M, Kehinde EO, Alsaleh M (2011) Selenium is significantly depleted among morbidly obese female patients seeking bariatric surgery. Obes Surg 21(11):1710–1713

Amsterdam JD, Kaplan M, Potter L, Bloom L, Rickels K (1986) Adinazolam, a new triazolobenzodiazepine, and imipramine in the treatment of major depressive disorder. Psychopharmacology 88(4):484–488

Barragan-Rodriguez L, Rodriguez-Moran M, Guerrero-Romero F (2008) Efficacy and safety of oral magnesium supplementation in the treatment of depression in the elderly with type 2 diabetes: a randomized, equivalent trial. Magnes Res 21(4):218–223

Berman RM, Cappiello A, Anand A et al (2000) Antidepressant effects of ketamine in depressed patients. Biol Psychiatry 47(4):351–354
Bikson M, Grossman P, Thomas C et al (2016) Safety of transcranial direct current stimulation: evidence based update 2016. Brain Stimul 9(5):641–661
Blier P, Zigman D, Blier J (2012) On the safety and benefits of repeated intravenous injections of ketamine for depression. Biol Psychiatry 72(4):e11–12
Blumberger DM, Vila-Rodriguez F, Thorpe KE et al (2018) Effectiveness of theta burst versus high-frequency repetitive transcranial magnetic stimulation in patients with depression (THREE-D): a randomised non-inferiority trial. Lancet 391(10131):1683–1692
Brunoni AR, Moffa AH, Sampaio-Junior B et al (2017) Trial of electrical direct-current therapy versus escitalopram for depression. N Engl J Med 376(26):2523–2533
Carhart-Harris RL, Goodwin GM (2017) The therapeutic potential of psychedelic drugs: past, present, and future. Neuropsychopharmacol: Off Publ Am Coll Neuropsychopharmacology 42(11):2105–2113
Carhart-Harris RL, Bolstridge M, Rucker J et al (2016) Psilocybin with psychological support for treatment-resistant depression: an open-label feasibility study. Lancet Psychiatry 3(7):619–627
Chilukuri H, Reddy NP, Pathapati RM, Manu AN, Jollu S, Shaik AB (2014) Acute antidepressant effects of intramuscular versus intravenous ketamine. Indian J Psychol Med 36(1):71–76
Dell'Osso B, Oldani L, Camuri G et al (2015) Augmentative repetitive transcranial magnetic stimulation (rTMS) in the acute treatment of poor responder depressed patients: a comparison study between high and low frequency stimulation. Eur Psychiatry: J Assoc Eur Psychiatrists 30(2):271–276
Diazgranados N, Ibrahim L, Brutsche NE et al (2010) A randomized add-on trial of an N-methyl-D-aspartate antagonist in treatment-resistant bipolar depression. Arch Gen Psychiatry 67(8):793–802
Donde C, Amad A, Nieto I et al (2017) Transcranial direct-current stimulation (tDCS) for bipolar depression: a systematic review and meta-analysis. Prog Neuro-Psychopharmacol Biol Psychiatry 78:123–131
Dos Santos RG, Balthazar FM, Bouso JC, Hallak JE (2016a) The current state of research on ayahuasca: a systematic review of human studies assessing psychiatric symptoms, neuropsychological functioning, and neuroimaging. J Psychopharmacol 30(12):1230–1247
Dos Santos RG, Osorio FL, Crippa JA, Riba J, Zuardi AW, Hallak JE (2016b) Antidepressive, anxiolytic, and antiaddictive effects of ayahuasca, psilocybin and lysergic acid diethylamide (LSD): a systematic review of clinical trials published in the last 25 years. Ther Adv Psychopharmacol 6(3):193–213
Drevets WC, Furey ML (2010) Replication of scopolamine's antidepressant efficacy in major depressive disorder: a randomized, placebo-controlled clinical trial. Biol Psychiatry 67(5):432–438
Emadi-Kouchak H, Mohammadinejad P, Asadollahi-Amin A et al (2016) Therapeutic effects of minocycline on mild-to-moderate depression in HIV patients: a double-blind, placebo-controlled, randomized trial. Int Clin Psychopharmacol 31(1):20–26
Fard FE, Mirghafourvand M, Mohammad-Alizadeh Charandabi S, Farshbaf-Khalili A, Javadzadeh Y, Asgharian H (2017) Effects of zinc and magnesium supplements on postpartum depression and anxiety: A randomized controlled clinical trial. Women Health 57(9):1115–1128
Fava M, McCall WV, Krystal A et al (2006) Eszopiclone co-administered with fluoxetine in patients with insomnia coexisting with major depressive disorder. Biol Psychiatry 59(11):1052–1060
Fond G, Hamdani N, Kapczinski F et al (2014) Effectiveness and tolerance of anti-inflammatory drugs' add-on therapy in major mental disorders: a systematic qualitative review. Acta Psychiatr Scand 129(3):163–179
Furey ML, Drevets WC (2006) Antidepressant efficacy of the antimuscarinic drug scopolamine: a randomized, placebo-controlled clinical trial. Arch Gen Psychiatry 63(10):1121–1129
Griffiths RR, Johnson MW, Carducci MA et al (2016) Psilocybin produces substantial and sustained decreases in depression and anxiety in patients with life-threatening cancer: a randomized double-blind trial. J Psychopharmacol 30(12):1181–1197

Grober U, Schmidt J, Kisters K (2015) Magnesium in prevention and therapy. Nutrients 7(9):8199–8226

Hallahan B, Ryan T, Hibbeln JR et al (2016) Efficacy of omega-3 highly unsaturated fatty acids in the treatment of depression. Br J Psychiatry: J Ment Sci 209(3):192–201

Hanusch KU, Janssen CH, Billheimer D et al (2013) Whole-body hyperthermia for the treatment of major depression: associations with thermoregulatory cooling. Am J Psychiatry 170(7):802–804

Hasler G, van der Veen JW, Tumonis T, Meyers N, Shen J, Drevets WC (2007) Reduced prefrontal glutamate/glutamine and gamma-aminobutyric acid levels in major depression determined using proton magnetic resonance spectroscopy. Arch Gen Psychiatry 64(2):193–200

Heresco-Levy U, Gelfin G, Bloch B et al (2013) A randomized add-on trial of high-dose D-cycloserine for treatment-resistant depression. Int J Neuropsychopharmacol 16(3):501–506

Huang CC, Wei IH, Huang CL et al (2013) Inhibition of glycine transporter-I as a novel mechanism for the treatment of depression. Biol Psychiatry 74(10):734–741

Husain MI, Chaudhry IB, Rahman RR et al (2015) Minocycline as an adjunct for treatment-resistant depressive symptoms: study protocol for a pilot randomised controlled trial. Trials 16:410

Husain MI, Chaudhry IB, Hamirani MM et al (2017) Minocycline and celecoxib as adjunctive treatments for bipolar depression: a study protocol for a multicenter factorial design randomized controlled trial. Neuropsychiatr Dis Treat 13:1–8

Ibrahim L, Diazgranados N, Luckenbaugh DA et al (2011) Rapid decrease in depressive symptoms with an N-methyl-d-aspartate antagonist in ECT-resistant major depression. Prog Neuro-Psychopharmacol Biol Psychiatry 35(4):1155–1159

Ibrahim L, Diaz Granados N, Jolkovsky L et al (2012) A Randomized, placebo-controlled, crossover pilot trial of the oral selective NR2B antagonist MK-0657 in patients with treatment-resistant major depressive disorder. J Clin Psychopharmacol 32(4):551–557

Iglesias P, Selgas R, Romero S, Diez JJ (2013) Selenium and kidney disease. J Nephrol 26(2):266–272

Jain FA, Hunter AM, Brooks JO 3rd, Leuchter AF (2013) Predictive socioeconomic and clinical profiles of antidepressant response and remission. Depress Anxiety 30(7):624–630

Janssen CW, Lowry CA, Mehl MR et al (2016) Whole-body hyperthermia for the treatment of major depressive disorder: a randomized clinical trial. JAMA Psychiatry 73(8):789–795

Kantrowitz JT, Halberstam B, Gangwisch J (2015) Single-dose ketamine followed by daily D-Cycloserine in treatment-resistant bipolar depression. J Clin Psychiatry 76(6):737–738

Kellner CH, Husain MM, Knapp RG et al (2016a) A novel strategy for continuation ECT in geriatric depression: phase 2 of the PRIDE study. Am J Psychiatry 173(11):1110–1118

Kellner CH, Husain MM, Knapp RG et al (2016b) Right unilateral ultrabrief pulse ECT in geriatric depression: phase 1 of the PRIDE study. Am J Psychiatry 173(11):1101–1109

Krystal JH, Madonick S, Perry E et al (2006) Potentiation of low dose ketamine effects by naltrexone: potential implications for the pharmacotherapy of alcoholism. Neuropsychopharmacol: Off Publ Am Coll Neuropsychopharmacology 31(8):1793–1800

Lai J, Moxey A, Nowak G, Vashum K, Bailey K, McEvoy M (2012) The efficacy of zinc supplementation in depression: systematic review of randomised controlled trials. J Affect Disord 136(1–2):e31–e39

Lapidus KA, Levitch CF, Perez AM et al (2014) A randomized controlled trial of intranasal ketamine in major depressive disorder. Biol Psychiatry 76(12):970–976

Larkin GL, Beautrais AL (2011) A preliminary naturalistic study of low-dose ketamine for depression and suicide ideation in the emergency department. Int J Neuropsychopharmacol 14(8):1 127–1131

Luscher B, Shen Q, Sahir N (2011) The GABAergic deficit hypothesis of major depressive disorder. Mol Psychiatry 16(4):383–406

Marks DM, Abramowitz JS, Spielmans GI (2012) Concerns about data reporting and interpretation in "Efficacy and tolerability of the novel triple reuptake inhibitor amitifadine in the treatment of patients with major depressive disorder: a randomized, double-blind, placebo-controlled trial". J Psychiatr Res 46(5):692–693; author reply 694–695

Martins JG (2009) EPA but not DHA appears to be responsible for the efficacy of omega-3 long chain polyunsaturated fatty acid supplementation in depression: evidence from a meta-analysis of randomized controlled trials. J Am Coll Nutr 28(5):525–542

Meron D, Hedger N, Garner M, Baldwin DS (2015) Transcranial direct current stimulation (tDCS) in the treatment of depression: systematic review and meta-analysis of efficacy and tolerability. Neurosci Biobehav Rev 57:46–62

Mion G, Villevieille T (2013) Ketamine pharmacology: an update (pharmacodynamics and molecular aspects, recent findings). CNS Neurosci & Ther 19(6):370–380

Miyaoka T, Wake R, Furuya M et al (2012) Minocycline as adjunctive therapy for patients with unipolar psychotic depression: an open-label study. Prog Neuro-Psychopharmacol Biol Psychiatry 37(2):222–226

Mokhber N, Namjoo M, Tara F et al (2011) Effect of supplementation with selenium on postpartum depression: a randomized double-blind placebo-controlled trial. J Matern-Fetal Neonatal Med: Off J Eur Assoc Perinat Med, Fed Asia Ocean Perinat Soc, Int Soc Perinat Obstet 24(1):104–108

Mrazek DA, Hornberger JC, Altar CA, Degtiar I (2014) A review of the clinical, economic, and societal burden of treatment-resistant depression: 1996–2013. Psychiatric Serv 65(8):977–987

Mullard A (2016) Deuterated drugs draw heavier backing. Nat Rev Drug Discov 15(4):219–221

Murrough JW, Perez AM, Pillemer S et al (2013) Rapid and longer-term antidepressant effects of repeated ketamine infusions in treatment-resistant major depression. Biol Psychiatry 74(4):250–256

Nemets H, Nemets B, Apter A, Bracha Z, Belmaker RH (2006) Omega-3 treatment of childhood depression: a controlled, double-blind pilot study. Am J Psychiatry 163(6):1098–1100

Nitsche MA, Paulus W (2000) Excitability changes induced in the human motor cortex by weak transcranial direct current stimulation. J Physiol 527 Pt 3:633–639

Osorio Fde L, Sanches RF, Macedo LR et al (2015) Antidepressant effects of a single dose of ayahuasca in patients with recurrent depression: a preliminary report. Revista brasileira de psiquiatria 37(1):13–20

Petty F, Trivedi MH, Fulton M, Rush AJ (1995) Benzodiazepines as antidepressants: does GABA play a role in depression? Biol Psychiatry 38(9):578–591

Practice guideline for the treatment of patients with major depressive disorder (revision) (2000) American Psychiatric Association. Am J Psychiatry 157(4 Suppl):1–45

Preskorn S, Macaluso M, Mehra DO et al (2015) Randomized proof of concept trial of GLYX-13, an N-methyl-D-aspartate receptor glycine site partial agonist, in major depressive disorder nonresponsive to a previous antidepressant agent. J Psychiatr Pract 21(2):140–149

Raison CL, Hale MW, Williams LE, Wager TD, Lowry CA (2014) Somatic influences on subjective well-being and affective disorders: the convergence of thermosensory and central serotonergic systems. Front Psychol 5:1580

Rasmussen KG, Lineberry TW, Galardy CW et al (2013) Serial infusions of low-dose ketamine for major depression. J Psychopharmacol 27(5):444–450

Rayman M, Thompson A, Warren-Perry M et al (2006) Impact of selenium on mood and quality of life: a randomized, controlled trial. Biol Psychiatry 59(2):147–154

Rice SM, Hickie IB, Yung AR et al (2016) Youth depression alleviation: the fish oil youth depression study (YoDA-F): a randomized, double-blind, placebo-controlled treatment trial. Early Interv Psychiatry 10(4):290–299

Rolls ET, Grabenhorst F, Parris BA (2008) Warm pleasant feelings in the brain. NeuroImage 41(4):1504–1513

Rush AJ, George MS, Sackeim HA et al (2000) Vagus nerve stimulation (VNS) for treatment-resistant depressions: a multicenter study. Biol Psychiatry 47(4):276–286

Rush AJ, Trivedi MH, Wisniewski SR et al (2006) Acute and longer-term outcomes in depressed outpatients requiring one or several treatment steps: a STAR*D report. Am J Psychiatry 163(11):1905–1917

Salardini E, Zeinoddini A, Mohammadinejad P et al (2016) Riluzole combination therapy for moderate-to-severe major depressive disorder: a randomized, double-blind, placebo-controlled trial. J Psychiatr Res 75:24–30

Sanacora G, Gueorguieva R, Epperson CN et al (2004) Subtype-specific alterations of gamma-aminobutyric acid and glutamate in patients with major depression. Arch Gen Psychiatry 61(7):705–713

Sawada T, Yokoi K (2010) Effect of zinc supplementation on mood states in young women: a pilot study. Eur J Clin Nutr 64(3):331–333

Segmiller F, Ruther T, Linhardt A et al (2013) Repeated S-ketamine infusions in therapy resistant depression: a case series. J Clin Pharmacol 53(9):996–998

Shiroma PR, Johns B, Kuskowski M et al (2014) Augmentation of response and remission to serial intravenous subanesthetic ketamine in treatment resistant depression. J Affect Disord 155:123–129

Short B, Fong J, Galvez V, Shelker W, Loo CK (2018) Side-effects associated with ketamine use in depression: a systematic review. Lancet Psychiatry 5(1):65–78

Siwek M, Dudek D, Paul IA et al (2009) Zinc supplementation augments efficacy of imipramine in treatment resistant patients: a double blind, placebo-controlled study. J Affect Disord 118(1–3):187–195

Steenbergen L, Sellaro R, van Hemert S, Bosch JA, Colzato LS (2015) A randomized controlled trial to test the effect of multispecies probiotics on cognitive reactivity to sad mood. Brain Behav Immun 48:258–264

Su KP, Lai HC, Yang HT et al (2014) Omega-3 fatty acids in the prevention of interferon-alpha-induced depression: results from a randomized, controlled trial. Biol Psychiatry 76(7):559–566

Szymkowicz SM, Finnegan N, Dale RM (2013) A 12-month naturalistic observation of three patients receiving repeat intravenous ketamine infusions for their treatment-resistant depression. J Affect Disord 147(1–3):416–420

Tarleton EK, Littenberg B, MacLean CD, Kennedy AG, Daley C (2017) Role of magnesium supplementation in the treatment of depression: a randomized clinical trial. PloS One 12(6):e0180067

Tiligada E, Kyriakidis K, Chazot PL, Passani MB (2011) Histamine pharmacology and new CNS drug targets. CNS Neurosci Ther 17(6):620–628

Tran P, Skolnick P, Czobor P et al (2012) Efficacy and tolerability of the novel triple reuptake inhibitor amitifadine in the treatment of patients with major depressive disorder: a randomized, double-blind, placebo-controlled trial. J Psychiatric Res 46(1):64–71

Viereck V, Gamper M, Re: Kirchin V, Page T, Keegan PE, Atiemo KO, Cody JD, McClinton S, Aluko P (2017) Urethral injection therapy for urinary incontinence in women. Cochrane Database Syst Rev 7:CD003881. https://doi.org/10.1002/14651858.cd003881.pub4. Neurourol Urodyn 37(7):2286–2287

Ward SE, Bax BD, Harries M (2010) Challenges for and current status of research into positive modulators of AMPA receptors. Br J Pharmacol1 60(2):181–190

Williams NR, Heifets BD, Blasey C et al (2018) Attenuation of antidepressant effects of ketamine by opioid receptor antagonism. Am J Psychiatry 175(12):1205–1215

Zarate CA Jr, Payne JL, Quiroz J et al (2004) An open-label trial of riluzole in patients with treatment-resistant major depression. Am J Psychiatry 161(1):171–174

Zarate CA Jr, Singh JB, Carlson PJ et al (2006) A randomized trial of an N-methyl-D-aspartate antagonist in treatment-resistant major depression. Arch Gen Psychiatry 63(8):856–864

Zarate CA Jr, Brutsche NE, Ibrahim L et al (2012) Replication of ketamine's antidepressant efficacy in bipolar depression: a randomized controlled add-on trial. Biol Psychiatry 71(11):939–946

Zhang XL, Sullivan JA, Moskal JR, Stanton PK (2008) A NMDA receptor glycine site partial agonist, GLYX-13, simultaneously enhances LTP and reduces LTD at Schaffer collateral-CA1 synapses in hippocampus. Neuropharmacology 55(7):1238–1250

Correction to: Neuroimmune Advance in Depressive Disorder

Guoqing Zhao and Xiaohua Liu

Correction to: Chapter 4 in: Y. Fang (ed.), *Depressive Disorders: Mechanisms, Measurement and Management*, Advances in Experimental Medicine and Biology 1180, https://doi.org/10.1007/978-981-32-9271-0_4

In the original version of the book, the following belated corrections have been incorporated:

"Schedlowski et al. 2016" has been replaced with "Walsh E, Eisenlohr-Moul T, Baer R (2016) Brief mindfulness training reduces salivary IL-6 and TNF-a in young women with depressive symptomatology. J Consult Clin Psychol 84(10):887–897. https://doi.org/10.1037/ccp0000122" in the reference list and citation in chapter 4. The chapter and book have been updated with the changes.

The updated version of this chapter can be found at https://doi.org/10.1007/978-981-32-9271-0_4

© Springer Nature Singapore Pte Ltd. 2020

Y. Fang (ed.), *Depressive Disorders: Mechanisms, Measurement and Management*, Advances in Experimental Medicine and Biology 1180,
https://doi.org/10.1007/978-981-32-9271-0_16

The manufacturer's authorised representative in the EU is Springer Nature Customer Service Centre GmbH, Europaplatz 3, 69115 Heidelberg, Germany. If you have any concerns regarding our products, please contact ProductSafety@springernature.com

Printed and bound by CPI Group (UK) Ltd, Croydon, CR0 4YY
24/07/2026
02175924-0008